AF806555

# Physiological, Molecular, and Genetic Perspectives of Wheat Improvement

Shabir H. Wani • Amita Mohan
Gyanendra Pratap Singh
Editors

# Physiological, Molecular, and Genetic Perspectives of Wheat Improvement

Springer

*Editors*
Shabir H. Wani
Mountain Research Centre For Field Crops
Sher e Kashmir University of Agricultural Sciences and Technology
Khudwani, Jammu and Kashmir, India

Amita Mohan
Department of Biology
University of Pennsylvania
Philadelphia, PA, USA

Gyanendra Pratap Singh
ICAR-Indian Institute of Wheat & Barley Research
Karnal, Haryana, India

ISBN 978-3-030-59576-0 ISBN 978-3-030-59577-7 (eBook)
https://doi.org/10.1007/978-3-030-59577-7

© The Editor(s) (if applicable) and The Author(s), under exclusive license to Springer Nature Switzerland AG 2021
This work is subject to copyright. All rights are solely and exclusively licensed by the Publisher, whether the whole or part of the material is concerned, specifically the rights of translation, reprinting, reuse of illustrations, recitation, broadcasting, reproduction on microfilms or in any other physical way, and transmission or information storage and retrieval, electronic adaptation, computer software, or by similar or dissimilar methodology now known or hereafter developed.
The use of general descriptive names, registered names, trademarks, service marks, etc. in this publication does not imply, even in the absence of a specific statement, that such names are exempt from the relevant protective laws and regulations and therefore free for general use.
The publisher, the authors, and the editors are safe to assume that the advice and information in this book are believed to be true and accurate at the date of publication. Neither the publisher nor the authors or the editors give a warranty, expressed or implied, with respect to the material contained herein or for any errors or omissions that may have been made. The publisher remains neutral with regard to jurisdictional claims in published maps and institutional affiliations.

This Springer imprint is published by the registered company Springer Nature Switzerland AG
The registered company address is: Gewerbestrasse 11, 6330 Cham, Switzerland

*This book is dedicated to Dr. Sanjaya Rajaram*

Dr. Sanjaya Rajaram, a globally decorated scientific laureate covered in various communications outlets like Wikipedia, news, and magazines. This dedication is additive to those for an exuding format and diction similarities as composed here. Sanjaya Rajaram was born in 1943 in a small farming village Raipur, District Varanasi, in the state of Uttar Pradesh in northern India. His family, including his parents, an older brother, and a younger sister, made a living on their five-hectare farm growing wheat, rice, and maize. Unlike most children in his socioeconomic position, he was encouraged to pursue an education by his parents and graduated from secondary school as the top-ranked student in the Varanasi District.

Rajaram went on to earn a BSc in agriculture from the University of Gorakhpur; an MSc in genetics and plant breeding from the Indian Agricultural Research Institute (IARI) in New Delhi, studying genetics and plant breeding under Dr. M. S. Swaminathan; and a PhD in plant breeding from the University of Sydney.

In 1969 Rajaram began working as a wheat breeder at the International Maize and Wheat Improvement Center in Mexico. He worked alongside scientist Norman Borlaug, in experimental wheat fields in El Batan (Texcoco), in Toluca and Ciudad Obregon, Sonora.

After Borlaug won the Nobel Peace Prize in 1970, he sought to address the growing agricultural needs across China and India. During this time, Rajaram was entrusted greater responsibility to execute the wheat breeding program, which he eventually inherited after Dr. Borlaug's retirement, thus taking over the work he had begun in Mexico.

Rajaram was instrumental in executing the unique "shuttle breeding" program and pioneering the crossing of the winter with the spring wheat type, which would usually never come into contact with one another, a strategy that revolutionized wheat varietal improvement across the world. His breeding techniques have resulted in enhanced nutrient-rich wheat product resistant to rusts, the major challenge of growing wheat in many parts of the world especially the Middle East and Asia.

Over his career, Dr. Sanjaya Rajaram has been instrumental in ushering in significantly enhanced production by breeding a series of wheat clusters initially from the famous winter/spring crosses that produce the VEERY, followed by Kauz and Attila wheat. Of this, Attila was adopted by many nations under various names due to its 15% higher yield over the rest. His plant-breeding accomplishments rendered the second push to the seeds of confidence that Borlaug developed, paving the way to the "Wheat Revolution."

After 33 years at CIMMYT, including the last seven as Director of the Global Wheat Program, Rajaram joined the International Center for Agricultural Research in the Dry Areas (ICARDA) as Director of Integrated Gene Management, before formally retiring in 2008.

Among several other international accolades, Rajaram is an elected Fellow of the National Academy of Agricultural Sciences and currently the owner and director of Resource Seeds International, a small private company specializing in wheat development and promotion based in Mexico.

In 2001, the Government of India awarded Rajaram the Padma Shri, the fourth highest civilian honor. In 2014, Rajaram received the prestigious World Food Prize for his scientific contributions and in developing 480 wheat varieties grown in 51 countries. His contribution has led to an increase in world wheat production, by more than 200 million tons, building up the Green Revolution a success.

"Rajaram's work serves as an inspiration to us all to do more, whether in the private or public sector," said US Secretary of State John Kerry at an event where he delivered the keynote address. "When you do the math, when our planet needs to support two billion more people in the next three decades, it's not hard to figure out: This is the time for a second green revolution," Kerry said. It is befitting to cap his career by the biographical assemblage of his immense contributions as done by Venkataramani Govindan in 2015 explicitly elucidates Sanjaya Rajaram "Mr. Golden Grain: The Life and Work of the Maharaja of Wheat Dr. Sanjaya Rajaram."

# Preface

Wheat is a staple crop of approximately 20% of the world populace. There is a dire need for significant yield advancement and improvement in the nutritional quality of wheat. Though wheat production has improved significantly since 1960, to keep pace with the growing demands of the projected human population, wheat productivity requires a 60% increase by 2050. While for global food security, we need increased yields, climate change is posing a severe threat to wheat productivity. The food insecurities due to the changing climate will negatively impact the socio-economic status, particularly in developing countries.

Traditional breeding methods and advanced crop production technologies have resulted in considerable augmentation of wheat production in Mexico, India and other Asian countries. However, due to increasing demands and projected threats to wheat productivity due to global climate change, it is indispensable to have a multi-disciplinary global effort to mitigate climate change and improve yields. This goal can be achieved by bringing together plant geneticists, molecular biologists, plant pathologists, and plant physiologists to develop wheat that yields better both in terms of quantity, quality, and resilience to environmental fluctuations.

Wheat is consumed in a variety of food products ranging from bread, cereals, pasta to cakes, and pastries. Thus, increasing the nutritional qualities of wheat will potentially contribute to reducing malnutrition and dietary mineral deficiencies. Nutritious wheat thus will aid in healthy growth and reduce mineral deficiency related ailments, particularly in children. Several chapters in the book summarize the efforts undertaken by scientists around the globe in developing better quality wheat along with reducing immunogenicity in wheat.

Climate change is posing a threat to food security. Wheat as a temperate crop is sensitive to heat stress. The Asian subcontinent, with more than half of the world population, is particularly vulnerable to changing climate. Recent advances in understanding the thermotolerance in wheat are summarized in two chapters in the book. Similarly, advances in molecular marker technologies, genome selection, and genome editing for improving wheat yield and quality are also presented in detail in the book.

The editors show appreciation to the contributors of different chapters. Authors of this book were chosen from a broad array of organizations, based on their expertise in the subject and to ensure the contributors are diverse. Author's profound perspective on specific topics has made this volume a state-of-the-art reference material. Our sincere gratitude goes to the young authors for their contributions and for sharing their current research. The book will serve as a ready reference for undergraduate and postgraduate students studying Wheat physiology, genetics, and genomics, in particular, and crop breeding, molecular genetics, and genomics. The book will be a good reference to wheat scientists, cereal researchers, and academicians. Libraries of institutions teaching crop science and molecular biology may need this book as a review material for students and teachers. The editors wish to show appreciation to all the contributors and the editorial workforce of Springer for their cooperation and speedy production of this book.

Srinagar, India — Shabir H. Wani
Pullman, USA — Amita Mohan
Karnal, India — Gyanendra Pratap Singh

# Contents

# Food Production: Global Challenges to Mitigate Climate Change

**Niaz Ali and Abdul Mujeeb-Kazi**

**Abstract** There is no simple solution to sustainably feeding a global population as large as 9.6 billion by 2050 while we focus to diminish the emission of greenhouse gases (GHG) and other crop productivity constraints that cumulatively can penalize outputs. Moreover, strong drifts in global climate change have already been recorded, indicative that prospects of further deterioration are inevitable. Sustaining future food security poses a serious challenge in the face of mounting population, climatic instability, and emergence of new crop uses such as biofuels. Bread wheat (*Triticum aestivum* L., AABBDD) is a major source of calories and protein, providing 20% of the total calories in human diet, and the importance to increase wheat production is widely acknowledged. It is unequivocally recognized as the major conduit toward food security. Nevertheless, yields of all major cereals have stagnated at farm levels with wheat showing the lowest rate of increase due to the emergence of various biotic and abiotic stress constraints. With almost no opportunity to expand agriculture on existing land, increasing genetic gains for grain yield and associated traits could significantly influence the number of people at hunger risk. However, yield is a complex trait, and obtaining higher yields under any situation is unlikely to be addressed with a single or uniform approach. While improvements in agronomy could boost the yield potential in some regions, yield gains in many other areas could be achieved with genetic improvements only. We argue that achieving increased crop adaptation is likely to be the key component of future food security, and this must be well integrated into climate change-related issues and sustainable agriculture. Public investments in agri-food infrastructure and adaptation of innovative farming practices will be crucial in developing resilient crops. Development of crops with a wide genetic base and adaptation to limited agricultural inputs are warranted. Thus dietary preferences could significantly reduce the emissions of GHG and are likely be necessary components of transition toward a low-carbon society. Further, application of the recently evolved high-throughput genotyping and phenomic tools in con-

N. Ali (✉)
Department of Botany, Hazara University, Mansehra, KP, Pakistan

A. Mujeeb-Kazi
Texas A&M University, College Station, TX, USA

© The Editor(s) (if applicable) and The Author(s), under exclusive license to Springer Nature Switzerland AG 2021

S. H. Wani et al. (eds.), *Physiological, Molecular, and Genetic Perspectives of Wheat Improvement*, https://doi.org/10.1007/978-3-030-59577-7_1

junction with genome editing tools could enable plant breeders to use the untapped genetic variability in crops that may ensure agricultural resilience, thereby increasing crop productivity.

**Keywords** Climate change · Genetic diversity · Synthetic hexaploid wheats · Food security · Alien introgression

## 1 Introduction

Despite doubling of the total population, the past half-century has witnessed a remarkable growth in food production, resulting in a dramatic decrease in global hunger (Godfray et al. 2010; FAO 2018). With no exaggeration, the rate of worldwide poverty today is lower than it has ever been in recorded human history. However, meeting the targets of "zero hunger" is still not over and requires substantially greater efforts (Liu et al. 2018; Mujeeb-Kazi et al. 2019). Even today more than one in seven individuals do not get balanced food, and an even greater number suffer from various forms of malnutrition (Godfray et al. 2010; Hawkesford et al. 2013; Mundial 2018). Since human population will continue to rise, this means that the worldwide demands for food will surge. Therefore, production of high-quality food must be doubled in ways that are environmentally and socially sustainable (Borlaug 2002; Ramírez-González et al. 2018; Hickey et al. 2019).

Spikes in food prices are signs of the tendencies of food availability. Although gross food values have curtailed, the rise in food prices more recently was driven primarily by the increasing demands from developing countries and to an extent by biofuel synthesis (Godfray et al. 2010; FAO 2018). The average cereal yield has increased from 1.35 t/ha in 1961 to 3.35 t/ha in 2007 and is likely to reach 4.8 t/ha by 2040. This indicates to the remarkable success of plant breeders during the last 60–70 years in increasing the yield potential of our important crops for human livelihood. If this increase was not attained, nearly three times more land would have been required to sustain the needs of the existing population (Smith and Gregory 2013). Nonetheless, man is perhaps confronted with one of the greatest challenges of this civilization, to sustainably feed over nine billion people by 2050 in sustainable ways and that the world's poorest people are no longer hungry (Borlaug 2002; Liu et al. 2018).

Recent studies have revealed that agricultural land is shrinking and areas that were once productive have been lost to urbanization. On the other hand, historical expansion of agriculture has significantly contributed to the loss of biodiversity (Mujeeb-Kazi et al. 2013; Hickey et al. 2019). Overarching is the major threat of climate adjustments where worries of how mitigation and adaptation procedures may affect food supply (Godfray et al. 2010). Slight changes in climate at the global level will impact plant distribution in both natural and managed ecosystems

(Coakley et al. 1999; Hawkesford et al. 2013). So while at the same time as we are to improve food production, we must take into account the needs to significantly decrease climatic impacts on agriculture as well as improve the resilience of our food production (Smith and Gregory 2013). The most probable scenario is that more food shall be required from the same or even less land and with fewer resources. Therefore, it is critical for us to identify approaches that integrate both the challenges of food security and climate change mitigation concerns while addressing the growing food demands set for 2050 (Godfray et al. 2010; Bevan et al. 2017).

Here we outline important strategies for addressing the challenge of feeding approximately 9.6 billion people and combat global challenges to mitigate climate change. Example of the wheat crop is explored because of its immense potential in feeding the global populace as well as of its being a major conduit toward future food security. Further, the prevalent current practices are unlikely to deliver the food and ecosystem services we require ahead; thus sustainable agriculture will require radical changes from sowing to harvesting supported by an astute dynamic policy approach for decades.

## 2 Global Climate Changes, Prospects, and Challenges for Sustainable Agriculture and Food Supply

Global food security is a major challenge of the twenty-first century to supply sufficient food while minimizing the climatic burdens of already stressed environment (Rasheed et al. 2017; Borrill et al. 2019). The recent visible changes in temperature and rainfall intensity have influenced agricultural production at the regional as well as global level and were associated with climate change phenomenon (Milus et al. 2009; Abberton et al. 2016). These changes have affected soil composition, living biota and agricultural production and threaten the household, national and global food securities (Nelson et al. 2009). Though contradictions exist in both causes as well as nature of climate changes, more recently the ecological consequences have been so evident that many of these disagreements have been resolved (Foley et al. 2011; Valizadeh et al. 2014).

Increasing efficiency of the agricultural system could galvanize sustainable agricultural production. Similarly, crop production is determined by optimum temperature and rainfall, therefore vulnerable to uneven climatic patterns (Milus et al. 2009; Campbell et al. 2016). Studies have shown that the average global temperature has increased and there would be considerable reduction in freshwater resources by the end of the twenty-first century. Likewise, variations in regional rainfall as well as snowfall have been reported, and these changes are likely to intensify in days ahead (Ewert et al. 2005; Stocker et al. 2013; Misra 2014).

Agriculture releases significant amounts of greenhouse gases (GHG) like $CO_2$, $CH_4$, $N_2O$, and halo-carbons to the atmosphere, playing an important role in absorbing the solar radiation (Valizadeh et al. 2014). Agriculture is responsible for the release of about 17–32% of all global anthropogenic GHG emissions (Bellarby et al. 2008). The potential to cut GHG released from agriculture and mitigate future climate changes offers a massive challenge while the focus is set for "zero hunger." Although climate change may benefit some crops in some regions (particularly the areas of the northern widths above 55°), the increased temperatures will eventually reduce crop yields at a large scale particularly in hot and dry areas (Milus et al. 2009; Smith and Gregory 2013; Valizadeh et al. 2014). Additionally, high temperatures will boost weed and pest attacks, thereby declining crop production (Ewert et al. 2005; Nelson et al. 2009).

Undoubtedly, the overall impacts of climate change on agriculture are expected to be harmful, will threaten global agricultural production systems, and ultimately influence food security. Food security is linked directly and indirectly to ecosystems through provisioning and supporting services; climate change will stretch this fine balance (Smith and Gregory 2013; Valizadeh et al. 2014). Although climate change will affect our overall ability to access food, marginalized populations in developing countries, like South and East Asia, are likely to be the worst affected by climate change (Misra 2014; Campbell et al. 2016). It is clear that from the food and ecosystem services, we anticipate that future demands offer more radical changes from production to consumption to policy making (Bellarby et al. 2008; Godfray et al. 2010; Bevan et al. 2017).

## 3 Strategies to Ensure Food Security and Mitigate Climate Change Impacts

### *Investments in Agri-Food Infrastructure and Adaptation of Innovative Farming Practices*

Investments in advance technologies and relevant infrastructure in conjunction with other cost-effective measures could make agricultural sector more productive and sustainable (Hazell and Wood 2007; Bevan et al. 2017; Rasheed et al. 2018). Roughly 30–40% of food is spoiled mainly due to the non-existence of agricultural infrastructure and the scarcity of financing in storage facilities. Moreover, poor transport facilities increase the prices of agricultural inputs and delivery of the harvest into markets (Godfray et al. 2010; Foley et al. 2011). The last few decades have seen remarkable technological innovations in agri-food industry. Robots and unmanned aerial vehicles (UAV) have been developed for farming purposes. The UAVs that are equipped with hyperspectral cameras could analyze crop status from remote centers (Walter et al. 2017). Similarly, nano-technology has emerged with the potential to minimize the adverse effects of agricultural practices on ecosystems, thereby promoting sustainable food security. The nanodevices could confer benefits to the agri-food sector by minimizing leaching, while improving nutrient uptake by plants. Furthermore, it is noteworthy to mention that nanomaterial could also be exploited to improve soil and waste management practices. Nanomaterial could improve pesticide efficiency by providing a more specific release toward target organisms (Mishra and Singh 2015; Fraceto et al. 2016).

Technological advancements and its judicious application hold immense potential, but there are potential risks if the benefits are embroidered alone. Similarly, investments in genetic modification technology and transgenic crop development are valuable, although the "fors and againsts" needed to be rigorously defined before it may substantially contribute toward global food security (Godfray et al. 2010; Walter et al. 2017; Mujeeb-Kazi et al. 2019). There is also a potential role for large-scale agricultural investments in poor countries and that are open to debate. No doubt, fair returns on investment are due, the rise of intellectual property rights raises concerns over the investments of the private sector, particularly for poor countries. Although the external investments in agriculture of developing countries may transform and bring on one side major benefits to crop production and processing, such investments may also be associated with "poverty traps" (Godfray et al. 2010).

## *Crops Adaptation to Limited Agricultural Inputs, Soil Nutrients, and Water Management*

Food production is carried out on almost 38% of earth's surface and having evident impacts on worldwide ecosystems. Still, food sufficiency can be satisfied by improvements in agriculture that is ecologically sustainable (Foley et al. 2011; Kitano et al. 2012). Climatic changes will modify temperature and soil moisture and increase $CO_2$ levels; therefore, agriculture is particularly vulnerable to climate change effects. Avoiding global hunger will require plant adaptation to the extreme conditions and land management to increase resilience of our agricultural systems (Borlaug 2002; Ogbonnaya et al. 2013). Best land management practices including judicious use of nutrients and multiple crops grown in rotation are a must to disrupt the life cycle of pests (Hawkesford et al. 2013; Borrill et al. 2019). Multiple cropping is an adjustment strategy to deal with rainfall, and it allows effective use of soil surface by reducing nutrient losses and erosion. The main objective of soil nutrient management is to enhance the yield and quality of crops and not compromising on environmental aspects (Agus et al. 2016; Walter et al. 2017).

Balance fertilizer applications are essential to increase crop yield as well as resilience to extreme events. Astonishingly, nitrogen production for agriculture accounts for 1.2% of global energy, and about 70% of freshwater supplies are used by agriculture (Foley et al. 2011; Kitano et al. 2012). The excessive use of chemicals is detrimental to crops, soil, and environment, and the ill impacts can be reduced if fertilizers are used in balance and in combination with organic matter to meet the crop needs (Misra 2014; Agus et al. 2016). Soil microbial communities carry out the bulk of decomposition and nutrient recycling activities. Further, changes in precipitation and temperature will modify soil microbial communities and thus the overall ecological niches. Also, plant endophytes (bacteria and fungi) have the potential to increase access of soil nutrients to plants and allow host plants to thrive on nutrient deficient soils, thus having a tremendous role in reducing the negative impacts of agriculture on environment (Ikram et al. 2018).

Similarly, freshwater supply is likely to be one of the main limiting factors for future agriculture; therefore, water efficient crops as well as investment in relevant infrastructure are required (Ogbonnaya et al. 2013; Misra 2014). Nonetheless, excessive amounts of water are also counterproductive and cause poor aeration, inhibiting plant growth. Exclusive of proper management, irrigated agriculture can be devastating to the environment and jeopardize agricultural sustainability. Breeding and selection for higher yield in stress-free conditions have indirectly improved yield in many water-limiting circumstances (Coakley et al. 1999; Godfray et al. 2010; Smith and Gregory 2013). Foremost pathways for enhancing water use efficiency in irrigated agriculture are to increase the output per unit of water. Moreover sprinkler irrigation is extremely useful and has shown promise in increasing the water use efficiency of crops (Ainsworth and Long 2005; Hawkesford et al. 2013; Walter et al. 2017). Yield potential of crops in water stress environments could be achieved by improving agronomy and plant physiology. The recent developments of traits affecting yield under drought have provided candidate genes to understand water use efficiency in much detail, thereby allowing to improve the quality and diversity of crops that are better adaptable to future human needs (Mujeeb-Kazi et al. 2013; Agus et al. 2016; Rasheed et al. 2017). Therefore, large-scale adaptation of crop breeding is essential to ensure food security. Plant breeders have pyramid traits sustaining yield under water-deficient conditions into future genotypes without yield penalties. This strategy will result in smart and better-adapted cultivars with high yield potential and stability under future climatic conditions (Borlaug 2002; Cattivelli et al. 2008; Hawkesford et al. 2013).

## *Improving Agricultural Resilience by Increasing Crop Productivity*

There is a wide range of variation in crop productivity, even across regions with similar agro-climates. To some extent, obtaining higher yields depends on the capacity of farmers to utilize the available resources (Godfray et al. 2010; Tariq et al. 2018). However, in comparing yields, it is important to also consider cropping systems, e.g., the highest wheat yields of over 15 t/ha are recorded for winter wheat (compared to the average of 3 t/ha) with a long growing season. In such cases, maximizing yield is not their sole objective; rather profitability is a more important criteria (Ainsworth and Long 2005; Hawkesford et al. 2013). Substantially more food could be produced with current crops if yield gaps are minimized. Similarly, trait stability, particularly yield and quality attributes, is essential that must be consistent across a range of environments (Masood et al. 2016; Rasheed et al. 2018; Borrill et al. 2019). Low yields may also occur due to technological constraints, for example, farmers may not have access to varieties or the technical skills required to maximize yields. Similarly, high costs to low returns ratio from increased production make it economically sub-optimal to raise production (Cattivelli et al. 2008; Hawkesford et al. 2013).

The tremendous success of the Green Revolution is attributed to the breeding efforts that resulted in the development of F1 maize hybrids and semi-dwarf varieties of wheat and rice with fungal resistance. These varieties were able to withstand more irrigation and fertilizer inputs without being vulnerable to lodging or disease epidemics (Borlaug 2002; Godfray et al. 2010; Mujeeb-Kazi et al. 2017). Augmented yield alone is although a major goal of food security, the importance of water and nutrient application, tolerance to stresses must be recognized. Indeed, ensuring sustainable food security is a multi-faceted challenge, involving much more than just increasing food production; it is also about protecting yield potential as well as increasing resource use efficiency (Hawkesford et al. 2013; Ali et al. 2016; Ikram et al. 2018).

The key to increased productivity is the ability of plants to harvest sunlight energy that regulates the ultimate yield (Hawkesford et al. 2013). The most productive crops like sugarcane convert solar energy into biomass with an efficiency of ~2% when grown in optimum conditions. Thus accelerating the rate of photosynthesis is the simplest way forward to increase yields (Ainsworth and Long 2005). Wheat converts 4.6% of the intercepted radiation into photosynthate where further improvement is at least theoretically possible (Zhu et al. 2010). Further, the $CO_2$-fixing enzyme Rubisco from different species has a good deal of variation in kinetic properties, and exploiting this pathway may deliver higher photosynthetic rates and in higher yields. Although theoretical yield limits for major crops vary greatly, there is clearly considerable scope for increasing production limits, and new models can predict more accurately these complex interactions (Godfray et al. 2010; Hawkesford et al. 2013; Agus et al. 2016).

## *Changing Lifestyles and Food Demand Patterns*

The recent tendencies of healthy life have considerably modified eating behaviors and food preferences. In addition rural migration toward cities for jobs and the "so-called" better lifestyles is shaping urban landscaping (hard infrastructure) and is associated with extension of the urban boundaries (Foley et al. 2011; FAO 2018). In addition to pressures on food web and sustainable ecosystem services, the changing standard of lifestyle and priorities is deeply rooted with the recent shifts in global climatic changes. Besides drastic reduction in net primary productivity, these big urban centers influence emissions of GHG from the transport and other building facilities (Creutzig et al. 2016; Ali and Abdullah 2017). This trend is threatening the socioeconomic stability of regions, especially in mega cities for transportation, social services, and residential settlements. To mitigate the impacts of urbanization on GHG and climate change, urban designing must integrate walking and bicycle lanes and discourage private motorized transport (Ainsworth and Long 2005; Creutzig et al. 2016; Walter et al. 2017).

The conversion efficiency of plant matter into animal is ~10%; thus, it is believed that more vegetarian people could be supported from the same arable land than if they were eating meat. Such dietary shifts alone could reduce emissions of anthropogenic GHG by more than 70% (Godfray et al. 2010; Creutzig et al. 2016). Thus eating behavior alone could reshape urban environment and is likely be a necessary component of transition toward a low-carbon society (Ali and Abdullah 2017). Additionally, consumers in the developed world are purchasing foods of the highest cosmetic standards and litigation on edible products safety; food fit for consumption is thrown away (Smith and Gregory 2013). It may be perceived that hunger is more likely a problem of income distribution rather than that of food shortages. While the hungry cannot afford to buy food, the rich have excessive food intake and suffer from obesity. Thus the global efforts to increase plant productivity may not address this problem (Hazell and Wood 2007). Many solutions of food shortages to climate change mitigation are aligned to changing habits, norms, and behavior, having immense potential for reducing GHG emissions and climate change effects (Godfray et al. 2010; Hawkesford et al. 2013; Walter et al. 2017).

## *Application of the Untapped Genetic Variability and Accelerated Domestication*

Modern agriculture is founded on the cultivation of only a few highly productive crop species that were domesticated from the wild (Mujeeb-Kazi et al. 1989; Tanksley and McCouch 1997; Tariq et al. 2018). Further, in almost all crop species including wheat, new varieties are virtually derived from crosses among genetically related modern varieties – excluding the ancestral species (Ali et al. 2016; Mujeeb-Kazi et al. 2019). Evidences suggest genetic diversity in all crop species has drastically declined during polyploid formation and domestication followed by intensive selection. This loss of useful genetic diversity has inspired maintenance of plant genetic resources in gene banks. International collections and gene banks provide precious repositories for alternative rare alleles of loci that have been exhaustively selected during domestication and modern breeding (Godfray et al. 2010; McCouch et al. 2013; FAO 2018). Virtually all crops including wheat need to be more tolerant to mitigate the impacts of future climate change hazards. Studies have suggested that each degree rise in temperature is associated with a 6% decrease in wheat production (Asseng et al. 2017; Borrill et al. 2019). Hence, it is appealing to increase the genetic base and breed stress-tolerant wheat genotypes deemed to address the

United Nations Millennium sustainability goals (Masood et al. 2016; Rasheed et al. 2017; Mujeeb-Kazi et al. 2019).

It is also documented that favorable introgressions from wild relatives could significantly improve grain yield, adaptability, end-use quality, and disease resistance of wheat as well as other important crops (Tanksley and McCouch 1997; Rasheed et al. 2018). Alien genes have long been recognized as excellent sources of allelic diversity against biotic and abiotic stresses, and high rates of alien introgressions in wheat cultivars indicate the global impact of wild relatives in farmer's fields (Ali et al. 2016). Successful gene transfers in wheat have been achieved from members of diverse genera including (but not limited to) *Triticum, Secale, Aegilops, Hordeum, Thinopyrum, Lophopyrum, Agropyrum, Psathyrostachys, Elymus, Leymus, Dasypyrum*, etc. (see Mujeeb-Kazi and Hettel 1995, Mujeeb-Kazi et al. 2017 for details). In addition to alien gene or chromosomal segments, introduction from related *Triticeae* species, the natural route of wheat polyploidization was exploited at CIMMYT, and synthetic hexaploid wheats (SHW's) were developed. D-genome synthetic wheats are prominent sources of unique and rare alleles for improving adaptation and yield potential in bread wheat. These SHW lines are distributed internationally for the introgression of suitable traits and represent one of the most effective breeding programs on the restoration of lost genetic diversity in wheat (Mujeeb-Kazi and Hettel 1995; Ali et al. 2016). In addition to bread making or end-use quality traits, SWHs undergo homologous pairing across all three genomes (ABD) to increase the genetic base of wheat (Mujeeb-Kazi et al. 1989; Ogbonnaya et al. 2013; Masood et al. 2016; Tariq et al. 2018).

Unlike when the first grain crops were domesticated 10,000 years ago, plant breeders today have an array of modern tools in their pursuit for crop improvement. The recent progress in high-throughput genotyping and phenotyping platforms and their affordability is instrumental to identification of genes in the quest of food security (Tanksley and McCouch 1997; Abberton et al. 2016; Mujeeb-Kazi et al. 2019). The development of high-throughput phenotyping enables appraisal of larger populations, thereby increasing selection accuracy. A key limiting factor in plant breeding was the long generation cycles of crops that has been reduced by "speed breeding," and this will significantly accelerate trait screening and discovery of favorable alleles (Rasheed et al. 2017, 2018).

Domestication has indeed resulted in the extensive appraisal of only a subset of useful genes available in ancestral species among crop cultivars (Ewert et al. 2005; Borrill et al. 2019). With the availability of the todays' inventions and remarkable biological developments, repeating the process of domestication through neo-domestication of wild species could be an alternative way to swiftly breed cultivars (Godfray et al. 2010; Bevan et al. 2017). Other routes to domestication of new species are possible via editing of known domestication genes with CRISPR–Cas9. The CRISPR system has major implications to produce climate-resilient crops (Abberton et al. 2016; Rasheed et al. 2017; Mujeeb-Kazi et al. 2019). Recently unbalanced expression and inheritance of the three wheat homoeologous genomes have been described; deciphering the mechanism may lead to breed improved wheat varieties (see Ramírez-González et al. 2018).

Persistent progresses in wheat and other important crops productivity will be ascertained by integrated approaches of combining genetic improvements supported by agronomy (McCouch et al. 2013; Hickey et al. 2019). Moreover, advances and cost-effectiveness of DNA sequencing and genomic prediction tools have shown incredible potential in plant breeding and improvement. Availability of reference genomes of wheat and other *Triticeae* species will allow insights into origin, evolution, and domestication of these species. Hence, it is likely that varieties of crop species with relevant phenotypic traits may be developed that will enable us to further capitalize on crop productivity and adaptation and address new challenges (King et al. 1997; Valizadeh et al. 2014; IWGSC 2018; Hickey et al. 2019).

## 4 Conclusions

Climate changes have started influencing agricultural production at the regional as well as global level. Preparing agricultural systems for climate change-related impacts would require more resilient agricultural system and investments in relevant infrastructure. Food security is already threatened, and yields of all major cereals have stagnated with wheat showing the lowest rate of increase. Increases in production will have an important part to play in food security; it will be constrained by the finite resources like land and freshwater availability. Noteworthy is to consider that there is no simple solution to sustainably feed ten billion people while aiming diminishing emission of GHGs at the same time. Sustainable food production is not simply to maximizing productivity but also is well connected to optimizing productivity across ecological landscape.

Addressing this challenge will require radical changes in the way food is produced, harvested, handled, and distributed. Unless urgent adaptive measures such as changes in crop growing patterns, eating habits, lifestyle, and innovative technologies are adopted, increases in global food production are likely to be non-sustainable and even counterproductive. Achieving increased crop productivity and adaptation are likely to be the key components of sustainable agriculture and food security. A multifaceted and well-integrated global strategy is demanded from producers to policy makers. The application of the recently evolved high-throughput platforms and genome editing as well as de novo domestication tools could enable plant breeders to identify and use the new genetic variations that may ensure future food security in ways that are environmentally and socially sustainable. We do need to be cognizant that to effectively combat the constraints that influence food productivity. The need is to carefully gauge the coverage on this treatise by the distinguished contributors, harness their conclusions in a well-integrated manner, and prioritize the best applicable modus operandi with an admixture blend of technologies that influences the multiple needs of a productive crop across diverse environments by recognizing the numerous micro- and macro-needs by establishing working teams that operate in unison, work in tandem, and make impacts that will deliver output gains in a timely manner that in recent years have fallen drastically below expectation targets.

## References


Abberton M, Batley J, Bentley A, Bryant J, Cai H, Cockram J, Costa de Oliveira A, Cseke LJ, Dempewolf H, De Pace C, Edwards D (2016) Global agricultural intensification during climate change: a role for genomics. Plant Biotechnol J 14(4):1095–1098

Agus F, Husnain H, Yustika RD (2016) Improving agricultural resilience to climate change through soil management. J Litbang Pert 34(4):147–158

Ainsworth EA, Long SP (2005) What have we learned from 15 years of free-air CO2 enrichment (FACE)? A meta-analytic review of the responses of photosynthesis, canopy properties and plant production to rising CO2. New Phytol 165(2):351–372

Ali N, Abdullah MA (2017) The food consumption and eating behaviour of Malaysian urbanites: issues and concerns. Geografia 8(6)

Ali N, Heslop-Harrison JP, Ahmad H, Graybosch RA, Hein GL, Schwarzacher T (2016) Introgression of chromosome segments from multiple alien species in wheat breeding lines with wheat streak mosaic virus resistance. Heredity 117(2):114–123

Asseng S, Cammarano D, Basso B, Chung U, Alderman PD, Sonder K, Reynolds M, Lobell DB (2017) Hot spots of wheat yield decline with rising temperatures. Glob Chang Biol 23(6):2464–2472

Bellarby J, Foereid B, Hastings A, Smith P (2008) Cool farming: climate impacts of agriculture and mitigation potential. Greenpeace International, Amsterdam

Bevan MW, Uauy C, Wulff BB, Zhou J, Krasileva K, Clark MD (2017) Genomic innovation for crop improvement. Nature 543(7645):346

Borlaug NE (2002) Feeding a world of 10 billion people: the miracle ahead. In Vitro Cell Dev Biol Plant 38(2):221–228

Borrill P, Harrington SA, Uauy C (2019) Applying the latest advances in genomics and phenomics for trait discovery in polyploid wheat. Plant J 97(1):56–72

Campbell BM, Vermeulen SJ, Aggarwal PK, Corner-Dolloff C, Girvetz E, Loboguerrero AM, Ramirez-Villegas J, Rosenstock T, Sebastian L, Thornton PK, Wollenberg E (2016) Reducing risks to food security from climate change. Glob Food Sec 11:34–43

Cattivelli L, Rizza F, Badeck FW, Mazzucotelli E, Mastrangelo AM, Francia E, Marè C, Tondelli A, Stanca AM (2008) Drought tolerance improvement in crop plants: an integrated view from breeding to genomics. Field Crop Res 105(1–2):1–14

Coakley SM, Scherm H, Chakraborty S (1999) Climate change and plant disease management. Annu Rev Phytopathol 37(1):399–426

Creutzig F, Fernandez B, Haberl H, Khosla R, Mulugetta Y, Seto KC (2016) Beyond technology: demand-side solutions for climate change mitigation. Annu Rev Environ Resour 41:173–198

Ewert F, Rounsevell MDA, Reginster I, Metzger MG, Leemans R (2005) Future scenarios of European agricultural land use. I. Estimating changes in crop productivity. Agric Ecosyst Environ 107:101–116

FAO STAT (2018) FAO online database. Available at: http://www.fao.org/faostat/en/#data. Last access: May 2019

Foley JA, Ramankutty N, Brauman KA, Cassidy ES, Gerber JS, Johnston M, Mueller ND, O'Connell C, Ray DK, West PC, Balzer C (2011) Solutions for a cultivated planet. Nature 478(7369):337

Fraceto LF, Grillo R, de Medeiros GA, Scognamiglio V, Rea G, Bartolucci C (2016) Nanotechnology in agriculture: which innovation potential does it have? Front Environ Sci 4:20

Godfray HCJ, Beddington JR, Crute IR, Haddad L, Lawrence D, Muir JF, Pretty J, Robinson S, Thomas SM, Toulmin C (2010) Food security: the challenge of feeding 9 billion people. Science 327(5967):812–818

Hawkesford MJ, Araus JL, Park R, Calderini D, Miralles D, Shen T, Zhang J, Parry MA (2013) Prospects of doubling global wheat yields. Food Energy Secur 2(1):34–48

Hazell P, Wood S (2007) Drivers of change in global agriculture. Philos Trans R Soc B Biol Sci 363(1491):495–515

Hickey LT, Hafeez AN, Robinson H, Jackson SA, Leal-Bertioli SC, Tester M, Gao C, Godwin ID, Hayes BJ, Wulff BB (2019) Breeding crops to feed 10 billion. Nat Biotechnol, p 1

Ikram M, Ali N, Jan G, Jan FG, Rahman IU, Iqbal A, Hamayun M (2018) IAA producing fungal endophyte Penicillium roqueforti Thom., enhances stress tolerance and nutrients uptake in wheat plants grown on heavy metal contaminated soils. PLoS One 13(11):e0208150

International Wheat Genome Sequencing Consortium (IWGSC) (2018) Shifting the limits in wheat research and breeding using a fully annotated reference genome. Science 361(6403):eaar7191

King IP, Forster BP, Law CC, Cant KA, Orford SE, Gorham J, Reader S, Miller TE (1997) Introgression of salt-tolerance genes from Thinopyrum bessarabicum into wheat. New Phytol 137(1):75–81

Kitano M, Inoue Y, Yamazaki Y, Hayashi F, Kanbara S, Matsuishi S, Yokoyama T, Kim SW, Hara M, Hosono H (2012) Ammonia synthesis using a stable electride as an electron donor and reversible hydrogen store. Nat Chem 4(11):934

Liu Y, Guo Y, Zhou Y (2018) Poverty alleviation in rural China: policy changes, future challenges and policy implications. China Agric Econ Rev 10(2):241–259

Masood R, Ali N, Jamil M, Bibi K, Rudd J, Mujeeb-Kazi A (2016) Novel genetic diversity of the alien D-genome synthetic hexaploid wheat (2n=6x=42, AABBDD) germplasm for various phenology traits. Pak J Bot 48:2017–2024

McCouch S, Baute GJ, Bradeen J, Bramel P, Bretting PK, Buckler E, Burke JM, Charest D, Cloutier S, Cole G, Dempewolf H (2013) Agriculture: feeding the future. Nature 499(7456):23

Milus EA, Kristensen K, Hovmøller MS (2009) Evidence for increased aggressiveness in a recent widespread strain of *Puccinia striiformis* f. sp. tritici causing stripe rust of wheat. Phytopathology 99:89–94

Mishra S, Singh HB (2015) Biosynthesized silver nanoparticles as a nanoweapon against phytopathogens: exploring their scope and potential in agriculture. Appl Microbiol Biotechnol 99(3):1097–1107

Misra AK (2014) Climate change and challenges of water and food security. Int J Sustain Built Environ 3(1):153–165

Mujeeb-Kazi A, Hettel GP (eds) (1995) Utilizing wild grass biodiversity in wheat improvement: 15 years of wide cross research at CIMMYT (No. 2). CIMMYT

Mujeeb-Kazi A, Roldan S, Suh DY, Ter-Kuile N, Farooq S (1989) Production and cytogenetics of *Triticum aestivum* L hybrids with some rhizomatous species. Theor Appl Genet 77:162–168

Mujeeb-Kazi A, Kazi AG, Dundas I, Rasheed A, Ogbonnaya F, Kishii M, Bonnett D, Wang RRC, Xu S, Chen P, Mahmood T (2013) Genetic diversity for wheat improvement as a conduit to food security. Adv Agron 122:179–257

Mujeeb-Kazi A, Ali N, Ibrahim A, Napar AA, Jamil M, Hussain S, Mahmood Z, Delgado R, Rosas V, Cortes A, Rajaram S (2017) Tissue culture mediated allelic diversification and genomic enrichment of wheat to combat production constraints and address food security. Plant Tissue Cult Biotechnol 27(1):89–140

Mujeeb-Kazi A, Munns R, Rasheed A, Ogbonnaya FC, Ali N, Hollington P, Dundas I, Saeed N, Wang R, Rengasamy P, Saddiq MS (2019) Breeding strategies for structuring salinity tolerance in wheat. Adv Agron 155:121–187

Mundial B (2018) Poverty and shared prosperity 2018: piecing together the poverty puzzle. Grupo Banco Mundial, Washington, DC

Nelson GC, Rosegrant MW, Koo J, Robertson R, Sulser T, Zhu T, Ringler C, Msangi S, Palazzo A, Batka M, Magalhaes M (2009) Climate change: Impact on agriculture and costs of adaptation. Intl Food Policy Res Inst

Ogbonnaya FC, Abdalla O, Mujeeb-Kazi A, Kazi AG, Xu SS, Gosman N, Lagudah ES, Bonnett D, Sorrells ME (2013) Synthetic hexaploid in wheat improvement. In: Janick J (ed) Plant breeding reviews, vol 37, pp 35–122

Ramírez-González RH, Borrill P, Lang D, Harrington SA, Brinton J, Venturini L, Davey M, Jacobs J, Van Ex F, Pasha A, Khedikar Y (2018) The transcriptional landscape of polyploid wheat. Science 361(6403):eaar6089

Rasheed A, Mujeeb-Kazi A, Ogbonnaya FC, He Z, Rajaram S (2017) Wheat genetic resources in the post-genomics era: promise and challenges. Ann Bot 121(4):603–616

Rasheed A, Ogbonnaya FC, Lagudah E, Appels R, He Z (2018) The goat grass genome's role in wheat improvement. Nat Plants 4(2):56–58

Smith P, Gregory PJ (2013) Climate change and sustainable food production. Proc Nutr Soc 72(1):21–28

Stocker TF, Qin D, Plattner GK, Tignor M, Allen SK, Boschung J, Nauels A, Xia Y, Bex V, Midgley PM (2013) Climate change 2013: the physical science basis. Cambridge University Press, Cambridge

Tanksley SD, McCouch SR (1997) Seed banks and molecular maps: unlocking genetic potential from the wild. Science 277(5329):1063–1066

Tariq MJ, Shah MKN, Hassan MU, Sajjad M, Jamil M, Ali N, Kazi AM (2018) Prevalence of higher glutenin variation in synthetic wheat germplasm. J Anim Plant Sci 28(2):568–575

Valizadeh J, Ziaei SM, Mazloumzadeh SM (2014) Assessing climate change impacts on wheat production (a case study). J Saudi Soc Agric Sci 13(2):107–115

Walter A, Finger R, Huber R, Buchmann N (2017) Opinion: smart farming is key to developing sustainable agriculture. Proc Natl Acad Sci 114(24):6148–6150

Zhu XG, Long SP, Ort DR (2010) Improving photosynthetic efficiency for greater yield. Annu Rev Plant Biol 61:235–261

# Reduced-Immunogenicity Wheat Now Coming to Age

**Sachin Rustgi, Samneet Kashyap, Lomme J. Deleu, and Jan A. Delcour**

**Abstract** This chapter focuses on the gluten-induced dietary disorders, conceived therapies, and hysteria associated with wheat/gluten consumption. Gluten proteins are one of the most widely consumed dietary proteins in the world and also the sole source of nutrition to many, especially those dwelling in developing countries. Prevalence of these disorders has compounded in the last couple of decades due to change in lifestyle, which includes an adaptation of the gluten-laden diet and excessive use of antibiotics in childhood with a suppressive effect on the development of the immune system and the improvements in diagnostics. Several therapies have been sought, but none of them has proven perfect. The issues associated with gluten-induced disorders and existing and possible therapies and prospects will be discussed under the following headings and subheadings.

**Keywords** Wheat · Celiac disease · Wheat allergy · Non-celiac wheat sensitivity · Reduced-immunogenicity wheat · Epitopes

## 1 Introduction

Wheat is a global staple and the second most-produced crop in the world after corn. In terms of calorific and nutritional output, wheat stands even before corn (Langridge 2017). It is the primary source of plant proteins in the most resource-deprived and

S. Rustgi (✉)
Department of Plant and Environmental Sciences, School of Health Research, Clemson University Pee Dee Research and Education Centre, Florence, SC, USA

Department of Crop and Soil Sciences, Washington State University, Pullman, WA, USA
e-mail: srustgi@clemson.edu

S. Kashyap
Department of Plant and Environmental Sciences, School of Health Research, Clemson University Pee Dee Research and Education Centre, Florence, SC, USA

L. J. Deleu · J. A. Delcour
Laboratory of Food Chemistry and Biochemistry, KU Leuven Food Science and Nutrition Research Centre (LFoRCe), Leuven, Belgium

© The Editor(s) (if applicable) and The Author(s), under exclusive license to Springer Nature Switzerland AG 2021

S. H. Wani et al. (eds.), *Physiological, Molecular, and Genetic Perspectives of Wheat Improvement*, https://doi.org/10.1007/978-3-030-59577-7_2

populated parts of the world (Langridge 2017). The common wheat is an outcome of the human selection of a natural hybrid of domesticated tetraploid wheat (emmer) and a wild diploid goat grass relative (Shewry 2019). Hence it is relatively a young species (Venske et al. 2019), which is evolving slowly under the intense selection pressure for enhanced yield and end-use quality. The intensive breeding as a consequence has narrowed the genetic base of elite wheat germplasm and also reduced the possibility to select for specific traits.

Gluten is a complex of seed storage proteins with unique structural and compositional properties (Rustgi et al. 2019). These proteins consist of repetitive tracts of proline and glutamine residues, which confer them unusual resistant to digestion. However, this unique composition of gluten proteins is inherently beneficial to the plant, as it allows dense packing of nitrogen in grains for use during germination and by making the grain less attractive to insect pest due to poor digestibility (Shewry 2019).

In the last few decades, a significant increase in the number of cases with gluten-associated disorders was reported. This increase in the number of cases with gluten-associated disorders could be attributed to many factors: (i) a dramatic change in the eating habits, which could be witnessed by the spread of celiac disease to areas where wheat is not grown or consumed historically; (ii) increasing adaptation of the plant-based diets and also fast foods enriched in gluten, due to affordability, convenience, durability in transport, etc.; (iii) better diagnostics and increasing public awareness; and (iv) however controversial, the underdeveloped immune system due to excessive use of antibiotics (Rustgi et al. 2019).

Since the gluten-associated disorders affect about 7–10% of the world population, and this number is increasing, a permanent and more affordable solution should be sought. Therefore, to promote research in this area, an effort has been made to summarize the current knowledge in this field of research.

## 2 Wheat Gluten Proteins

Gluten proteins contain two major fractions, the monomeric gliadins (30–80 kDa) and the polymeric glutenins (up to 20 MDa) (Delcour et al. 2012). Gliadins are monomeric and have a more or less globular shape (Veraverbeke and Delcour 2002). They are soluble in aqueous ethanol and thus classified as prolamins (Osborne 1907). An important difference between glutenins and gliadins is that the latter have no free sulfhydryl (SH) groups (Shewry et al. 1986).

Gliadins can be further subdivided in three groups: α-, γ-, and ω-gliadins (Shewry et al. 1986; Balakireva and Zamyatnin 2016; Shewry 2019). Only the first two types have intramolecular disulfide (SS) bonds (respectively, 6 and 8 in α- and γ-gliadins). The intramolecular SS bonds are found in highly conserved regions, which makes them unaccessible for SH/SS exchange reactions at room temperature (e.g., during dough mixing) (Muller and Wieser 1995, 1997). However, during heat treatments, they can become involved in intermolecular SS bonds (see below). Omega-gliadins have no cysteine amino acid residues and are believed to have a stiff coil structure (Shewry et al. 2009). In general, gliadins are rich in glutamine, proline, asparagine, and arginine (Muller and Wieser 1997).

Glutenins, due to their large size, are not soluble in mild media. They consist of different glutenin subunits (GS), the structures and solubility of which are comparable to those of gliadins, but they do contain free SH groups. With their SH groups, the GS form intermolecular SS bonds which are at the basis of the polymeric glutenin structure in mature wheat. There are two types of GS: low molecular weight-GS (LMW-GS) and high molecular weight-GS (HMW-GS).The LMW-GS show high similarities with α- and γ-gliadins (but as stated above, they do have free SH groups) and can be further subdivided in different subcategories (Delcour et al. 2012).

HMW-GS also have free SH groups. They are important contributors to the elasticity of gluten networks, even if they only occur in small numbers (Gianibelli et al. 2001). These subunits are rich in glutamine, proline, and glycine (Shewry et al. 1992).

The huge variation in both the amount and the occurrence of the different types of gliadins and of GSs is an important element at the base of the distinction between good and poor bread making quality wheat (Veraverbeke and Delcour 2002).

A wide range of proteins with similarities to gluten proteins at sequence or structure levels were identified. These proteins are collectively grouped under the "prolamin-superfamily". These proteins generally show homology to gluten proteins in the non-repetitive cysteine-rich N- and C-terminal domains and perform diverse metabolic or structural roles in grains or other plant parts. Small but some effect of these proteins on the processing quality was also reported (Shewry 2019). Among these proteins, amylase trypsin inhibitors (ATIs) and lipid transfer proteins (LTPs) were shown to be involved in gluten-associated disorders (Juhász et al. 2018).

## 3 Gluten-Associated Disorders

A large number of epitopes belonging to all families of gluten proteins have been shown to elicit various reactions in different individuals, which correspond with their genetic constitutions. In other words, different celiac patients are sensitive to different gluten proteins (Koning 2012). Despite extensive efforts, the repertoire of epitopes is still incomplete. So far, 356 genes with known epitopes and an additional 472 potential allergen genes were assigned to the wheat genome. Of these 356 genes, 226 belong to the prolamin gene superfamily (Juhász et al. 2018). Of all the epitopes with a known immune response (determined based on the IFNγ-ELISpot assay), 25 mapped to the HMW glutenin subunits, and only 1 of these 25 epitopes was shown to trigger a medium immune response (SFU value between 10 and 20). The rest of the epitopes were reported to have weak immune reactions (SFU values of less than ten) (Juhász et al. 2018).

Similarly, all epitopes that mapped to sequences of the LMW glutenin subunits were known to have a weak immune response. It suggests that all families of gliadins (α-, γ-, and ω-gliadins) are highly immunoreactive and especially the one mapping to D- and A-subgenomes of wheat and related species. The epitopes that map to the repetitive domain in the gliadin sequences were more immunoreactive than the one mapping to the C-terminal non-repetitive domain. The epitopes rarely mapped to the N-terminal non-repetitive domain of prolamin sequences (Tye-Din et al. 2010; Juhász et al. 2018).

As mentioned earlier, gluten intake in sensitive individuals could manifest diverse symptoms – cutaneous, gastrointestinal, or neurological – and these reactions could be from mild to fatal (Brouns et al. 2019). The symptoms can be widely classified into celiac disease, wheat allergy, and wheat sensitivity (Sapone et al. 2012) (Fig. 1). The manifestation of celiac disease in an individual depends primarily on the three factors: (i) the environmental trigger, which is exposure to gluten and related proteins of the prolamin superfamily (Rustgi et al. 2019; Shewry 2019); (ii) gut abnormalities, i.e., leaky intestine (Fasano 2009); and iii) genetic predisposition, i.e., the presence of susceptibility alleles (Fig. 2) (Brouns et al. 2019).

The adaptive immune system mediates celiac disease (gluten intolerance). If left untreated, it induces the production of antibodies against the indigestible gluten peptides and also against a housekeeping enzyme, tissue transglutaminase 2 (tTG2) (Brouns et al. 2019). The tTG2 is also responsible for chemical modification of gluten peptides, which facilitates their recognition as foreign entities by the immune system. But the faulty immune system in genetically predisposed individuals recognizes tTG2 as an enemy and triggers an autoimmune response (Osorio et al. 2012).

Given the parallelism between the gluten peptides and living (bacteria) or non-living (prions and viruses) pathogens, Dr. Chaitan Khosla of Stanford University, a pioneer in the oral enzyme therapy for celiac disease considered gluten peptides as the non-replicating pathogens. Since the gluten peptides like pathogens evade

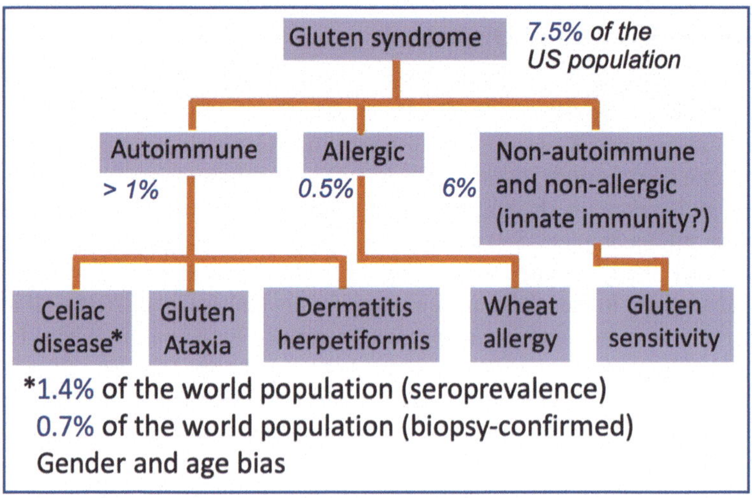


**Fig. 1** Gluten-associated dietary disorders and the present US population affected by these disorders. (Modified from Sapone et al. 2012)

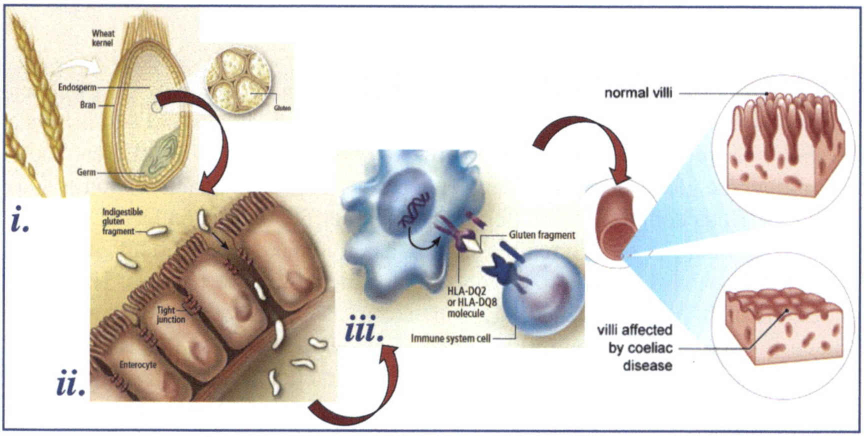


**Fig. 2** A trio of factors responsible for celiac disease. (Modified from Fasano 2009)

"host" defenses by escaping digestion through gastrointestinal enzymes, invade intestinal epithelium, take a more aggressive form after chemical modification by tTG2, and trigger the cascade of reaction leading to the intestinal and extraintestinal symptoms (Bethune and Khosla 2008). As stated above, the first reaction initiated by gluten peptides gets amplified to take a more aggressive form of an autoimmune disorder upon recognition of tTG2 by the immune system as antigen to develop autoantibodies against it, which cause damage to the intestine and other tissues. The second kind of reaction is a wheat allergy, which involves both innate and adaptive immune systems. It is a quick reaction against the external allergen within a couple of minutes to hours after ingestion, which results in various symptoms including dermatitis, anaphylaxis, and various other symptoms (Tatham and Shewry 2008). The third kind of reaction known as gluten sensitivity is very complex and least understood. It involves the innate immune system, and the symptoms associated with this reaction are quite diverse, ranging from fatigue, distress, depression, migraines to gastrointestinal symptoms (Sapone et al. 2012). The trigger to the latter reaction is yet unknown and has been recently suggested to be fermentable oligo-, di-, monosaccharides, and polyols (FODMAPs) that coexist with gluten in wheat grains (Skodje et al. 2018; Verbeke 2018; Brouns et al. 2019).

To sum up, wheat and derived products elicit many diet-induced health issues in more than 7.5–10% of the population in some countries (Rosella et al. 2014; Aziz et al. 2015; Golley et al. 2015). In particular, the celiac disease alone affects more than 71 million individuals around the globe (i.e., ~1% of the world population), which makes it one of the most devastating disorders of the gastrointestinal tract (Bai et al. 2013). There is no known therapy for these disorders other than the strict lifelong adherence to wheat (gluten)-exclusion diet, which has associated side effects (Rustgi et al. 2019). Given the high prevalence of gluten-induced disorders in all studied populations throughout the globe, a large number of studies have been dedicated to finding more effective therapies for these disorders.

## 4 Gluten Threshold

A gluten-free diet does not necessarily signify "zero gluten" as low levels of gluten are generally tolerated by gluten intolerant and sensitive individuals. Establishing a threshold for gluten intake is of high interest to regulatory bodies of different countries around the globe and also to develop methods of precise quantification of gluten from various commodities. After a large number of studies conducted globally and the meta-analysis of the credible studies, a daily intake of less 50 mg gluten for an extended period was found to be generally tolerated by celiac patients. Therefore, a threshold of 20 ppm (20 mg in a kg), which restricts the daily intake of gluten from "gluten-free" food far below 50 mg, was considered safe. This decision on the threshold depended not only on the maximum tolerable dose of gluten in food but also on the amount of "gluten-free" product(s) consumed daily in different parts of the world. In this respect, the current limit of 20 ppm allows a safety margin for variation in the gluten sensitivities and dietary habits of different patients. Therefore, now, most of the countries around the world have adopted the ≤20 ppm limit recommended by the Codex Alimentarius Commission (Brouns et al. 2019).

## 5 Gluten Detection Methods

Over the years, several gluten detection and quantification methods have been developed and tested using the gluten-containing and/or spiked samples. These methods can be grossly partitioned into immunological and non-immunological methods. The non-immunological methods rely on the physical and biochemical properties of the gluten proteins and involve several different methods including the Kjeldahl and the Dumas combustion method, which are very restrictive and can only be applied to test the wheat starches used in the preparation of the gluten-free products. These methods rely on the determination of nitrogen content, which should stay below 0.05% on the dry matter basis. Other assays include the polymerase chain reaction (PCR), which relies on the determination of specific DNA and is more sensitive by several orders of magnitude in comparison with protein assays. Some research groups suggested that PCR shows 10–30 times more sensitivity than ELISA (Koppel et al. 1998; Henterich et al. 2003). Albeit PCR-based assays are a highly sensitive tool for gluten analysis in comparison with ELISA and/or Western blotting, these cannot be applied to the hydrolyzed products such as beer, syrup, and malt extracts for determination of their gluten content.

The relatively more direct and precise method for gluten detection and quantification is matrix-assisted laser desorption/ionization time-of-flight mass spectrometry (MALDI-TOF MS), which can simultaneously measure protein and protein hydrolysate ranging in size from 1000 to 100,000 Daltons without a need of chromatographic purification. In addition, this technique allows reliable determination of protein levels as low as 0.01 mg/ml in the food samples (Camafeita et al. 1997,

1998; Iametti et al. 2005, 2006). Although MALDI-TOF MS is a highly sensitive non-immunological approach for detection and quantification of gluten contamination in food samples, its routine application is constrained by the considerable sample processing cost and the requirement of the specialized equipment.

Another approach that has extensively been used for characterization, separation, and quantification of the cereal protein fractions is column chromatography. Among chromatographic methods, gel permeation (GP) chromatography, which separates proteins based on their molecular weights, and reversed-phase (RP) chromatography that separates proteins according to their hydrophobicities have been used often. These procedures have advantages in terms of speed (usually 30 min) and detection capacity, which is as low as 1–2 μg for gluten (Weiser and Seilmeier 2003). Although this method can be used to determine gluten contamination reliably, it has the disadvantage of being unable to differentiate between gluten and non-gluten proteins in the analysis of complex food products.

The more versatile and commonly accepted assays are immunological assays in particular ELISA. Owing to the sensitivity and speed of detection, the Codex Committee on Methods of Analysis and Sampling has endorsed these methods. Several variations of these methods have been developed over the years (extensively reviewed in Scherf and Poms 2016). A number of antibodies (monoclonal and polyclonal) and a variety of commercial kits are available in the market to perform these assays. The commonly used ELISA systems can be grossly divided into two categories: the sandwich ELISA and the competitive ELISA. The sandwich ELISA is only suitable for large antigens because the antigen should have at least two spatially separated epitopes to bind both of the antibodies. Thus, this ELISA system is not an appropriate choice when working with partially hydrolyzed gluten samples like in the sourdough products, malt, and beer, whereas the competitive ELISA is suitable for the detection of small-sized antigens with a single epitope. The major problem associated with both of the ELISA systems is the determination of gluten contamination in heat-processed food samples, which cause conformational changes to the antigen masking or modifying the antibody recognition site(s). It has been documented that the α/β- and γ-gliadins by the heat treatment lose 49–67% of the original reactivity, while the ω-gliadins remain mostly unaffected, i.e., they only lose reactivity by 7% (Ellis et al. 1994; Rumbo et al. 2001). A detailed list of commercially available prolamin detection kits and specifications can be found in Scherf and Poms (2016) and Osorio et al. (2019a).

Recently, aptamers have emerged as an alternative to antibodies because these molecules can overcome the limitations of using antibodies for the detection, identification, and quantification of specific targets (Song et al. 2012). The aptamers are "single-stranded oligonucleotides that can bind proteins, small-molecules, and living cells with high affinity and specificity" (Berezovski et al. 2006). In the later years, aptamers against the immunodominant 33-mer peptide of α2-gliadin have been developed and successfully used in a variety of assays for gluten quantification. Specifically, the 33-mer peptide-specific aptamers dubbed "Gli4" showed a gluten detection limit of 0.5 ppm, but it failed to detect gluten in heat-treated and hydrolyzed food samples, whereas "Gli1" worked better on such samples, but exhibited a detection limit of 5 ppm (Amaya-Gonzalez et al. 2014, 2015; Pinto et al. 2014; López-López et al. 2017; Malvano et al. 2017).

## 6 Approaches to Reduce Gluten-Exposure in Sensitive Individuals

So far, the only approved prescription for the gluten-associated disorders is adherence to a gluten-free diet. However, following a gluten-free lifestyle is challenging. And as mentioned earlier, it is not without penalties. For instance, (i) strict adherence to a diet devoid of gluten-containing grains deteriorates gut health by its negative influence on the gut microbiota, and (ii) long-term adherence to carbohydrate-rich gluten-free diet results in multiple deficiencies and change in the patient's body mass index (BMI). Therefore, a significant effort has been put in developing therapies for these disorders. The treatments in development for gluten-associated disorders can be grossly divided into dietary and non-dietary approaches, which are discussed below.

### *Dietary Procedures*

The approaches which are preventive or prophylactic are grouped under this category. These approaches include the use of reduced-gluten wheat genotypes or gluten detoxification methods. And each of these approaches is elaborated in the following headings.

#### Screening of Wheat Germplasm

A body of research has suggested that any gluten peptide larger than nine amino acids can elicit an immune reaction in the susceptible individuals (Osorio et al. 2012). Therefore, no wheat genotype either new or old wheat varieties, landraces, or diploid/tetraploid wheat progenitors could be considered safe for celiac patients (Mitea et al. 2010; Goryunova et al. 2012; Brouns et al. 2013; Shewry 2018). The wide genetic screens performed on wheat and related species using immunological and non-immunological methods to study allergenicity and antigenicity of these genotypes supported this conclusion. The immunological methods used were ELISA and the T-cell assays, whereas the non-immunological methods used were based on sequence analysis, gene/transcript sequencing, and gluten profiling (cf. Rosella et al. 2014; Gilissen et al. 2014).

These studies conclusively revealed that gliadins are ubiquitously present in all wheat lines and related wild species. Also, seeds of certain ancient tetraploid wheat types like Graziella Ra, Khorasan, or Kamut have shown to have even higher amounts of total gliadin than modern accessions (Colomba and Gregorini 2012; Brouns et al. 2013), therefore deemed unsuitable for celiac patients (Gregorini et al. 2009; Shewry 2018). However, based on limited data, Pizzuti et al. (2006) proposed that the diploid Einkorn wheat (*Triticum monococcum*) is non-toxic for celiac patients, but later studies revealed its unsuitability for consumption by celiac patients (Kasarda 2007; Vaccino et al. 2009; Gianfrani et al. 2012). Similarly, none of the tetraploid durum wheat (van den Broeck et al. 2010a; Salentijn et al. 2013) and hexaploid wheat genotypes (Molberg et al. 2005; van Herpen et al. 2006; van den Broeck et al. 2010b; Gilissen et al. 2014) were found suitable for general use by celiac patients. To sum up, after careful scrutiny of the facts, it would be safe to say that all wheat and related species such as barley, rye, triticale, tritordeum, and their hybrids are immunogenic and should be avoided by celiac patients (Rustgi et al. 2019).

## Screening of the Genetic Stocks of Wheat and Related Cereals

Wheat cultivar 'Chinese Spring'-derived nulli-tetrasomic and deletion lines lacking a specific chromosome or chromosome segment were screened for their immunogenic potential. As expected, these genotypes showed low toxicity with gliadin-specific antibodies and under the T-cell-based assays, due to the lack of particular gliadin loci (Ciclitira et al. 1980a, b; Frisoni et al. 1995; van den Broeck et al. 2009, 2011). However, concerning the technological properties of these lines, mixed results were obtained. The results showed that deleting the α-gliadin locus from the short arm of chromosome 6 of the D genome leads to substantial loss in dough mixing and rheological properties. However, deleting the ω-gliadin, γ-gliadin, and LMW glutenin subunit loci from the short arm of chromosome 1D showed little to no effect on the technological properties (van den Broeck et al. 2009).

A large number of wheat genotypes in both winter and spring backgrounds and different market classes (hard, soft, red, and white) were bred to carry a reciprocal chromosome translocation involving wheat chromosome arm 1BS [with loci for ω- and γ-gliadins (*Gli1*) and LMW glutenin subunits (*Glu3*)] and rye chromosome arm 1RS (Lukaszewski 2015). The rye chromosome arm carries genes for resistance to three major rust diseases of wheat, grain yield, and the *Sec1* locus that encodes ω-secalins. This translocation was primarily bred in wheat for the agronomical advantage, but it was later realized to damage the technological properties (Lukaszewski 2015). Specifically, the dough made from some 1BL/1RS hard wheat lines was found unacceptable for breadmaking purposes because of excessive stickiness and mixing intolerance (Schwarzlaff et al. 2001). The inheritance of secalin proteins from rye and absence of glutenin subunits in these genotypes was suggested as a possible explanation for the sticky dough phenotype (Barbeau et al. 2003). However, higher amounts and/or differences in the composition of cell wall polysaccharides, β-glucans, and pentosans and/or the presence of a ferulic acid ester were later suggested as other possible explanations (Barbeau et al. 2003). Besides the sticky dough phenotype in hard wheat lines, the 1BL/1RS translocation has been shown to reduce cookie diameter in soft wheat lines.

Upon 2D-PAGE gel analysis of 1BL/1RS translocation lines, eight protein spots were explicitly found in these genotypes; at the same time, 16 other spots were found missing. And another 12 protein spots, which were present in both regular wheat and the translocation lines showed either up- or downregulation. Out of these 12 spots, a highly overexpressed spot in translocated genotypes was identified as a γ-gliadin. It suggested that overexpression of a γ-gliadin compensates for the lack of LMW subunits in translocation lines. Also, a spot that was absent from the translocation line was identified as an α-amylase inhibitor, which was also proposed as a candidate for the sticky dough phenotype observed in the translocation lines (Gobaa et al. 2007).

Recent studies have revealed that all ω-secalins are enriched in tetrapeptide, PQQP, commonly present in celiac disease-associated epitopes. It suggested that ω-secalins can potential have celiac toxicity. A more recent study suggested that besides immunodominant and toxic epitopes, ω-secalin encodes a decapeptide QQPQRPQQPF that prevents K562(S) cell agglutination and celiac mucosa immune activation induced by toxic gliadins (De Vita et al. 2012). Therefore, identification of this immunomodulatory gliadin sequence, naturally occurring in wheat cultivars toxic for celiac patients, might offer new therapeutic strategies for celiac disease.

Wheat mutants lacking α/β-, γ-, and/or ω-gliadins and/or showing reduced accumulation to complete elimination of specific gliadins and/or LMW glutenin subunits were identified (Rustgi et al. 2019). Among these genotypes, the ω-gliadins-free genotype "3xN" (*Gli-B1*, *Gli-A1*, and *Gli-D1* null) developed by intercrossing of mutant lines lacking particular ω-gliadin groups and a genotype lacking almost all gliadins "TeM1" (*Gli-B1*, *Gli-D1*, *Gli-A2*, and *Gli-D2* null) deserve specific mention. These genotypes are not glute-free, albeit when 3xN was tested with the sera derived from the patents with wheat allergy showed a significant reduction in allergenicity (Waga and Skoczowski 2014; Skoczowski et al. 2017). Similarly, when peptic-tryptic digest of prolamins from TeM1 was tested for toxicity in celiac disease via monitoring the agglutinating activity against human myelogenous leukemia K562(s) cells, 3.5-fold more (572.5 mg/L) prolamin digest in comparison to the single mutants (161.5 mg/L) was tolerated (Pogna et al. 1998). Albeit the observed reductions in allergenicity and antigenicity of these genotypes are remarkable, these genotypes are still unsafe for consumption by celiac patients (Rustgi et al. 2019).

Similar reduced-gluten (hordein) mutants were also identified in barley. However, these mutants were initially selected for their high lysine content, which is an important trait in feed barley (Rustgi et al. 2019). One such low hordein barley mutant is *Risø 1508*; it is also known as *sex3c* (shrunken endosperm xenia) due to its shrunken endosperm and altered carbohydrate profile (Munck 1992). The mutant *Risø 1508* completely lacks class C hordeins and accumulates considerably reduced amount of class B hordeins (200 ppm hordein, 100-fold less than control). When this mutant was fed to gluten-sensitive rhesus macaques (*Macaca mulatta*), remission of the anti-gliadin antibody serum responses and improvement of clinical diarrhea were observed. However, the subjects never showed complete recovery. Hence the authors of the study concluded that "the reduced gluten barley diet might be used for the partial improvement of gluten-induced disease, but its therapeutic value still requires upgrading" (Sestak et al. 2015). Recently, Tanner and co-workers developed ultra-low gluten (ULG) barley genotype using this mutant in a cross-breeding approach with another reduced-gluten barley mutant (Tanner et al. 2016) and achieved almost zero gluten status. However, given the large number and complexity of the gliadin genes in wheat and their inheritance in blocks, the possibility of pyramiding all low toxicity gliadin genes in a single wheat variety seems remote (Koning 2012).

## Other Cereals and Non-cereals as an Alternative

Other than wheat, some individuals show sensitivity to oat gluten proteins (avenins) and, in rare cases, to even maize gluten proteins dubbed zeins (Comino et al. 2013; Rosella et al. 2014; Ortiz-Sánchez et al. 2013). However, all oat varieties are not immunogenic. So far, two cereals, which are unequivocally accepted for celiac patients' consumption, are rice and sorghum (Rosella et al. 2014; Pontieri et al. 2013). But, the rice kernels have low protein and fiber content and are highly enriched in easily digestible carbohydrates that may contribute to the high glycemic index. The rice kernels also tend to sequester large quantities of arsenic (Rosella et al. 2014; Da Sacco et al. 2013; Munera-Picazo et al. 2014), and its grain storage proteins (other than prolamins and glutelins) are reported to trigger a variety of allergic reactions (asthma, atopic dermatitis, diarrhea, and anaphylaxis) in different individuals (Matsuda et al. 2006; Nambu 2006; Trcka et al. 2012; Gilissen et al. 2014). Therefore, rice is not the best choice for consumption by celiac patients. Sorghum is primarily used as animal feed in the Western countries, albeit in many parts of Africa and Asia, it is used for human food. Therefore, the main issue hampering its acceptance in the West is the lack of research into the end-uses of sorghum. There are some alternatives available for celiac patients, in particular, the minor cereals like fonio, tef, millet, teosinte, and Job's tears. However, these cereals are less common and have been cultivated regionally; for instance, tef is a crop in Ethiopia. All tef varieties examined so far are free of stimulatory epitopes (Hopman et al. 2008; Spaenij-Dekking et al. 2005), but the primary concern about its use is the possible cross-contamination with other gluten-containing grains like wheat (Saturni et al. 2010). There are other crops that are processed similarly to cereals and hence called pseudocereals. The most popular of these is the nutritionally dense quinoa, which unfortunately is controversial due to the immunotoxicity of some varieties (Zevallos et al. 2012, 2014). Similarly, for buckwheat, there have been reports of allergies (Panda et al. 2010; Stember 2006).

## Engineered Celiac-Safe Wheat Genotypes

In the wake of the difficulties associated with breeding "celiac-safe" genotypes, many research groups adapted to the genetic engineering procedures. Two kinds of wheat genotypes were developed, one where gluten proteins were eliminated, and the other where the gluten-detoxification enzymes were expressed. Following the former lead, Becker and co-workers produced a series of transgenic lines where α-gliadin genes were downregulated using RNA interference (RNAi). In these lines, α-gliadins were reduced by over 60% compared to the control cultivar (Becker et al. 2006, 2012; Becker and Folck 2006; Wieser et al. 2006). Using a similar approach silencing of the ω5-gliadins (Altenbach and Allen 2011; Altenbach et al. 2014) and ω1,2-gliadins was achieved by Altenbach and co-workers (Altenbach et al. 2019). However, in the attempt to silence the ω1,2-gliadins, the authors identified a transgenic line almost completely lacking gliadins and LMW glutenin subunits. When tested, the flour proteins from this genotype showed a stark decline in reactivity with serum IgG and IgA antibodies from a cohort of celiac disease patients (Altenbach et al. 2019). But the line suffered from the diminished mixing properties (Altenbach et al. 2019). Similarly, downregulation of γ-gliadins was also achieved using RNAi, and genotypes showing 65–97% reduction in the target proteins were identified (Gil-Humanes et al. 2008; Piston et al. 2011). More recently, following this lead, Smulders and co-workers developed CRISPR-Cas9-based constructs to specifically induce mutations in the genes encoding α- and γ-gliadin genes (Jouanin et al. 2018, 2019) and Barro and co-workers in the α2-gliadin genes (Sánchez-León et al. 2018).

The studies mentioned in the paragraph above were focused on the elimination of the specific gluten proteins using an RNA interference approach or genome editing. On the other hand, the studies mentioned below have either utilized a chimeric hairpin construct to target all gliadin (α/β-, γ-, and ω-) genes together (Gil-Humanes et al. 2010, 2011, 2012a, b, 2014a, b) or used RNAi to silence the master regulator (*DEMETER*) of the prolamin transcription (Wen et al. 2012; Rustgi et al. 2014). The lines showing 60–88% reductions in the gliadin content were identified using the chimeric hairpin construct. Tests of these genotypes with the intestinal T-cell clones derived from the biopsy samples of celiac patients showed almost complete suppression of disease-related T-cell epitopes (Gil-Humanes et al. 2010). When tested, these lines also showed reasonable baking characteristics and organoleptic properties as well as exhibited increased lysine content (Gil-Humanes et al. 2012a, b, 2014a, b). Two kinds of transgenic lines were produced to achieve DME suppression, one with DME-specific hairpin RNA and the other with DME-specific artificial micro RNA (amiRNA). The lines expressing DME-specific hairpin construct showed 45–76% reductions in the content of immunogenic prolamins (Wen et al. 2012; Rustgi et al. 2014) (Fig. 3). And the lines expressing one of the three amiRNAs exhibited 54–88% reductions in their respective prolamin contents (Rustgi et al. 2014).

Following the latter (gluten detoxification) approach, the Rustgi and co-workers expressed "glutenases" in wheat endosperm. Based on the parameters like target specificity, substrate length, optimum pH, and site of action, two prolyl endopeptidases one from *Flavobacterium meningosepticum* and the other from a thermophilic bacterium, *Pyrococcus furiosus*, and a glutamine-specific endoprotease from barley (EP-B2) were selected for expression in wheat endosperm (Osorio et al. 2012, 2019b). Wheat transformants expressing a FmPEP-EPB2 combination with up to

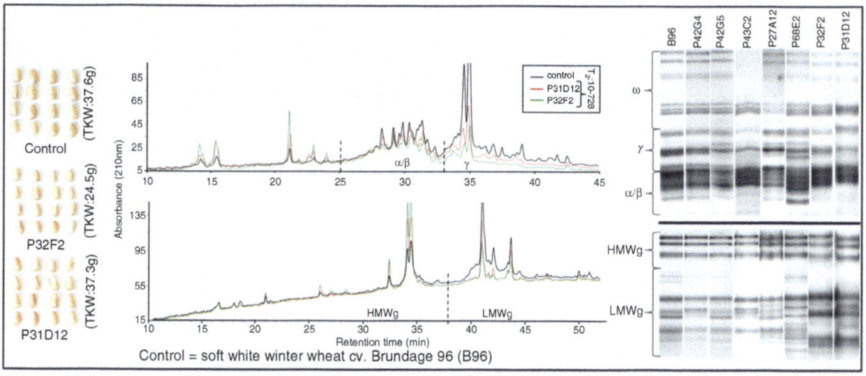


**Fig. 3** Liquid chromatography (middle) and polyacrylamide gel electrophoresis (right) of gliadins (top) and glutenin (bottom) fractions extracted from the grains of the two progeny plants (P31D12 and P32F2) of a genetically engineered wheat line. A random sample of grains from the selected lines with their respective thousand kernel weights (TKWs) is shown on the left. (Modified from Wen et al. 2012)

58% reduction and a PfuPEP-EP-B2 combination with up to 68% reduction in the content of the immunogenic gluten peptides were obtained (Osorio et al. 2019b) (Fig. 4).

This latter approach has specific advantages: (1) Some celiac patients show sensitivity to the HMW-GSs peptides (Dewar et al. 2006). Therefore, the formerly discussed transformants, which lack specific gliadins and/or LMW-GSs, are unsuitable for such patients. (2) The combination of enzymes used in this approach prevents degradation of gluten proteins within grains; therefore avoid the distraction of the end-use quality. The glutamine-specific endoprotease used in this study is encoded as a proenzyme, where the propeptide serves as both inhibitor and chaperone to respectively facilitate spatiotemporal regulation of the proteolytic activity and proper folding of the proteases (Bethune et al. 2006; Cappetta et al. 2002; Schilling et al. 2009; Cambra et al. 2012). These properties are of immense importance, as it avoids degradation of the prolamins in the protein bodies within grains and also in flour during the dough-making process. In addition, the prolyl endopeptidase due to its strict preference for substrates with ≤33 amino acids in size can only degrade peptides generated by the endoprotease (Gass and Khosla 2007), therefore permitting both of these enzymes to accumulate within the protein bodies containing the gluten proteins without degrading them and affecting the baking properties of the flour. (3) Intake of foods prepared from wheat engineered to express glutenases in grains does not require consumers to intake dietary supplements (none of these supplements are yet available in the market) before or with each meal. (4) The proposed therapy is expected to reach the general public without specific efforts and/or adding to the daily expenses of the consumers, as the remedy to wheat allergy and gluten intolerance is packed in the grain. (5) Contamination of regular wheat in the glutenases expressing wheat, at any level from farm to shelf, is less likely to make

| Parameter | EP-B2 | Fm-PEP |
|---|---|---|
| Substrate specificity | Glutamine-specific endoprotease | Post-proline cleaving endopeptidase |
| Substrate length | Protein | Peptide (~30 aa) |
| pH | 4-7 | 7 |
| Site of action | Stomach | Small intestine |

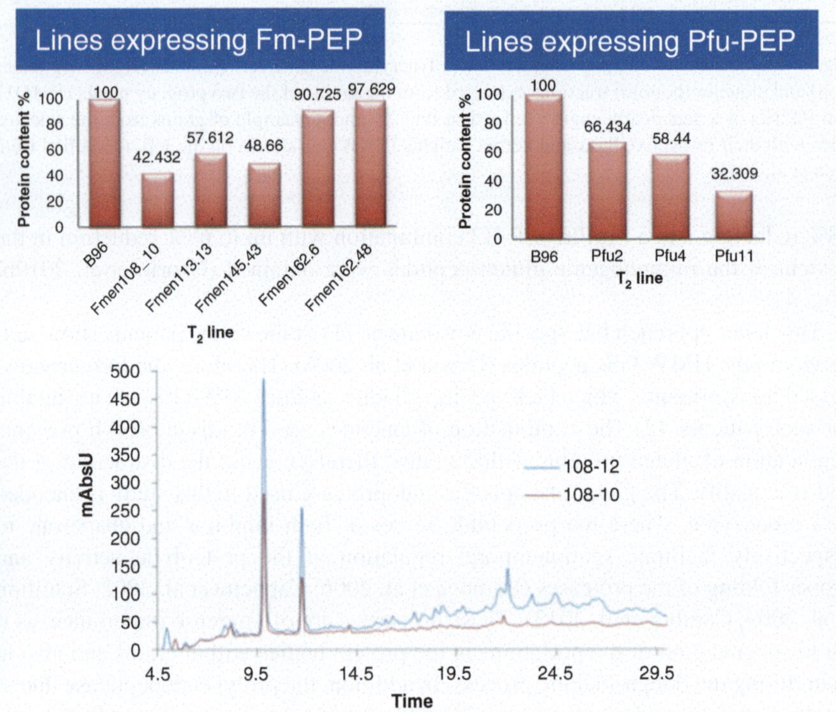


**Fig. 4** List of parameters used to select the two peptidases, one from barley (barley endoprotease B2; EP-B2) and another from *Flavobacterium* (*F. meningosepticum* prolyl endopeptidase; Fm-PEP). The total reduction in the content of the immunogenic gluten peptides in wheat transformants expressing a FmPEP-EPB2 combination (left) or a PfuPEP (*Pyrococcus furiosus* prolyl endopeptidase)-EP-B2 combination. Total protein was extracted from the grains of genetically engineered lines, treated with gastric enzymes and quantified by liquid chromatography (an example profile shown below, where 108–12 = control and 108–10 = a transgenic line). Proteins were measured in μg/mL and expressed as percent amount of immunogenic peptides remaining in each line in relation to the control line (B96). (Modified from Osorio et al. 2019b)

it unsuitable for celiac patients, as the glutenases expressing in the grains will degrade the contaminating gluten protein.

## Management Practices and Processing Procedures

Other than using genetic alterations, the reduced-immunogenicity wheat can be achieved by modulating growth conditions of wild-type wheat genotypes or by changing the processing parameters of the whole grains or the wheat flour. In fact, a correspondence was observed between nitrogen and sulfur dose and the amount as well as the composition of proteins accumulated in wheat grains (Godfrey et al. 2010; Shewry 2011) (see Table 1 for examples). An increase in nitrogen supply results in a significant increase in the content of gliadins and glutenins, but not of albumins and globulins (Johansson et al. 2001). Specifically, the effect on gliadins was more pronounced than on glutenins. High levels of nitrogen increased the proportions of hydrophilic proteins (ω-gliadins and HMW subunits), and those of hydrophobic proteins (γ-gliadins and LMW subunits) were decreased (Wieser and Seilmeier 1998). In a separate study, the majority of HMW subunits and ω-gliadins and some α-gliadins showed increased accumulation, while two LMW subunits and a minor γ-gliadin exhibited decreased accumulation in response to fertilizer or high temperature, whereas fertilizer did not influence gluten protein accumulation under high-temperature conditions (Hurkman et al. 2013). More recently, two commercial spelt wheat varieties evaluated through seven nitrogen fertilization modalities did not influence the epitope expression of the first variety, whereas it had a slight effect on the epitope expression of the second variety (Dubois et al. 2018). Similar effects of nitrogen fertilizer on hordein, specifically C-hordein biosynthesis during early stages of grain development, were reported in barley (Giese and Hopp 1984; Müller and Knudsen 1993).

Much like nitrogen fertilizers, sulfur fertilization showed influence on the amount of total gluten as well as the crude protein content of flour. In the case of sulfur deficiency, the amount of S-free ω-gliadins increased drastically and that of S-poor HMW subunits increased moderately. In contrast, the amounts of S-rich γ-gliadins and LMW subunits decreased significantly, whereas the amount of α-gliadins was reduced only slightly. Sulfur deficiency results in a remarkable shift in protein proportions, such that the gliadin to glutenin ratio increases distinctly, and among gliadins, the ω-gliadins become significant components and γ-gliadins minor elements (Wieser et al. 2004).

Other than nutrient status, temperature regime during grain development was reported to have a significant influence on the amount and type of proteins accumulation. For instance, the low-temperature conditions during grain development were shown to decrease the level of protein fractions primarily associated with celiac disease but increase the content of protein families related to WDEIA or barker's asthma, such as LTPs, hydrolases, peroxidases, and ATIs. On the other hand, under the high temperature conditions, the changes in seed storage protein accumulation were shown to result in slightly increased accumulation of ω-gliadins (3–26%) and

**Table 1** Effect of different environmental factors and management regimens on the gluten content and composition

| Type | Genotype | Experiment duration (years) | Locations | Factor(s) | Comment | References |
|---|---|---|---|---|---|---|
| Winter wheat | Ana Morava, Toplica, Vizija, Takovcanka, and Lazarica | 3 | 3 | Genotype x environment effect on gluten content | Analysis of variance suggested highly significant differences among genotypes, years, and locations for gluten content | Zecevic et al. (2009) |
| Durum wheat | Canadian varieties – Kyle, AC Avonlea, AC Navigator, AC Pathfinder, and AC Morse; breeding lines, DT674, DT 672, DT 675, and DT 666; U.S. cultivar, Durex | 3 | 2 | Effect of nitrogen on gluten content | Increase in nitrogen fertilizer resulted in increased protein content in all cultivars across environments | Ames et al. (2003) |
| Bread wheat | Eagle, Oxley, Egret, Halberd, and Cook | 11 | 3 | Effect of heat stress on gluten | Gliadin synthesis continues at a greater rate than glutenin synthesis during a period of heat stress. Consequently, the mature grains show a higher ratio of gliadins to glutenins and produce weaker dough. These results now provide a basis for formulating strategies to minimize variations in dough properties due to growing conditions | Blumenthal et al. (1993) |
| Bread wheat | Pandas, Spada, Maestra, and Oderzo | – | 4 | Effect of heat stress on gluten | During heat stress, there is a continuous synthesis of both gliadin and LMW glutenin subunits, but synthesis of HMW glutenin subunits is suppressed | Ciaffi et al. (1996) |
| Semi-hard and soft common wheat | 40 wheat genotypes | 2 | 3 | Genotype x environment effect on gluten content | Significant effects of both genotype and environment were onserved on the gluten content | Jing et al. (2003) |

| Type | Genotype | Experiment duration (years) | Locations | Factor(s) | Comment | References |
|---|---|---|---|---|---|---|
| Bread wheat | 140 cultivars and experimental lines | 4 | – | Genotype x environment effect on gluten content | Both cultivar and cultivar by environment interaction had significant effects on all quality traits. Variances of quality traits associated with genetic factors (cultivar) were generally larger than those for cultivar by environmental interaction effects | Denčić et al. (2011) |
| Winter wheat | Lars | 4 | 2 | Effect of sulfur on gluten proteins | Applied sulfur positively affected the quality of wheat flour by increasing gluten and protein concentration. Furthermore, sulfur positively affected the development, stability, softening, and quality of dough as well as bread volume | Järvan et al. (2008) |
| Bread wheat | Ningmai9 | – | 2 | Effect of nitrogen and potassium on gluten content | N and K fertilization significantly increased grain yield, protein content, and wet gluten content | Zou et al. (2006) |
| Spring weak-gluten wheat | Yangmai 9 (YM9), Ningmai 9 (NM9), and Yangmai 13 (YM13) | 2 | – | Effect of phosphorus fertilizer on grain proteins | Optimum P fertilizer increased grain yield and improved grain quality of weak-gluten wheat | Zhu et al. (2012) |
| Winter wheat | Erzurum (Turkey) | – | – | Effect of water stress at various growth stages on quality | Late water stress caused an increase of 8.3% in grain protein content, 8.7% in sedimentation volume, 10.8% in wet gluten content, and a reduction in 1000-kernel weight | Ozturk and Aydin (2004) |

(continued)

**Table 1** (continued)

| Type | Genotype | Experiment duration (years) | Locations | Factor(s) | Comment | References |
|---|---|---|---|---|---|---|
| Winter wheat | S + D23ulamit (E), Samanta (A), apache (B), Meritto (B), Mladka (C), and Rapsodia (C) | 2 | 2 | The effect of organic and conventional growing systems on gluten | Conventionally grown wheat varieties showed more HMW glutenins than organically grown wheat varieties. However, organically grown varieties showed more albumin and globulin content | Krejčířová et al. (2008) |
| Bread wheat | PBW 343 and C306 | 2 | 1 | Effect of sowing time and drought on gluten content | Late sown conditions lead to high grain protein content by increasing albumins/globulins but decreasing glutenins. Rain-fed conditions decreased trypsin inhibitors' activity, and exposure to heat and drought together led to a decline in the glutenin-to-gliadin ratio. | Singh et al. (2012) |
| Weak- or medium-gluten cultivar | Yangmai 9 (weak-gluten) and Yangmai 12 (medium-gluten) | – | – | Effect of planting density on the protein content | Increasing planting density from $105 \times 10^4$ to $240 \times 10^4$ plants/ha increased the protein content | Liu et al. (2006) |
| Spring wheat | – | 21 | – | Effect of temperature on the gluten content | The gluten content decreased when daily minimum and maximum temperatures exceeded 11–12 °C and 21–22 °C, respectively | Peltonen et al. (1990) |
| Winter wheat | Spark, rialto, Soissons, and beaver | 1 | 1 | Effect of temperature and moisture on gluten | Good processing quality varieties showed significant differences related to environmental conditions than the variety with poor processing qualities | Georget et al. (2008) |

| Type | Genotype | Experiment duration (years) | Locations | Factor(s) | Comment | References |
|---|---|---|---|---|---|---|
| Red spring wheat | Butte 86 | 1 | 1 | Effect of temperature, water and fertilizer on gluten content | Under high-temperature regimen, transcripts from all gene families appeared slightly earlier, but the time frame of accumulation was shorter as compared to plants under moderate daytime temperature with optimum water and fertilizer | Altenbach et al. (2002) |
| Spring wheat | Star | – | 2 | Effect of sulfur on gluten properties | Overall, gluten content was affected little by different doses of the sulfur fertilizer, but the individual single protein sub-units were influenced strongly. With a sulfur deficiency, the gliadins: glutenins ratio increased, whereas the ratio of HMW glutenins to LMW glutenins stayed in balance | Wieser et al. (2004) |
| Spring wheat | Sport, Dacke, Dragon, Thasos | 3 | 3 | Effect of nitrogen application rate on gluten strength | High nitrogen fertilizer rate increased protein concentration, decreased gluten strength, and increased the total amount of glutenins and gliadins. The timing of fertilization (early application) increases gluten strength | Johansson et al. (2004) |
| Red spring wheat | Butte 86 | 1 | 1 | Effect of temperature and fertilizer on gluten composition | At high temperature, HMW glutenins, omega-gliadins, and some alpha gliadins increased, while LMW glutenins and gamma gliadins decreased. Fertilizer did not influence gluten protein accumulation under high-temperature conditions | Hurkman et al. (2013) |

(continued)

α-gliadins (25–33%), more specifically in the content of 33-mer containing α-gliadins (Juhász et al. 2018). Collectively, these studies suggest that it is possible to alter the amount and type of proteins accumulated in wheat grains by modulating with the growing condition of wheat plants.

Besides, it is possible to obtain reduced-immunogenicity flour from regular wheat genotypes by applying specific processing procedures such as milling techniques or twin-screw extrusion techniques. More recently, the use of the microwaves to remove antigenic properties of the wheat gluten proteins was proposed (Landriscina et al. 2017). Additionally, the use of wheat, barley, and rye sprouts (germinated grains) as a safe food for celiac patients or to be used as an ingredient for other products was proposed. Although cereal endopeptidases synthesized during sprouting can efficiently hydrolyze gluten (Hartmann et al. 2006), other research showed that using peptidases from sprouted wheat to digest gliadin did not result in food safe for celiac patients (Stenman et al. 2009). Another proposed method was the use of sourdough fermentation to produce bakery products suitable for celiac patients (Zannini et al. 2012). However, no conclusive data exist on the use of any of the methods mentioned above [for details the readers are recommended to consult Rustgi et al. (2019) and references cited therein].

### *Non-dietary Procedures*

In parallel to the efforts to develop dietary therapies for the celiac disease, extensive research was performed to developing non-dietary therapies. These therapies can be largely classified into (1) luminal therapies which are based on the detoxification of gluten proteins and can be further classified into enzyme therapy, probiotic therapy, flour/dough pretreatment, and gluten inactivation by polymeric binding; (2) intestinal barrier enhancing therapies, which focus on reducing the permeability of intestinal epithelial barrier; and (3) immune-targeted therapies, which target either celiac disease-specific pathways or inflammatory mediators common in gastrointestinal inflammation. These non-dietary therapies to treat the celiac disease has been extensively reviewed in the past by Schuppan et al. (2009), Sollid and Khosla (2011), Osorio et al. (2012), Rashtak and Murray (2012), McCarville et al. (2015), and Ribeiro et al. (2018) and, therefore, have not been discussed here.

## 7 Conclusion

Outstanding genetic resources, such as conventionally produced reduced-gluten mutations in each one of the gliadin and glutenin loci and the cytogenetically, as well as genetically engineered reduced-gluten lines, are available to researchers to breed wheat genotypes for celiac patients. Specifically, a vast collection of well-characterized chromosome substitution and alien introgression lines developed in the background of elite hexaploid and tetraploid wheat genotypes exist today, which could be screened for their gluten composition, antigenicity, and allergenicity as well as technological properties. These lines carry alien introgression spanning almost all parts of the wheat genome, exist in the elite background, and carry many desirable exotic attributes such as insect pest, fungal, or abiotic stress tolerance. Besides, remarkable genomic resources and approaches such as genomic selection are available, which could facilitate the selection process of desirable lines from the interbreeding program. These new genomic prediction methods also reduce the dependence on the expensive phenotyping for technological properties, allergenicity, and antigenicity tests. Therefore, we believe that desired resources are available to the breeders today to make sturdy progress in the direction of developing "celiac-safe" wheat genotypes.

**Acknowledgments** This work was supported by the Clemson Faculty Succeeds Grant and NIFA Hatch/Multi-state grant (S009).

## References

Altenbach SB, Allen PV (2011) Transformation of the US bread wheat "Butte 86" and silencing of omega-5 gliadin genes. GM Crops 2:66–73

Altenbach SB, Kothari KM, Dao L (2002) Environmental conditions during wheat grain development alter temporal regulation of major gluten protein genes. J Cereal Chem 79:279–285

Altenbach SB, Tanaka CK, Seabourn BW (2014) Silencing of omega-5 gliadins in transgenic wheat eliminates a major source of environmental variability and improves dough mixing properties of flour. BMC Plant Biol 14:1

Altenbach SB, Chang H-C, Yu XB, Seabourn BW, Green PH, Alaedini A (2019) Elimination of omega-1,2 gliadins from bread wheat (*Triticum aestivum*) flour: effects on immunogenic potential and end-use quality. Front Plant Sci 10:580

Amaya-Gonzalez S, De-Los-Santos-Alvarez N, Miranda-Ordieres AJ, Lobo-Castanon MJ (2014) Aptamer binding to celiac disease-triggering hydrophobic proteins: a sensitive gluten detection approach. Anal Chem 86:2733–2739

Amaya-Gonzalez S, De-Los-Santos-Alvarez N, Miranda-Ordieres AJ, Lobo-Castanon MJ (2015) Sensitive gluten determination in gluten-free foods by an electrochemical aptamer-based assay. Anal Bioanal Chem 407:6021–6029

Ames NP, Clarke JM, Dexter JE, Woods SM, Selles F, Marchylo B (2003) Effects of nitrogen fertilizer on protein quantity and gluten strength parameters in durum wheat (Triticum turgidum L. var.durum) cultivars of variable gluten strength. J Cereal Chem 80:203–211

Aziz I, Hadjivassiliou M, Sanders DS (2015) The spectrum of noncoeliac gluten sensitivity. Nat Rev Gastroenterol Hepatol 12:516–526

Bai JC, Fried M, Corazza GR, Schuppan D, Farthing M et al (2013) World gastroenterology organisation global guidelines on celiac disease. J Clin Gastroenterol 47:121–126

Balakireva AV, Zamyatnin AA (2016) Properties of gluten intolerance: gluten structure, evolution, pathogenicity and detoxification capabilities. Nutrients 8:E644

Barbeau WE, Schwarzlaff SS, Uriyo MG, Johnson JM, Harris CH, Griffey CA (2003) Origin and practical significance of the sticky dough factor in 1BL/1RS wheats. J Sci Food Agric 83:29–38

Becker D, Folck A (2006) Inhibierung der α-Gliadingeneexpression in hexaploiden Brotweizen. Getreidetechnologie 30:153–156

Becker D, Folck A, Knies P, Lörz H, Wieser H (2006) Silencing the a-gliadins in hexaploid bread wheat. In: Lookhart LG, Ng WPK (eds) Gluten proteins. AACC International, St Paul, pp 86–89
Becker D, Wieser H, Koehler P, Folck A, Mühling KH, Zörb C (2012) Protein composition and techno-functional properties of transgenic wheat with reduced α-gliadin content obtained by RNA interference. J Appl Bot Food Qual 85:23–33
Berezovski M, Musheev M, Drabovich A, Jitkova J, Krylov S (2006) Non-SELEX: selection of aptamers without intermediate amplification of candidate oligonucleotides. Nat Protoc 1:1359–1369
Bethune MT, Khosla C (2008) Parallels between pathogens and gluten peptides in celiac sprue. PLoS Pathog 4:e34
Bethune MT, Strop P, Tang Y, Sollid LM, Khosla C (2006) Heterologous expression, purification, refolding, and structural-functional characterization of EP-B2, a self-activating barley cysteine endoprotease. Chem Biol 13:637–647
Blumenthal CS, Barlow EWR, Wrigley CW (1993) Growth environment and wheat quality: the effect of heat stress on dough properties and gluten proteins. J Cereal Sci 18:3–21
Brouns F, van Buul VJ, Shewry PR (2013) Does wheat make us fat and sick? J Cereal Sci 58:209–215
Brouns F, Van Rooy G, Shewry P, Rustgi S, Jonkers D (2019) Adverse reactions to wheat or wheat components. Compr Rev Food Sci Food Saf. https://doi.org/10.1111/1541-4337.12475
Camafeita E, Alfonso P, Lopez JA, Sorell L, Mendez E (1997) Sample preparation optimization for the analysis of gliadins in food by matrix-assisted laser desorption/ionization time-of-flight mass spectrometry. J Mass Spectrom 32:444–449
Camafeita E, Solis J, Alfonso P, Lopez JA, Sorell L, Mendez E (1998) Selective identification by matrix-assisted laser desorption/ionization time-of-flight mass spectrometry of different types of gluten in foods made with cereal mixtures. J Chromatogr A 823:299–306
Cambra I, Hernández D, Diaz I, Martinez M (2012) Structural basis for specificity of propeptide-enzyme interaction in barley C1A cysteine peptidases. PLoS One 7:e37234
Cappetta M, Roth I, Díaz A, Tort J, Roche L (2002) Role of the prosegment of Fasciola hepatica cathepsin L1 in folding of the catalytic domain. Biol Chem 383:1215–1221
Ciaffi M, Tozzi L, Borghi B, Corbellini M, Lafiandra D (1996) Effect of heat shock during grain filling on the gluten protein composition of bread wheat. J Cereal Sci 24:91–100
Ciclitira PJ, Lennox ES, Hunter JO (1980a) Clinical testing of bread made from nullisomic-6A wheats in celiac patients. Lancet 2:234–236
Ciclitira PJ, Hunter JO, Lennox ES (1980b) Clinical testing in celiac patients of bread made from wheats deficient in some alpha-gliadins. Clin Sci 59:25
Colomba MS, Gregorini A (2012) Are ancient durum wheats less toxic to celiac patients? A study of α-gliadin from Graziella Ra and Kamut. Sci World J 2012:837416
Comino I, Moreno M, Real A, Rodríguez-Herrera A, Barro F, Sousa C (2013) The gluten-free diet: testing alternative cereals tolerated by celiac patients. Nutrients 5:4250–4268
Da Sacco L, Baldassarre A, Masotti A (2013) Diet's role in the toxicity of inorganic arsenic (iAs): a journey from soil to children's mouth. J Geochem Explor 131:45–51
De Vita P, Ficco DBM, Luciani A, Vincentini O, Pettoello-Mantovani M, Silano M, Maiuri L, Cattivelli L (2012) A ω-secalin contained decamer shows a celiac disease prevention activity. J Cereal Sci 55:234–242
Delcour JA, Joye IJ, Pareyt B, Wilderjans E, Brijs K, Lagrain B (2012) Wheat gluten functionality as a quality determinant in cereal-based food products. Annu Rev Food Sci Technol 3:469–492
Denčić S, Mladenov N, Kobiljski B (2011) Effects of genotype and environment on breadmaking quality in wheat. Int J Plant Prod 5:71–82
Dewar DH, Amato M, Ellis HJ, Pollock EL, Gonzalez-Cinca N, Wieser H, Ciclitira PJ (2006) The toxicity of high molecular weight glutenin subunits of wheat to patients with coeliac disease. Eur J Gastroenterol Hepatol 18:483–491
Dubois B, Bertin P, Hautier L, Muhovski Y, Escarnot E, Mingeot D (2018) Genetic and environmental factors affecting the expression of α-gliadin canonical epitopes involved in celiac

disease in a wide collection of spelt (*Triticum aestivum* ssp. *spelta*) cultivars and landraces. BMC Plant Biol 18:262

Ellis HJ, Doyle AP, Wieser H, Sturgress RP, Day P, Ciclitira PJ (1994) Measurement of gluten using a monoclonal antibody to a sequenced peptide of α-gladian from the coeliac-activating domain I. J Biochem Biophys Methods 28:77–82

Fasano A (2009) Surprises from celiac disease. Sci Am 301:54–61

Frisoni M, Corazza GR, Lafiandra D, De Ambroggio E, Filipponi C et al (1995) Wheat deficient in gliadins: promising tool for coeliac disease. Gut 36:375–378

Gass J, Khosla C (2007) Prolyl endopeptidases. Cell Mol Life Sci 64:345–355

Georget DMR, Underwood-Toscano C, Powers SJ, Shewry PR, Belton PS (2008) Effect of variety and environmental factors on gluten proteins: an analytical, spectroscopic and rheological study. J Agric Food Chem 56:1172–1179

Gianfrani C, Maglio M, Rotondi Aufiero V, Camarca A, Vocca I et al (2012) Immunogenicity of monococcum wheat in celiac patients. Am J Clin Nutr 96:1339–1345

Gianibelli MC, Larroque OR, MacRitchie F, Wrigley CW (2001) Biochemical, genetic, and molecular characterization of wheat glutenin and its component subunits. Cereal Chem 78:635–646

Giese H, Hopp HE (1984) Influence of nitrogen nutrition on the amount of hordein, protein z and α-amylase messenger RNA in developing endosperms of barley. Carlsberg Res Commun 49:365–383

Gil-Humanes J, Piston F, Hernando A, Alvarez JB, Shewry PR, Barro F (2008) Silencing of γ-gliadins by RNA interference (RNAi) in bread wheat. J Cereal Sci 48:565–568

Gil-Humanes J, Piston F, Tollefsen S, Sollid LM, Barro F (2010) Effective shutdown in the expression of celiac disease-related wheat gliadin T-cell epitopes by RNA interference. Proc Natl Acad Sci USA 107:17023–17028

Gil-Humanes J, Piston F, Shewry PR, Tosi P, Barro F (2011) Suppression of gliadins results in altered protein body morphology in wheat. J Exp Bot 62:4203–4213

Gil-Humanes J, Piston F, Gimenez MJ, Martin A, Barro F (2012a) The introgression of RNAi silencing of γ-gliadins into commercial lines of bread wheat changes the mixing and technological properties of the dough. PLoS One 9:e45937

Gil-Humanes J, Piston F, Rosell CM, Barro F (2012b) Significant down-regulation of γ-gliadins has minor effect on gluten and starch properties of bread wheat. J Cereal Sci 56:161–170

Gil-Humanes J, Piston F, Altamirano-Fortoul R, Real A, Comino I et al (2014a) Reduced-gliadin wheat bread: an alternative to the gluten-free diet for consumers suffering gluten-related pathologies. PLoS One 9:e90898

Gil-Humanes J, Piston F, Barro F, Rosell CM (2014b) The shutdown of celiac disease-related gliadin epitopes in bread wheat by RNAi provides flours with increased stability and better tolerance to over-mixing. PLoS One 9:e91931

Gilissen LJWJ, van der Meer IM, Smulders MJM (2014) Reducing the incidence of allergy and intolerance to cereals. J Cereal Sci 59:337–353

Gobaa S, Bancel E, Kleijer G, Stamp P, Branlard G (2007) Effect of the 1BL.1RS translocation on the wheat endosperm, as revealed by proteomic analysis. Proteomics 7:4349–4357

Godfrey D, Hawkesford MJ, Powers SJ, Millar S, Shewry PR (2010) Effects of crop nutrition on wheat grain composition and end use quality. J Agric Food Chem 58:3012–3021

Golley S, Corsini N, Topping D, Morell M, Mohr P (2015) Motivations for avoiding wheat consumption in Australia: results from a population survey. Public Health Nutr 18:490–499

Goryunova SV, Salentijn EM, Chikida NN, Kochieva EZ, van der Meer IM, Gilissen LJ, Smulders MJM (2012) Expansion of the gamma-gliadin family in *Aegilops* and *Triticum*. BMC Evol Biol 12:215

Gregorini A, Colomba M, Ellis HJ, Ciclitira PJ (2009) Immunogenicity characterization of two ancient wheat α-gliadin peptides related to coeliac disease. Nutrients 1:276–290

Hartmann G, Koehler P, Wieser H (2006) Rapid degradation of gliadin peptides toxic for coeliac disease patients by proteases from germinating cereals. J Cereal Sci 44:368–371

Henterich N, Osman AA, Mendez E, Mothes T (2003) Assay of gliadins by real time immunopolymerase chain reaction. Nahrung 47:345–348

Hopman E, Dekking L, Blokland ML, Wuisman M, Zuijderduin W, Koning F, Schweizer J (2008) Tef in the diet of celiac patients in The Netherlands. Scand J Gastroenterol 43:277–282

Hurkman WJ, Tanaka CK, Vensel WH, Thilmony R, Altenbach SB (2013) Comparative proteomic analysis of the effect of temperature and fertilizer on gliadin and glutenin accumulation in the developing endosperm and flour from Triticum aestivum L. cv. Butte 86. J Proteome Sci 11:8

Iametti S, Bonomi F, Ferranti P, Picariello G, Gabrovska D (2005) Characterization of gliadin content in beer by using different approaches. In: Stern M (ed) Proceedings of the 19th meeting of working group on Prolamin analysis and toxicity. Verlag Wissenschaftliche Scripten, Zwickau, pp 73–78.

Iametti S, Bonomi F, Ferranti P, de Martino A, Picariello G (2006) Characterization of peptides and proteins in beer by using different approaches. In Stern M (ed) Proceedings of the 20th meeting of working group on Prolamin analysis and toxicity. Verlag Wissenschaftliche Scripten, Zwickau, pp 47–52

Järvan M, Edesi L, Adamson A, Lukme L, Akk A (2008) The effect of Sulphur fertilization on yield, quality of protein and baking properties of winter wheat. Agronomy Res 6:459–469

Jing Q, Jiang D, Dai T, Cao W (2003) Effects of genotype and environment on wheat grain quality and protein components. J Appl Ecol 14:1649–1653

Johansson E, Prieto-Linde ML, Jönsson JÖ (2001) Effects of wheat cultivar and nitrogen application on storage protein composition and bread-making quality. Cereal Chem 78:19–25

Johansson E, Prieto-Linde ML, Svensson G (2004) Influence of nitrogen application rate and timing on grain protein composition and gluten strength in Swedish wheat cultivars. J Sci Food Agri 167:345–350

Jouanin A, Boyd L, Visser RGF, Smulders MJM (2018) Development of wheat with hypoimmunogenic gluten obstructed by the gene editing policy in Europe. Front Plant Sci 9:1523

Jouanin A, Schaart JG, Boyd LA, Cockram J, Leigh FJ, Bates R, Wallington EJ, Visser RGF, Smulders MJM (2019) Outlook for coeliac disease patients: towards bread wheat with hypoimmunogenic gluten by gene editing of α- and γ-gliadin gene families. BMC Plant Biol 19:333

Juhász A, Belova T, Florides CG, Maulis C, Fischer I et al (2018) Genome mapping of seed-borne allergens and immunoresponsive proteins in wheat. Sci Adv 4:eaar8602

Kasarda DD (2007) Letter to the editor: *Triticum monococcum* and celiac disease. Scand J Gastroenterol 42:1141–1142

Koning F (2012) Celiac disease: quantity matters. Seminars Immunopathol 34:541–549

Koppel E, Stadler M, Luthy J, Hubner P (1998) Detection of wheat contamination in oats by polymerase chain reaction (PCR) and enzyme-linked immumosorbent assay (ELISA). Z Lebensm Untersuch Forsch 206:399–403

Krejčířová L, Capouchová I, Petr J, Bicanová E, Faměra O (2008) The effect of organic and conventional growing systems on quality and storage protein composition of winter wheat. Plant Soil Environ 53:499–505

Landriscina L, D'Agnello P, Bevilacqua A, Corbo MR, Sinigaglia M, Lamacchia C (2017) Impact of gluten-friendly™ technology on wheat kernel endosperm and gluten protein structure in seeds by light and electron microscopy. Food Chem 221:1258–1268

Langridge P (2017) Achieving sustainable cultivation of wheat: breeding, quality traits, pests and diseases. Burleigh Dodds Science Publishing, Cambridge

Liu P, Guo WS, Xu YM, Feng CN, Zhu XK, Peng YX (2006) Effect of planting density on grain yield and quality of weak-gluten and medium-gluten. Wheat J Triticecae S512:1

López-López L, Miranda-Castro R, De-los-Santos-Alvarez N, Miranda-Ordieres AJ, Lobo-Castanon MJ (2017) Disposable electrochemical aptasensor for gluten determination in food. Sensors Actuators B Chem 241:522–527

Lukaszewski AJ (2015) Introgressions between wheat and rye. In: Molnár-Láng M, Ceoloni C, Doležel J (eds) Alien introgression in wheat. Springer, Cham, pp 163–190

Malvano F, Albanese D, Pilloton R, Matteo MD (2017) A new label-free impedimetric aptasensor for gluten detection. Food Control 79:200–206
Matsuda T, Nakase M, Alvarez AM, Izumi H, Kato T, Tada Y (2006) Rice-seed allergenic proteins and hypoallergenic rice. In: Mine Y, Shahidi F (eds) Nutraceutical proteins and peptides in health and disease. CRC Press, Hoboken, pp 493–511
McCarville JL, Caminero A, Verdu EF (2015) Celiac treatments, adjuvant therapies and alternatives to the gluten-free diet. In: Arranz E, Fernández-Bañares F, Rosell CM, Rodrigo L, Peña AS (eds) Advances in the understanding of gluten related pathology and the evolution of gluten-free foods. OmniaScience, Barcelona, pp 223–253
Mitea C, Salentijn EM, van Veelen P, Goryunova SV, van der Meer IM et al (2010) A universal approach to eliminate antigenic properties of alpha-gliadin peptides in celiac disease. PloS One 5:e15637
Molberg Ø, Uhlen AK, Jensen T, Flæte NS, Fleckenstein B et al (2005) Mapping of gluten T-cell epitopes in the bread wheat ancestors: implications for celiac disease. Gasteroenterology 128:393–401
Müller M, Knudsen S (1993) The nitrogen response of a barley C-hordein promoter is controlled by positive and negative regulation of the GCN4 and endosperm box. Plant J4:343–355
Muller S, Wieser H (1995) The location of disulfide bonds in alpha-type gliadins. J Cereal Sci 22:21–27
Muller S, Wieser H (1997) The location of disulphide bonds in monomeric gamma-type gliadins. J Cereal Sci 26:169–176
Munck L (1992) The case of high-lysine barley breeding. In: Shewry PR (ed) Barley: genetics, biochemistry, molecular biology and biotechnology. CAB International, Wallingford, pp 573–601
Munera-Picazo S, Ramirez-Gandolfo A, Burlo F, Carbonell-Barrachina AA (2014) Inorganic and total arsenic contents in rice-based foods for children with celiac disease. J Food Sci 79:T122–T128
Nambu M (2006) Rice allergy. Pediatrics 117:2331
Ortiz-Sánchez JP, Cabrera-Chávez F, de la Barca AM (2013) Maize prolamins could induce a gluten-like cellular immune response in some celiac disease patients. Nutrients 5:4174–4183
Osborne TB (1907) The proteins of the wheat kernel. Carnegie Institute, Washington
Osorio C, Wen N, Gemini R, Zemetra R, von Wettstein D, Rustgi S (2012) Targeted modification of wheat grain protein to reduce the content of coeliac causing epitopes. Funct Integr Genomics 12:417–438
Osorio CE, Mejías JH, Rustgi S (2019a) Gluten detection methods and their critical role in assuring safe diets for celiac patients. Nutrients 11:2920
Osorio C, Wen N, Mejias JH, Liu B, Reinbothe S, von Wettstein D, Rustgi S (2019b) Development of wheat genotypes expressing a glutamine-specific endoprotease from barley and a prolyl endopeptidase from *Flavobacterium meningosepticum* or *Pyrococcus furiosus* as a potential remedy to coeliac disease. Funct Integr Genomics 19:123–136
Ozturk A, Aydin F (2004) Effect of water stress at various growth stages on some quality characteristics of winter wheat. J Agron Crop Sci 190:93–99
Panda R, Taylor SL, Goodman RE (2010) Development of a sandwich enzyme-linked immunosorbent assay (ELISA) for detection of buckwheat residues in food. J Food Sci 75:T110–T117
Peltonen J, Karvonen T, Kivi E (1990) The effect of climatic factors on production of spring wheat quantity to quality ratio in southern Finland. Agric Food Sci 62:227–236
Pinto A, Polo PN, Henry O, Redondo MCB, Svobodova M, O'Sullivan CK (2014) Label-free detection of gliadin food allergen mediated by real-time apta-PCR. Anal Bioanal Chem 406:515–524
Piston F, Gil-Humanes J, Rodríguez-Quijano M, Barro F (2011) Down-regulating γ-gliadins in bread wheat leads to non-specific increases in other gluten proteins and has no major effect on dough gluten strength. PLoS One 6:e24754

Pizzuti D, Buda A, D'Odorico A, D'Incà R, Chiarelli S, Curioni A, Martines D (2006) Lack of intestinal mucosal toxicity of *Triticum monococcum* in coeliac disease patients. Scand J Gastroenterol 41:1305–1311

Pogna NE, Monari AM, Cacciatori P, Redaelli R, Ng PKW (1998) Development and characterization of bread wheat lines lacking chromosome 1B-, 1D-, 6A- and 6D-encoded prolamins. In: Proceedings of the IXth international wheat genetics symposium. Saskatoon, Saskatchewan, Canada, pp 265–268

Pontieri P, Mamone G, De Caro S, Tuinstra MR, Roemer E et al (2013) Sorghum, a healthy and gluten-free food for celiac patients as demonstrated by genome, biochemical, and immunochemical analyses. J Agric Food Chem 61:2565–2571

Rashtak S, Murray JA (2012) Review article: coeliac disease, new approaches to therapy. Aliment Pharmacol Ther 35:768–781

Ribeiro M, Nunes FM, Rodriguez-Quijano M, Carrillo JM, Branlard G, Igrejas G (2018) Next-generation therapies for celiac disease: the gluten-targeted approaches. Trends Food Sci Technol 75:56–71

Rosella CM, Barro F, Sousa C, Mena MC (2014) Cereals for developing gluten-free products and analytical tools for gluten detection. J Cereal Sci 59:354–364

Rumbo M, Chirdo FG, Fossati CA, Anon MC (2001) Analysis of the effect of heat treatment on gliadin immunochemical qualification using a panel of anti-prolamin antibodies. J Agric Food Chem 49:5719–5726

Rustgi S, Wen N, Osorio C, Brew-Appiah RAT, Wen S et al (2014) Natural dietary therapies for the 'gluten syndrome'. Scientia Danica Series B Biol 3:1–87

Rustgi S, Shewry P, Brouns F, Deleu LJ, Delcour JA (2019) Wheat seed proteins – factors influencing their content, composition, and technological properties, and strategies to reduce adverse reactions. Compr Rev Food Sci Food Saf 18:1751–1769

Salentijn EMJ, Esselink DG, Goryunova SV, van der Meer IM, Gilissen L, Smulders MJM (2013) Quantitative and qualitative differences in celiac disease epitopes among durum wheat varieties identified through deep RNA-amplicon sequencing. BMC Genomics 14:16

Sánchez-León S, Gil-Humanes J, Ozuna CV, Giménez MJ, Sousa C, Voytas DF, Barro F (2018) Low-gluten, nontransgenic wheat engineered with CRISPR/Cas9. Plant Biotechnol J 16:902–910

Sapone A, Bai JC, Ciacci C, Dolinsek J, Green PH, Hadjivassiliou M, Kaukinen K, Rostami K, Sanders DS, Schumann M, Ullrich R, Villalta D, Volta U, Catassi C, Fasano A (2012) Spectrum of gluten-related disorders: consensus on new nomenclature and classification. BMC Med 10:13

Saturni L, Ferretti G, Bacchetti T (2010) The gluten-free diet: safety and nutritional quality. Nutrients 2:16–34

Scherf KA, Poms RE (2016) Recent developments in analytical methods for tracing gluten. J Cereal Sci 67:112–122

Schilling K, Körner A, Sehmisch S, Kreusch A, Kleint R et al (2009) Selectivity of propeptide-enzyme interaction in cathepsin L-like cysteine proteases. Biol Chem 390:167–174

Schuppan D, Junker Y, Barisani D (2009) Celiac disease: from pathogenesis to novel therapies. Gastroenterology 137:1912–1933

Schwarzlaff SS, Uriyo MG, Johnson JM, Barbeau WE, Griffey CA (2001) Apparent dough stickiness of selected 1BL/1RS translocated soft wheat flours. Cereal Chem 78:93–96

Sestak K, Thwin H, Dufour J, Aye PP, Liu DX, Moehs CP (2015) The effects of reduced gluten barley diet on humoral and cell-mediated systemic immune responses of gluten-sensitive *Rhesus macaques*. Nutrients 7:1657–1671

Shewry PR (2011) Effects of nitrogen and sulfur nutrition on grain composition and properties of wheat and related cereals. In: Hawkesford MJ, Barraclough P (eds) The molecular and physiological basis of nutrient use efficiency in crops. Wiley-Blackwell, Chichester, pp 103–120

Shewry PR (2018) Do ancient types of wheat have health benefits compared with modern bread wheat? J Cereal Sci 79:469–476

Shewry PR (2019) What is gluten – why is it special? Front Nutr 6:101

Shewry PR, Tatham AS, Forde J, Kreis M, Miflin BJ (1986) The classification and nomenclature of wheat gluten proteins: a reassessment. J Cereal Sci 4:97–106

Shewry PR, Halford NG, Tatham AS (1992) High-molecular-weight subunits of wheat glutenin. J Cereal Sci 15:105–120

Shewry PR, D'Ovidio R, Lafiandra D, Jenkins JA, Mills ENC, Bekes F (2009) Wheat grain proteins. In: Khan K, Shewry PR (eds) Wheat: chemistry and technology, 4th edn. AACC International, St. Paul, pp 223–298

Singh S, Gupta A, Kaur N (2012) Influence of drought and sowing time on protein composition, antinutrients, and mineral contents of wheat. Sci World 2012:1–9

Skoczowski A, Obtułowicz K, Czarnobilska E, Dyga W, Mazur M, Stawoska I, Waga J (2017) Antibody reactivity in patients with IgE-mediated wheat allergy to various subunits and fractions of gluten and non-gluten proteins from ω-gliadin-free wheat genotypes. Ann Agric Environ Med 24:229–236

Skodje GI, Sarna VK, Minelle IH, Rolfsen KL, Muir JG et al (2018) Fructan, rather than gluten, induces symptoms in patients with self-reported non-celiac gluten sensitivity. Gastroenterology 154:529–539

Sollid LM, Khosla C (2011) Novel therapies for coeliac disease. J Intern Med 269:604–613

Song K, Lee S, Ban C (2012) Aptamers and their biological applications. Sensors 12:612–631

Spaenij-Dekking L, Kooy-Winkelaar Y, Koning F (2005) The Ethiopian cereal tef in celiac disease. N Engl J Med 353:1748–1749

Stember RH (2006) Buckwheat allergy. Allergy Asthma Proc 27:393–395

Stenman SM, Venalainen JI, Lindfors K, Auriola S, Mauriala T et al (2009) Enzymatic detoxification of gluten by germinating wheat proteases: implications for new treatment of celiac disease. Ann Med 41:390–400

Tanner GJ, Blundell MJ, Colgrave ML, Howitt CA (2016) Creation of the first ultra-low gluten barley (*Hordeum vulgare* L.) for coeliac and gluten-intolerant populations. Plant Biotechnol J 14:1139–1150

Tatham AS, Shewry PR (2008) Allergens to wheat and related cereals. Clin Exp Allergy 38:1712–1726

Trcka J, Schäd SG, Scheurer S, Conti A, Vieths S, Gross G, Trautmann A (2012) Rice-induced anaphylaxis: IgE-mediated allergy against a 56-kDa glycoprotein. Int Arch Allergy Immunol 158:9–17

Tye-Din JA, Stewart JA, Dromey JA, Beissbarth T, van Heel DA, Tatham A, Henderson K, Mannering SI, Gianfrani C, Jewell DP, Hill AV, McCluskey J, Rossjohn J, Anderson RP (2010) Comprehensive, quantitative mapping of T cell epitopes in gluten in celiac disease. Sci Transl Med 2:41ra51

Vaccino P, Becker HA, Brandolini A, Salamini F, Kilian B (2009) A catalogue of *Triticum monococcum* genes encoding toxic and immunogenic peptides for celiac disease patients. Mol Genet Genomics 281:289–300

van den Broeck HC, van Herpen TW, Schuit C, Salentijn EM, Dekking L et al (2009) Removing celiac disease-related gluten proteins from bread wheat while retaining technological properties: a study with Chinese spring deletion lines. BMC Plant Biol 9:41

van den Broeck HC, Hongbing C, Lacaze X, Dusautoir J-C, Gilissen L, Smulders M, van der Meer I (2010a) In search of tetraploid wheat accessions reduced in celiac disease-related gluten epitopes. Mol BioSyst 6:2206–2213

van den Broeck HC, de Jong HC, Salentijn EMJ, Dekking L, Bosch D et al (2010b) Presence of celiac disease epitopes in modern and old hexaploid wheat varieties: wheat breeding may have contributed to increased prevalence of celiac disease. Theor Appl Genet 121:527–1539

van den Broeck HC, Gilissen LJWJ, Smulders MJM, van der Meer IM, Hamer RJ (2011) Dough quality of bread wheat lacking alpha-gliadins with celiac disease epitopes and addition of celiac-safe avenins to improve dough quality. J Cereal Sci 53:206–216

Van Herpen TWJM, Goryunova S, van der Schoot J, Mitreva M, Salentijn EMJ et al (2006) Alpha-gliadin genes from the A, B, and D genomes of wheat contain different sets of celiac disease epitopes. BMC Genomics 7:1
Venske E, dos Santos RS, Busanello C, Gustafson P, de Oliveira AC (2019) Bread wheat: a role model for plant domestication and breeding. Hereditas 156:16
Veraverbeke WS, Delcour JA (2002) Wheat protein composition and properties of wheat glutenin in relation to breadmaking functionality. Crit Rev Food Sci Nutr 42:179–208
Verbeke K (2018) Non-celiac gluten sensitivity: What is the culprit? Gastroenterology 154:471–473
Waga J, Skoczowski A (2014) Development and characteristics of ω-gliadin-free wheat genotypes. Euphytica 195:105–116
Weiser H, Seilmeier W (2003) Determination of gliadin and gluten in wheat starch by means of alcohol extraction and gel permeation chromatography. In: Stern M (ed) Proceedings of the 17th meeting of working group on prolamin analysis and toxicity. Verlag Wissenschaftliche Scripten, Zwickau, pp 53–57
Wen S, Wen N, Pang J, Langen G, Brew-Appiah RAT et al (2012) Structural genes of wheat and barley 5-methylcytosine DNA glycosylases and their potential applications for human health. Proc Natl Acad Sci USA 109:20543–20548
Wieser H, Seilmeier S (1998) The influence of nitrogen fertilisation on quantities and proportions of different protein types in wheat flour. J Sci Food Agric 76:49–55
Wieser H, Gutser R, Tucher SV (2004) Influence of Sulphur fertilization on quantities and proportions of gluten protein types in wheat flour. J Cereal Sci 40:239–244
Wieser H, Koehler P, Folck A, Becker D (2006) Characterization of wheat with strongly reduced α-gliadin content. In: Lookhart GL (ed) Gluten proteins 2006. C.H.I.P.S., Weimar, pp 13–16
Zannini E, Pontonio E, Waters DM, Arendt EK (2012) Applications of microbial fermentations for production of gluten-free products and perspectives. Appl Microbiol Biotechnol 93:473–485
Zecevic V, Knezevic D, Boskovic J, Madic M (2009) Effect of genotype and environment on wheat quality. Genetika 41:247–253
Zevallos VF, Ellis HJ, Suligoj T, Herencia LI, Ciclitira PJ (2012) Variable activation of immune response by quinoa (*Chenopodium quinoa* Willd.) prolamins in celiac disease. Am J Clin Nutr 96:337–344
Zevallos VF, Herencia LI, Chang FJ, Donnelly S, Ellis HJ, Ciclitira PJ (2014) Gastrointestinal effects of eating quinoa (*Chenopodium quinoa* Willd.) in celiac patients. Am J Gastroenterol 109:270–278
Zhu X, Li C, Jiang Z, Huang L, Feng C, Guo W, Peng Y (2012) Responses of phosphorus use efficiency, grain yield, and quality to phosphorus application amount of weak-gluten wheat. J Integr Agric 11:1103–1110
Zou TX, Dai TB, Jiang D, Qi J, Cao WX (2006) Effects of nitrogen and potassium supply on grain yield and quality in weak gluten wheat. J Triticeae Crops S512:1

# Wheat Quality Improvement for Micronutrients

**Ashita Bisht, Satveer Kaur, Shivani Sharma, Abhishek Bhandawat, Shubham Bhardwaj, Monika Garg, Ajay Kumar Pandey, Mahendra Bishnoi, Tilak Raj Sharma, and Joy K. Roy**

**Abstract** Micronutrients are crucial for plant growth and human health. Up to two billion people worldwide suffer from iron and zinc deficiencies, particularly in regions with predominantly cereal-based diets. Wheat is recognized as a widely consumed staple food and a rich source of protein and dietary fibre, which contains a low level of the essential micronutrients. There are many possible strategies to improve micronutrient availability in staple food, including dietary diversification, mineral supplementation, and post-harvest food fortification. Biofortification is the most convenient, faster, and cost-effective method to combat malnutrition by increasing the density and bioavailability of critical essential micronutrients. This book chapter describes the promising sustainable approaches for biofortification of wheat, which includes breeding, agronomy, and genetic engineering to improve Fe and Zn content in wheat grains. Enhanced micronutrient density in grain destined for human consumption may alleviate hidden hunger or micronutrient malnutrition among the malnourished population of the world. Therefore, the biofortification of wheat with increased nutritional quality is expected to gain more attention in the future.

**Keywords** Biofortification · Iron · Zinc · Micronutrient · Wheat

## 1 Introduction

Wheat is a widely consumed staple cereal as it holds a significant proportion of total global acreage (14%, 21.8 million hectares) and production (13.64%, 771.7 million tonnes) with average productivity of 3531.2 kg/ha (FAOSTAT Database 2019). It is readily available to the majority of the population, but it lacks essential

A. Bisht · S. Kaur · S. Sharma · A. Bhandawat · S. Bhardwaj · M. Garg · A. K. Pandey
M. Bishnoi · T. R. Sharma · J. K. Roy (✉)
National Agri-Food Biotechnology Institute, Mohali, Punjab, India
e-mail: joykroy@nabi.res.in

© The Editor(s) (if applicable) and The Author(s), under exclusive license to Springer Nature Switzerland AG 2021

S. H. Wani et al. (eds.), *Physiological, Molecular, and Genetic Perspectives of Wheat Improvement*, https://doi.org/10.1007/978-3-030-59577-7_3

micronutrients. Essential micronutrients are trace elements like zinc (Zn), copper (Cu), iron (Fe), manganese (Mn), boron (B), molybdenum (Mo), chlorine (Cl), nickel (Ni), selenium (Se), and cobalt (Co) that play a crucial role in various plant biosynthetic pathways. However, if the supply of one or more nutrients is inadequate, it would negatively affect crop production and human health (Shukla et al. 2018). Micronutrient malnutrition is considered to be a severe global health problem, which results in low productivity and poor standard of nutritional quality of products, which in turn causes long-term health consequences in humans. About two billion people in the world are confronting this considerable problem (Fig. 1, data adapted from WHO 2002). Deficiencies of iron, iodine, vitamin A, and zinc cast an adverse effect on human health such as anaemia, growth retardation, weakening of immune system, severity of infectious diseases, rickets, osteoporosis, and blindness of young children which further increases the risk of mortality and burden of death (Muller and Krawinkel 2005). In pregnant women, deficiencies of these nutrients increase the possibility of low birth weight, congenital disabilities, and pregnancy-related complications. Leaching, liming of soils, insignificant utilization of manures, and accelerated exploitation of micronutrient-pure chemical fertilizers are the major factors which are responsible for the loss of micronutrients. In addition to these factors, an inappropriate supply of micronutrients through fertilization is the primary cause of rampant deficiencies of micronutrients in the soil. Each micronutrient plays a precise role in plant, animal, and human metabolism, and their deficiency cannot be mitigated by the replacement of other elements (Ahmed et al. 2012).

Over the past half a century, the constant selection for increased yield has led to a severe decrease in micronutrient content (Velu et al. 2014). Common symptoms observed in plants due to nutrient deficiencies are stunted growth, chlorosis, twisted

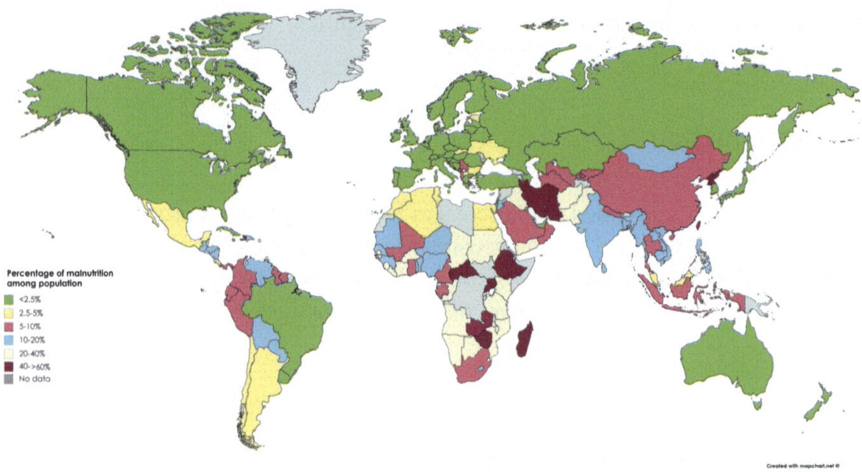


**Fig. 1** Country-wise depiction of the population (%) suffering from malnutrition. (Drawn using Map Chart. https://mapchart.net/)

leaves, purplish-red colouring, and loss of turgor pressure (Hodges and Constable 2010). Among them, iron and zinc deficiency are the most ubiquitous micronutrient deficiency, which may affect major cereals such as wheat, rice, and maize. The nutrient deficiencies and toxicities in the soil can be diagnosed through soil testing, plant investigation, and visual inspection (Beede et al. 2005). Various effective approaches exist that help to alleviate micronutrient deficiency such as biofortification and dietary diversification having supplementation with multiple micronutrients. Biofortification is considered as the best strategy to enhance the nutritional quality of various crops and to develop crop varieties that possess specific end product and nutritional requirements. It helps in minimizing the nutritional gap by offering low affordability and accessibility in reaching low-income households of developing regions such as Asia and Africa (Beal et al. 2017).

The development of micronutrient-rich genotypes poses a great challenge due to its complex genetics. Therefore, the identification of genome-wide distributed QTLs and genes is vital to design an effective biofortification strategy. Extensive consideration of five key factors, viz. efficient root uptake of the mineral from the soil, mobilization of micronutrients towards the grain, exclusive storage in the endosperm, enhanced bioavailability, and suppressing the level of anti-nutrients, is required. The molecular components controlling each of these areas have been used individually or in combinations to increase the nutrition concentration in grains (Ludwig and Slamet-Loedin 2019). This chapter aims at understanding various strategies such as agronomic biofortification, conventional breeding, molecular breeding, and transgenic approaches to provide a holistic system (Fig. 2) for increasing the genetic potential for enhanced micronutrient content. The effect of environment, factors affecting bioavailability, and worldwide programmes on mineral biofortification have also been discussed. All the components, as mentioned earlier, could help a conceptual increase of 25 and 10 mg/g in Fe and Zn concentrations, respectively, above the baseline. Thus, bringing the targeted concentration to 60 mg/g in Fe and 45 mg/g in Zn (Ortiz-Monasterio et al. 2007) is elucidated in Fig. 2.

## 2 Importance of Wheat Quality from Health Perspectives

Wheat is a good source of nutrition and calories that are essential for healthy growth and development. Consumption of wheat-based foods associated with both positive and adverse health effects, depending on the quantity and quality of the whole grain wheat that is included in the diet (Shewry et al. 2012). Wheat grain is also known as a caryopsis, is usually oval in shape, and is comprised of 2–3% wheat germ, 13–17% bran, and 80–85% endosperm (Belderok et al. 2000). The outer layer of the grain is bran, which contains fibres and is also rich in vitamin B and minerals. The chemical composition of wheat contains a significant amount of various nutrients such as polysaccharides, proteins, lipids, vitamins, minerals, dietary fibres, and phytochemicals, which may contribute to a healthy diet (Shewry and Hey 2015) as given in

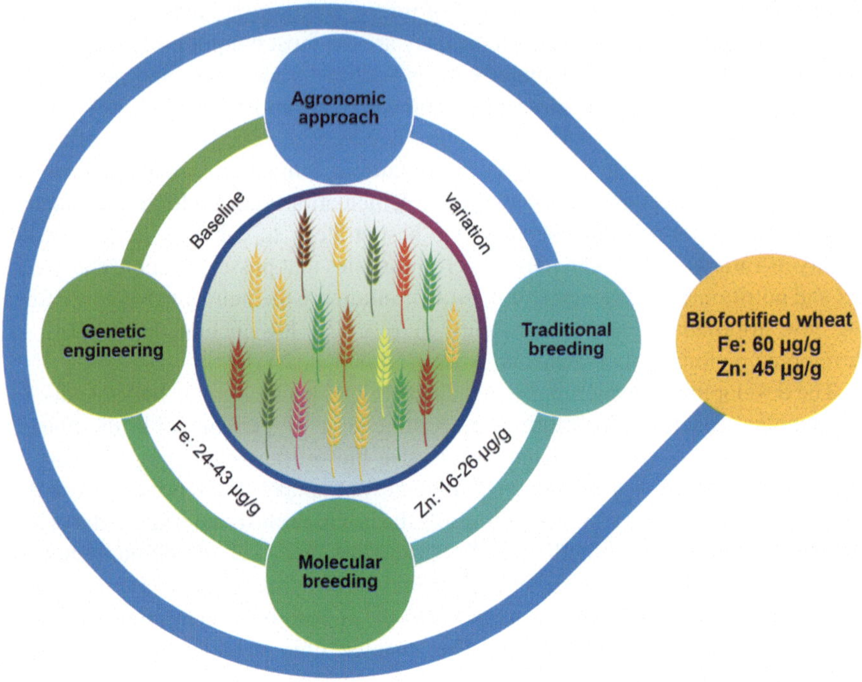


**Fig. 2** Various approaches used in wheat to increase mineral (Fe and Zn) concentration above baseline

**Table 1** Chemical composition of wheat grain

| Part of the wheat grain | Chemical components | References |
|---|---|---|
| Bran | Dietary fibres (53%), vitamins, and minerals (7.2%) | Cornell and Cauvain (2003) |
| Endosperm | Carbohydrates, fats (1.5%), and proteins (13%) | Belderok et al. (2000) and Šramková et al. (2010) |
| Wheat germ | Protein (25%), lipids (8–13%), vitamins, and minerals (4.5%) | Cornell and Cauvain (2003) |

Table 1. The major accumulation of minerals, including Fe and Zn, is found in the embryo and aleurone layer. Still, the aleurone layer comparatively stores more Fe fraction, while zinc is more concentrated in the embryo (Balk et al. 2019).

The consumption of wheat-based foods such as bread, pasta, noodles, breakfast cereals, cookies, and crackers contribute to body weight maintenance as well as to diet quality, providing a more balanced intake of nutrients. The nutritional value of wheat helps to fight against diet and lifestyle-related diseases (Table 2), including obesity, cardiovascular diseases, and type 2 diabetes. Antioxidant phytochemicals found in wheat bran fractions may modulate cellular oxidative status and prevent biologically important molecules such as DNA, proteins, and membrane lipids from

**Table 2** Potential health benefits of whole-grain wheat components

| Whole grain wheat components | Health benefits | References |
|---|---|---|
| *Dietary fibres* | Improve laxation, reduce blood cholesterol and glucose level | Slavin et al. (2009) |
| *Vitamins and provitamins* (tocopherols and tocotrienols) | Acts as a potent antioxidant, diminish the risk of cancer, CVD, and type 2 diabetes | Atkinson et al. (2010), Traber et al. (2008) and Milman et al. (2008) |
| *Minerals* | | |
| (a) Magnesium | Improves insulin metabolism and reduce endothelial dysfunction | Fung et al. (2002) |
| (b) Selenium | Helps in redox reaction and thyroid metabolism | Chacko et al. (2009) |
| *Other bioactive compounds* | | |
| (a) Polyphenols | Decreases oxidative stress and reduce the risk of cancer, heart, and neurodegenerative diseases | Aviram et al. (2005), Barone et al. (2009) and Duffy and Vita (2003) |
| (b) Phenolic acids | Hypocholesterolemic and anti-atherogenic | Graf (1992) |
| (c) Ferulic acids | Antimutagenic and anti-inflammatory | Birošová et al. (2005) and Ozaki (1992) |

**Table 3** Recommendations for micronutrients

| Gender | Fe (mg/day) | Zn (mg/day) | Cr (ug/day) | Cu (mg/day) | Se (mg/day) |
|---|---|---|---|---|---|
| Women | 14.8 | 7.0 | 25.0 | 1.2 | 0.060 |
| Men | 8.7 | 9.5 | 25.0 | 1.2 | 0.075 |

oxidative damage and that this consequently involved in diminishing the risk of chronic diseases such as CVD and cancer (Stevenson et al. 2012). So, there is an instant need to integrate measures of nutritional quality in wheat end-use. Increase in global demand for wheat is attributed to the physical and viscoelastic properties of the flour and dough which help in the processing of wheat to produce various unique wheat products such as bread, noodles, pasta, cakes, and biscuits which are not possible from other staple crops (Shewry and Hey 2015). It has high grain protein content; therefore, the demand for wheat-based convenience foods has extensively increased due to urbanization and industrialization (Kong 2013).

## 3 Bioavailability

Daily recommendations for micronutrients vary across gender and age (Table 3, NHS 2019). The Fe uptake in women is higher than that of men. In general, the bioavailability of micronutrients is very little, i.e., Fe (5%) and Zn (25%). Thus, to provide the daily requirement of 14.8 mg and 8.7 mg Fe to women and men, respectively would, in turn, require an intake of 296 mg/day and 174 mg/day Fe. Similarly, to fulfil the daily demand of 7 mg and 9.5 mg Zn in women and men, respectively, would involve the consumption of 28 mg/day and 38 mg/day of Zn. Thus, apart from increasing the micronutrient content, the efforts to decrease the level of anti-nutrients/inhibitors and increasing the extent of the substances which promotes nutrient absorption (Bouis 2003; Bouis and Welch 2010) for bioavailability are equally important.

Phytic acid is a common anti-nutrient that accounts for 75% of the total storage fraction of phosphorus in wheat seed. It is abundant in the germ and aleurone layers of cereal grains (Lott and Spitzer 1980). It comprises inositol hexaphosphates and pentaphosphates which readily binds with cations like Ca, Fe, Mn, magnesium (Mg), potassium (K), and Zn. However, it is also known to chelate Fe and Zn forming insoluble complexes with them, which make these nutrients poorly soluble in the gastrointestinal tract causing a reduction in their bioavailability (Iqbal et al. 1994). On the contrary, it is also known to be vital for seed germination and growth, along with providing defence against oxidative stress (Doria et al. 2009; Balk et al. 2019). It also possesses antioxidant or anticarcinogen properties (Schlemmer et al. 2009). So, it is vital to achieving a balanced concentration of phytic acid in seed grain (Garcia-Oliveira et al. 2018).

Ferritin protein is ubiquitous in nature and acts as an iron reserve. It is known to promote bioavailability (Bouis and Welch 2010). For both phytic acid and ferritin, the number of genes involved in biosynthesis and metabolism is lesser in comparison with those associated with uptake, transport, and deposition of Fe and Zn. Consequently, theoretically, improving bioavailability should be less challenging than micronutrient enrichment (Bouis and Welch 2010).

## 4 Types of Wheat-Based on Colour

Colour wheat exists in various forms (Fig. 3) such as purple, blue, black, red, and yellow coloured depending upon the types and occurrence of the phenolics/anthocyanins/xanthophylls in wheat seed coat layers (Ficco et al. 2014). Purple, blue, and black colours are due to anthocyanins that are well-known antioxidants that remove harmful free radicals from body. In purple wheat, anthocyanins are present in pericarp layer, while blue wheat has anthocyanins in the aleurone layer, and the black wheat is a combination of both, i.e., anthocyanins which is present in pericarp as well as in aleurone layer (Abdel-Aal et al. 2006, 2008; Garg et al. 2016). The blue and purple colours are developed naturally at the time of grain filling. Apart from anthocyanins, coloured wheat has been reported to have higher accumulation of minerals in the grain. Guo et al. (2012) and his colleagues observed higher contents of Fe, Zn, Mn, Cu, Se, Mg, K, and P in black wheat. The high concentration of iron, zinc, and magnesium has been reported in purple wheat. High organic chromium has been reported from coloured wheat (Guo et al. 2012). Coloured wheat

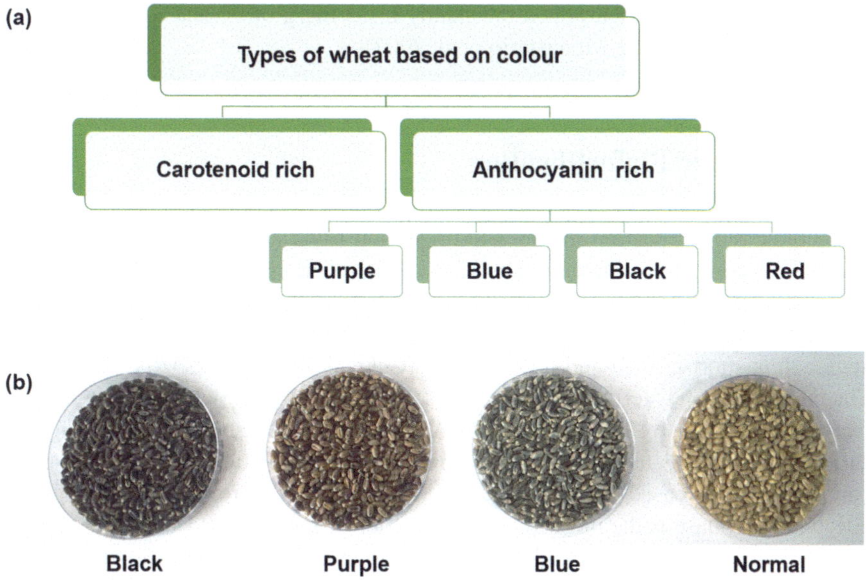


**Fig. 3** Classification of wheat. (**a**) Types of nutritionally rich wheat based on colour. (**b**) Anthocyanin-rich black, purple, and blue wheat in comparison with normal wheat

anthocyanins are sugar or phenolic group derivatives of six major anthocyanidins, i.e. cyanidin, delphinidin, malvidin, pelargonidin, peonidin, and petunidin (Guo et al. 2012). Another class of wheat based on colour is red wheat, which does not contain anthocyanins but have catechins (Havrlentová et al. 2014). On the other hand, the yellow-type wheat predominantly has colour in the endosperm of durum wheat. The yellow pigment is mostly associated with carotenoid (carotene and xanthophyll), mainly, lutein and zeaxanthin, yet the latter two do not display any provitamin A activity but have been reported to be good for eye health (Yeum and Russell 2002; Leenhardt et al. 2006; Ortiz-Monasterio et al. 2007).

## 5 Effect of Environment on Micronutrient Content

The environmental conditions, chiefly the composition of the soil, impart variation in the determination of grain micronutrient (Garvin et al. 2006; Trethowan 2007; Ortiz-Monasterio et al. 2007). Even after breeding for efficient uptake and mobilization of micronutrients, their limited supply in the soil acts as a limiting factor. A century-long study conducted on 14 US hard red winter wheat varieties showed that different locations have a significant effect on the seed mineral content viz., Fe, Zn, Cu, and Se (Garvin et al. 2006). Studies on Indian landrace 'C306' showed contrasting Fe and Zn concentrations in hydroponic and field conditions. Under hydroponics, the amount of Fe and Zn was 220 mg/g and 130 mg/g, respectively, while in the field setting the quantity dropped drastically to 33 mg/g of Fe and 31 mg/g of Zn (Welch et al. 2005; Ortiz-Monasterio et al. 2007).

# 6 Agronomic Biofortification

Agronomic biofortification includes fertilizer and foliar application of micronutrients, which could be sufficiently absorbed by the plant system determined by factors such as leaf anatomy, stage of the application, soil, pH, and climate (Alshaal and El-Ramady 2017). However, in comparison to breeding methods, the agronomic approach is short-termed and less efficient (Cakmak 2008; Aciksoz et al. 2011b). The most common is fertigation with zinc salt, such as, $ZnSO_4$, which has reports of increasing grain Zn by around 60% (Zhang et al. 2012b). The timing of foliar application further affects its efficiency. Applying Zn during the grain development stage contributes to increased grain Zn concentration. Studies have revealed that the grain development phase is the most important for foliar application (Ozturk et al. 2006; Zhang et al. 2010). A recent study also corroborates the foliar application of Fe and Zn to enhance the concentration of these micronutrients in wheat (Niyigaba et al. 2019).

Another innovative method for nutrient delivery is nanofertilizer. These fertilizers are either encapsulated or covered with nanoparticles with the high surface-to-volume ratio, which helps in precise and slow delivery of single or combination of nutrients. These are more efficient in nutrient use and environment safe. It can be absorbed via both roots and leaves which is an added advantage (Zuverza-Mena et al. 2017; Feregrino-Pérez et al. 2018; Adisa et al. 2019; Zulfiqar et al. 2019). Different types of nanomaterials are either single or multiwalled nanotubes and magnetized iron nanoparticles such as Cu, Zn, zinc oxide (ZnO), silica, gold (Au), silver (Ag), and aluminium (Al). Several studies have reported its successful application for N, P, K, Cu, Fe, Mn, Mo, and Zn (Liu and Lal 2015; Solanki et al. 2016). Foliar application of Zn-chitosan at a concentration of 20 mg/g (w/w) revealed increased grain Zn content by 27% and 42% in durum wheat cultivars MACS 3125 and UC1114, respectively (Deshpande et al. 2017).

Furthermore, iron-chelating compounds like phytosiderophores are also known to be released from roots to facilitate Fe and Zn mobilization under deficiency conditions (Cakmak et al. 1994). These are instrumental in Fe uptake and are known to be promoted by enhancing nitrogen (N) concentration in wheat (Murata et al. 2006; Suzuki et al. 2006; Aciksoz et al. 2011a). Other chelating proteins of Fe and Zn, such as nicotianamine, are also positively influenced by improved N status. Hence, the application of N could also influence uptake and translocation for other micronutrients. So, the method of agronomic biofortification is mainly successful in nutrient-deficient soils. The soil and foliar application become even more practical when jointly applied with herbicides, pesticides, and fungicides, thereby reducing cost and time (Velu et al. 2014).

# 7 Conventional Methods of Breeding for Quality Improvement

Conventional breeding is the genetic manipulation of crops using traditional methods (mass selection, pure line, pedigree, bulk population, backcrossing, single seed descent, multiline, and composite) within restricted gene pools (Acquaah 2016). Therefore genetic biofortification using conventional breeding aims for a simultaneous increase in nutritional quality (Bouis 2003) along with yield. For improvement of a trait, identification and utilization of genetic variation are of prime importance. Screening of germplasm in several studies indicated higher levels of Fe and Zn (2–3 times) in progenitors (*Aegilops speltoides*, and *Ae. tauschii*), wild relatives (*Triticum monococcum, T. dicoccoides, T. boeticum, Ae. kotschyi, Ae. longissima, Ae. peregrina, Ae. cylindrica, Ae. ventricosa*, and *Ae. geniculata*), landraces, and synthetic amphiploids than commercial cultivars (Cakmak et al. 2000; Monasterio and Graham 2000; Calderini and Ortiz-monasterio 2003; Chhuneja et al. 2006; Ortiz-Monasterio et al. 2007; Genc et al. 2009; Velu et al. 2011, 2014; Xu et al. 2011). Higher concentrations were also observed in disomic hexaploid lines of bread wheat and monosomic and disomic addition lines of *Ae. peregrina*, *Ae. longissima*, and *Ae. umbellulata* (Neelam et al. 2012; Velu et al. 2014; Goudia and Hash 2015; Kumar et al. 2019). Still, *T. dicoccoides* (Cakmak et al. 2000; Ortiz-Monasterio et al. 2007; Garcia-Oliveira et al. 2018) and *T. turgidum ssp. dicoccum* (Velu et al. 2014; Cheema et al. 2018) were among the most promising sources for improved mineral levels. Further, *T. dicoccoides* substitution lines indicated the chromosomes 6A, 6B, and 5B as putative regions for genes responsible for improved concentrations of Fe and Zn (Cakmak et al. 2004). Similarly, screening of *A. tauschii* and rye revealed a higher concentration of 42% and 35% for Se. Significantly higher contents for Se, Li, and Mg were also reported in *T. dicoccon* and *T. spelta* (Piergiovanni et al. 1997).

Once a suitable genetic variation is present, traditional breeding relies on effective selection dependent upon additive genetic effects, the phenomenon of heterosis in $F_1$ progeny, and transgressive segregation in later generations (Garcia-Oliveira et al. 2018). As depicted in Fig. 4, the crossing of genotypes with distant ancestry with intermediate values to produce superior transgressive segregants and introgression of genes from wild relatives with higher micronutrient content are common practices in wheat biofortification for micronutrients (Ortiz-Monasterio et al. 2007).

Conventional breeding for quality aspects such as carotenoid, anthocyanin, Fe, Zn, and protein is a slow process as these traits are polygenic in nature with complex inheritance, low heritability, and affected mainly by the environment (Trethowan et al. 2005; Trethowan 2007). To overcome the challenge of slow generation time, a recent advancement called 'speed breeding' can be employed. This is a promising method to shorten the breeding cycle by providing an extended photoperiod ensuring six generations per year. The optimum conditions involves light of wavelengths 400–700 (photosynthetically active radiation (PAR)) with quality of photosynthetic photon flux density (PPFD) or Lux of ~450–500 μmol/m$^2$/s at plant canopy height, photoperiod of 22 h/2 h (light/dark), temperature (22 °C/17 °C), and humidity

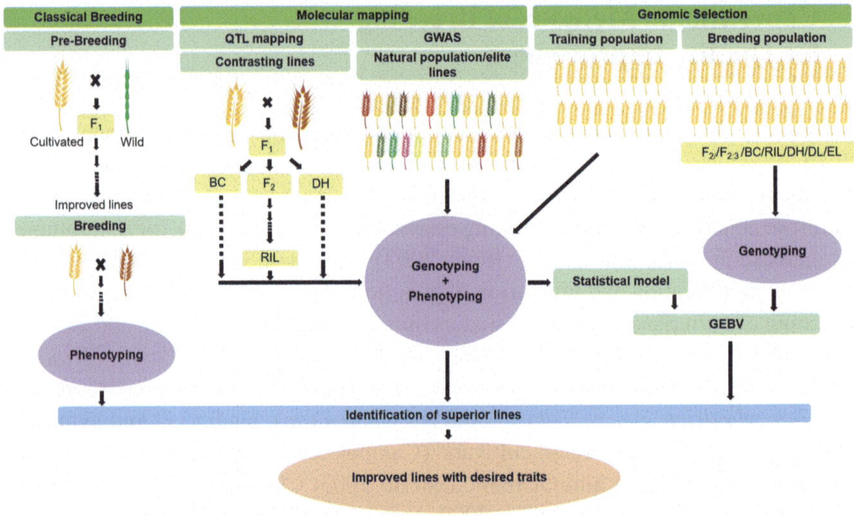


**Fig. 4** Different breeding approaches for micronutrient biofortification. Abbreviations: *F1* first filial generation, *BC* back cross, *F2* second filial generation, *DH* double haploid, *RIL* recombinant inbred line, *F2:3* F3 derived from F2 generation, *DL* diverse line, *EL* elite lines, *GEBV* genomic estimated breeding values

60–70% (Ghosh et al. 2018; Watson et al. 2018). Another obstacle in breeding for improved quality is the negative correlation between yield and Fe, Zn, and grain protein concentration (Kibite and Evans 1984; Graham et al. 1999; Monasterio and Graham 2000; Ortiz-Monasterio et al. 2007; Laidig et al. 2017; Kondić-Špika et al. 2019). Nevertheless, a positive correlation between Fe and Zn (Cakmak et al. 2004; Morgounov et al. 2007; Genc et al. 2009; Peleg et al. 2009; Gomez-Becerra et al. 2010; Zhang et al. 2010; Velu et al. 2011, 2012) suggests colocalization of these elements and the genetic behaviour of the alleles responsible for their accumulation and storage in grains which are either pleiotropic or co-segregate (Velu et al. 2014; Cheema et al. 2018) which facilitates the concurrent improvement of both micronutrients (Monasterio and Graham 2000; Badakhshan et al. 2013; Heidari et al. 2016). In comparison to the total number of released wheat varieties, the number of biofortified varieties for Fe and Zn is quite small; moreover, only a few known cultivars are released for high Fe content (Table 4). Still, continuous efforts are being made across the globe to identify and develop superior breeding lines.

## 8 Molecular Breeding for Quality Improvement

Molecular breeding or marker-assisted breeding (MAB) utilizes molecular markers that are tightly linked to the trait of interest. The genetic analysis of markers associated with the target QTL (quantitative trait loci) is accomplished using QTL

**Table 4** List of biofortified wheat varieties for various micronutrients

| Variety | Biofortified for | Year of release | Institute | References |
|---|---|---|---|---|
| HI 8627 (Malav Kirti) | Carotene | 2005 | Indian Agricultural Research Institute (IARI), India | IARI Database (2019) |
| HD 2932 (Pusa Wheat 111) | Zn | 2007 | IARI, India | IARI Database (2019) |
| BHU 1, Akshai (BHU3), BHU 5, BHU 6, BHU 17, BHU 18 | Zn | 2014 | CIAT, CIMMYT, Harvest Plus | Velu et al. (2015) and Harvest Plus (2019) |
| Abhay (Zinc Shakthi) | Zn | 2015 | Nirmal seeds and HarvestPlus | Velu et al. (2015, 2018) |
| Zincol | Zn | 2015 | CIMMYT<br>National Agricultural Research Center, Pakistan | Singh and Velu (2017) |
| NABIMG-9, ABIMG-10, NABIMG-11 | Anthocyanin | 2016 | National Agri-Food Biotechnology Institute, India | Garg et al. (2016) |
| Zinc Shakti (Chitra) | Zn | 2016 | Harvest Plus | Singh and Velu (2017) |
| HPBW-01 (PBW 1 Zn) | Fe and Zn | 2017 | Punjab Agricultural University, India | Singh and Velu (2017) and Yadava et al. 2017 |
| WB02 | Fe and Zn | 2017 | Indian Institute of Wheat and Barley Research, India | Singh and Velu (2017) and Yadava et al. (2017) |
| BARI Gom 33 | Zn | 2017 | Bangladesh Agricultural Research Institute (BARI) collaborated with CIMMYT | CIMMYT (2019) |

mapping. For QTL mapping, various mapping populations are used, which are either mortal (segregating) or immortal (non-segregating) lines (Fig. 3). The mortal lines consist of $F_2$, $F_{2:3}$, and back cross (BC), while double haploid (DH), recombinant inbred lines (RIL) attained after 6–8 cycles of single seed descent method (SSD) are covered under immortal lines (Singh and Singh 2015). Many QTLs have been identified in diverse germplasm and advanced breeding lines for Fe and Zn; some selected studies are presented in Table 5. However, QTLs are not stable across different environments and show additive and epistatic nature (Garcia-Oliveira et al. 2018). The method of meta-QTL has been proposed to identify a few robust and reproducible markers, which will be present across diverse environments. Moreover, the latest development in genome-wide association studies (GWAS), which involves variation present in naturally diverse lines (DL) or elite lines (EL), helps in the production of dense linkage maps.

**Table 5** List of QTLs identified for iron and zinc

| Trait | Cross/parents | QTLs | References |
|---|---|---|---|
| Fe and Zn | *Triticum dicoccoides* | *GPC-B1* (6 7 BS) | Joppa et al. (1997), Uauy et al. (2006) and Distelfeld et al. (2007) |
| Fe and Zn | *Triticum dicoccoides* | *TtNAM-B1* | Distelfeld et al. (2007) |
| Fe | RIL (*Triticum boeoticum* × *Triticum monococcum*) | *QFe.pau-7A, QFe.pau-2A* | Tiwari et al. (2009) |
| Fe and Zn | RIL (Xiaoyan × 54 Jing 411) | *QZn-5A, QFe-5A2, QGpc-5A1, QGpc-6A* | Xu et al. (2012) |
| Zn | RIL (PBW343 × Kenya Swara) | *QGzncpk.cimmyt-1BS, QGzncpk.cimmyt-2Bc, QGzncpk.cimmyt-3AL* | Hao et al. (2014) |
| Fe and Zn | RIL (*T. spelta* (H+ 26 (PI348449) × *T. aestivum* cv. HUW 234) | *QZn.bhu-2B, QZn.bhu-6A, QFe.bhu-3B* | Srinivasa et al. (2014) |
| Fe and Zn | DH (Berkut × Krichauff) Hexaploid (Adana99 × 70711) | *QGfe.ada-2B, QGfe.ada-2B, QGZn.ada-2B, QGfe.ada-2B, QFe.bhu-2B* | Tiwari et al. (2016) and Velu et al. (2016) |
| Fe | Tetraploid (Saricanak98 × MM5/4) | *QGfe.sar-5B* | Velu et al. (2016) |
| Zn | Tetraploid (Saricanak98 × MM5/4) | *Qzneff.sar-6A, Qzneff.sar-6B* | Velu et al. (2016) |
| Zn | DH (Berkut × Krichauff) | *QZn.bhu-1B, QZn.bhu-2* | Tiwari et al. (2016) and Velu et al. (2016) |
| Zn | Tetraploid (Saricanak98 × MM5/4) | *QGzn.sar-1B, QGzn.sar-6B, QGZn.sar-1B* | Velu et al. (2016) |
| Zn | Hexaploid (Adana99 × 70711) | *QGzn.ada-6B, QGzn.ada-1D QGzn.ada-7B* | Velu et al. (2016) |
| Fe and Zn | RIL (Synthetic hexaploid wheat × *Triticum spelta*) | *QGZn.cimmyt-7B_1P2, QGFe.cimmyt-4A_P2, QGZn.cimmyt-7B_1P2, QGZn.cimmyt-7B_1P1* | Crespo-Herrera et al. (2017) |
| Fe and Zn | *Triticum dicoccon* PI94624/*Aegilops squarrosa* [409] × BCN | *QGFe.iari-2A, QGFe.iari-5A, QGFe.iari- 7A and QGFe.iari-7B, QGZn.iari-2A, QGZn.iari-4A, QGZn.iari-5A, QGZn. iari-7A and QGZn.iari-7B* | Krishnappa et al. (2017) |

An essential gene, GPC-B1 (high grain protein content), was identified on six BS chromosome (Joppa et al. 1997) of wild emmer wheat (*Triticum turgidum* ssp. *dicoccoides*) and plays an integral part in the simultaneous improvement of grain Fe, Zn, and protein content. The introgression of GPC-B1 clone in the recombinant chromosome substitution lines (RSLs) of *Triticum dicoccoides* exhibited multiple positive effects. These lines accumulated more Fe (18%), Zn (12%), Mn (28%), and protein (38%) in the grains as compared to RSLs carrying the same allele from *Triticum durum* (Distelfeld et al. 2007). Most of the chromosomal regions are associated with QTL for Fe and Zn. Still, majority are clustered within A followed by B and D (Shi et al. 2008; Genc et al. 2009; Peleg et al. 2009; Xu et al. 2012; Rong-li et al. 2013; Zhi-en et al. 2014; Crespo-Herrera et al. 2016; Crespo-Herrera et al. 2017; Velu et al. 2017). Moreover, these studies pointed chromosome 7A as a critical region linked to grain Fe and Zn. Additionally some of the targeted QTLs responsible for both Fe and Zn are colocalized and can be manipulated together.

GWAS, along with genomic selection (GS), has greatly facilitated the mining of specific chromosomal regions/alleles responsible for mineral concentration enrichment. These approaches can be integrated for quicker identification of underlying candidate genes for micronutrient biofortification. Initial GWAS performed on grain Zn accumulation employing data utilizing 90K iSelect SNP data on wheat (Wang et al. 2014b) identified weak and varying marker associations (Guttieri et al. 2015). Recent findings on GWAS revealed several marker-trait associations (MTAs) related to micronutrients. In a study conducted for Fe, Zn, β-carotene, and grain protein content led to the identification of 136, 587, 28, and 33 MTAs using four different methods viz., single locus trait analysis, multi-locus mixed model, multi-trait mixed model, and matrix variate linear mixed model, respectively (Kumar et al. 2018). Bhatta et al. (2018) identified 13 stable MTAs in *Triticum turgidum* × *Aegilops tauschii* for a higher grain quantity of Cu, Fe, Mg, Mn, Ni, and Zn. Also, applying Illumina iSelect 90 K Infinitum SNP array platform revealed 39 MTAs for Zn concentration. Additionally, two significant QTLs were also detected on chromosomes 2 and 7 (Velu et al. 2018). A similar application of 90k iSELECT Infinium and 35 k Affymetrix arrays platforms helped in the detection of 40 MTAs for zinc with two significant effect associations on chromosomes 3B (723,504,241–723,611,488 bp) and 5A (462,763,758–466,582,184 bp, Alomari et al. 2018). The latest study reports significant SNPs (single-nucleotide polymorphism) and MTAs on chromosomes 2A, 3B, and 5A for grain Fe content. It further indicated the presence of candidate genes (transcription factors and transmembrane proteins) on chromosome 2A (763,689,738–765,710,113 bp, Alomari et al. 2019).

## 9 Transgenics

A complementary approach to conventional biofortification involves manipulation of the plant genome through the application of biotechnological approaches, commonly referred to as genetic modification or genetic engineering. The approach has been successful in creating several biofortified crops such as rice, maize, and sorghum over the past years (Masuda et al. 2012; Wei et al. 2012; Kappara et al. 2018) and can be utilized in increasing Fe and Zn concentrations in grain (Balk et al. 2019). Since wheat is a hexaploid with a large and complex genome (17 Gb), these attributes hinder the process of genetic engineering, making it recalcitrant to transformation. Moreover, Fe and Zn regulation are under rigid genetic control, and proper methods are required to decipher the underlying genes. Also, a limited number of studies are available for transgenic wheat.

Nonetheless, significant progress has been made in designing and optimizing *Agrobacterium*-mediated and biolistic wheat transformation methods raising the bar of transformation efficiency from as low as 5% to a greater than tenfold increase (Borisjuk et al. 2019). Recently emerging genome editing technologies such as CRISPR-Cas have greatly facilitated in generating a loss of function mutants, which would assist in mining gene functions in polyploid species such as wheat. Moreover, multiple genes can be targeted at once through this approach via a single construct, thus allowing multiplexing (Čermák et al. 2017).

The first transgenic approach in wheat for increased Fe came from endosperm-specific overexpression of wheat FERRITIN gene under the control of Ta-GLU-D1 promoter in wheat cultivar Bobwhite (Borg et al. 2012). As FERRITIN protein is involved in sequestering and storing iron atoms, it consequently helps in increasing iron accumulation reaching up to 50–85% in the wheat grains. Another study focused on incorporating the FERRITIN gene from *Phaseolus vulgaris* (common bean) under the rice endosperm-specific promoter, OsGLOBULIN, and rice Nicotianamine synthase gene (*OsNAS2*) driven by a constitutive promoter from *Zea mays*, ZmUBIQUITIN, individually or in combination (Singh et al. 2017a). This resulted in grains harbouring as high as 93.1 μg/g of Fe in greenhouse conditions. Most of the gene families targeted for Fe and Zn are illustrated in Fig. 5. Another successful example is the increase in Fe content by overexpressing wheat vacuolar transporter *TaVIT2* gene and using endosperm-specific high molecular weight (HMW) Glutenin D1 as promoter. This resulted in 21.7 ± 2.7 μg/g grain Fe content in lines with single-copy insertion and a fourfold increase (41.5 ± 8.2 μg/g) in lines with multiple copy insertion (Connorton et al. 2017). A recent report described the constitutive expression of *OsNAS2* driven by maize ubiquitin 1 promoter in wheat that resulted in 80 ppm of grain Fe in field trial conditions (Beasley et al. 2019). Several studies have (Table 6) focused on enriched for micronutrient content in wheat and major cereals.

Numerous efforts have been deployed to reduce the level of phytic acid in cereal grains using two commonly used approaches that include mutagenesis and transgenic expression of phytic acid degrading enzyme, i.e., phytase or targeted knockdown of genes involved in phytic acid synthesis and transport such as *MIPs* (Myo inositol 3 phosphate synthase) and IPK1 (inositol penta*kis*phosphate kinase). Success for generating low phytate was also obtained by targeting the vacuolar transporter *MRP/ABCC* (multidrug resistance-associated proteins/ATP-binding cassette type-C transporters, Sparvoli and Cominelli 2015; Bhati et al. 2016). Some of the important cereal crops generated with reduced phytic acid content are enlisted in Table 7.

Therefore, transgenics serve as an alternative method when none other seems to be effective. With technologies like RNAi and CRISPR-Cas, an undesirable vector backbone can be removed, making biotech crops more acceptable to consumers as traditional crops. Such approaches mediated by zinc finger nuclease were used for generating low phytate in maize (Shukla et al. 2009). Therefore, by completely understanding the mechanisms for Fe and Zn homeostasis, availability of these micronutrients could be improved.

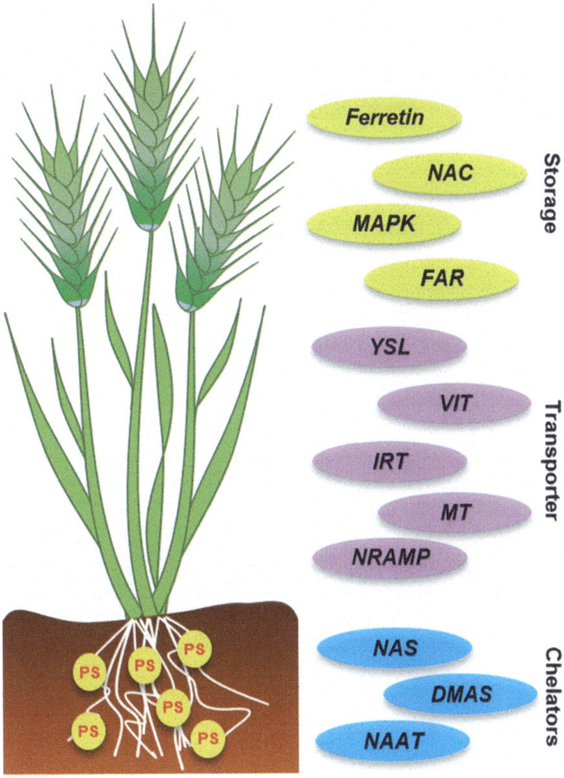


**Fig. 5** Major gene families involved with Fe and Zn regulation. Abbreviations: *NAC*, (*NAM*, *ATAF1*, *ATAF2* and *CUC2*); *MAPK* mitogen-activated protein kinase, *FAR* fatty acyl-CoA reductase, *YSL* yellow stripe-like, *VIT* vacuolar iron transporter, *IRT* iron regulated transporter, *MT* metal transporter, *NRAMP* natural resistance-associated macrophage proteins, *NAS* nicotianamine synthase, *DMAS* deoxymugineic acid synthase, *NAAT* nicotianamine aminotransferase, *PS* phytosiderophores. PS are produced by wheat plant under Fe- and Zn-deficient condition which helps in increasing the uptake of these metal ions

**Table 6** List of genes targeted for the increment in micronutrient content

| Micronutrient | Crop | Gene expression | Increment (μg/g) | References |
|---|---|---|---|---|
| Pro-vitamin A | Wheat | Maize *Psy1* and bacterial *CrtI* | 4.96 | Cong et al. (2009) |
| | | *CrtB* or *CrtI* | 3.21 | Wang et al. (2014a) |
| | Rice | Phytoene synthase (*Psy*) from daffodil and phytoene desaturase (*CrtI*) from *Erwinia uredovora* | 1.6 | Ye et al. (2000) |
| | | *Psy* from maize and *CrtI* from *Erwinia* | 37 | Paine et al. (2005) |
| | Maize | Bacterial *CrtB* or *CrtI* | 9.8 | Aluru et al. (2008) |
| | | *Psy1* (maize) | 59.32 | Naqvi et al. (2009) |
| Zinc | Rice | *HvNAS1* | 35 | Masuda et al. (2009) |
| | | *Ferritin* from soybean, *Aspergillus flavus phytase*, and *OsNAS1* | 35 | Wirth et al. (2009) |
| | | *OsNAS2* overexpression | 76 | Johnson et al. (2011) |

(continued)

**Table 6** (continued)

| Micronutrient | Crop | Gene expression | Increment (µg/g) | References |
|---|---|---|---|---|
| Iron | Wheat | *TaFERRITIN* | ~88.5 | Borg et al. (2012) |
| | | *TaVIT2* | ~40–55 | Connorton et al. (2017) |
| | Rice | *OsNAS2 + PvFerritin* | 93.1 | Singh et al. (2017b) |
| | | *OsIRT1* | 12 | Lee and An (2009) |
| | | *TOM1* | 18 | Nozoye et al. (2011) |
| | | *OsYSL13* | 15 | Zhang et al. (2018) |
| | | *OsNAS1, OsNAS2, OsNAS3* | 19 | Johnson et al. (2011) |
| | | *OsFER2* | 15.9 | Lee and An (2009) |
| | | *OsIRO2* | 15.5 | Ogo et al. (2011) |
| | | *OsVIT2 knockdown* | 28 | Zhang et al. (2012a) |
| | | *PvFerritin + rgMT + phyA* | 22 | Lucca et al. (2001) |
| | | *AtIRT1, PvFerritin, AtNAS1* | 10.46 | Boonyaves et al. (2016) |
| | | *GmFerritin, OsNAS2* | 15 | Trijatmiko et al. (2016) |
| | | *AtNAS1, AtFRD3, PvFer* | 11.08 | Wu et al. (2018) |
| | | *AtNAS1, PvFer, AtNRAMP3* | 13.65 | Wu et al. (2019) |
| | | *OsNAS2* | 80 | Beasley et al. (2019) |

Abbreviations: *Psy* phytoene synthase, *Hv Hordeum vulgare*, *NAS* nicotianamine synthase, *Os Oryza sativa*, *Ta Triticum aestivum*, *Pv Phaseolus vulgaris*, *IRT* iron regulated transporter, *TOM* transporter of mugineic acid, *YSL* yellow stripe-like, *FER* ferritin, *IRO* iron oxidase, *rgMT* metallothionein-like clone of rice, *phyA* phytochrome A, *Gm Glycine max*, *At Arabidopsis thaliana*, *NRAMP* natural resistance-associated macrophage proteins

## 10 Methods of Estimation of Quality

Micronutrients such as carotenoids, Fe, and Zn are present in trace levels, which make their estimation quite tricky. For high-throughput screening, simple, rapid, accurate, and inexpensive approaches are required. Methods like high-performance liquid chromatography (HPLC) and ultra-performance liquid chromatography

**Table 7** List of crops and the important targeted genes to generate low phytate in grains

| Crop | Gene | References |
|---|---|---|
| Wheat | Mutagenesis-based approach | Guttieri et al. (2004) |
| | Phytase of *Aspergillus fumigatus* driven by high molecular weight glutenin 1 DX5 promoter | Brinch-Pedersen et al. (2006) |
| | RNAi for genes involved in biosynthesis and transport of phytic acid (IPK1 and ABCC13) | Aggarwal et al. (2018) and Bhati et al. (2016) |
| | *Aspergillus japonicus* phytase gene (*phy*A) in wheat endosperm | Abid et al. (2017) |
| Rice | RNAi construct of *MIP* gene driven by seed-specific oleosin18 promoter | Ali et al. (2013) |
| | *OsMRP5* artificial microRNA driven by Ole18 promoter | Li et al. (2014) |
| | Phytase genes from ruminal bacterium *Selenomonas ruminantium* and *E. coli* | Hong et al. (2004) |
| | Phytase of *A. fumigatus*, ferritin of *Phaseolus vulgaris* and endogenous cysteine-rich metallothionein like protein | Lucca et al. (2002) |
| Maize | Zinc finger nucleases mediated genome editing of IPK1 | Shukla et al. (2009) |
| | Phytase of *A. fumigatus* and ferritin were driven by seed-specific rice glutelin promoter | Chen et al. (2008) |

(UPLC) have been reported for carotenoid determination in grain, but it is time-consuming (Weber 1987; Kurilich and Juvik 1999; Gama and Sylos 2005; Eagling et al. 2014). A variety of semi-quantitative colourimetric methods are available for Fe and Zn like Perl's Prussian blue and diphenylthiocarbazone-based dithizone (Ozturk et al. 2006; Velu et al. 2006, 2008; Choi et al. 2007). However, the screening at large scale becomes a laborious process while using these colorimetric methods. For the determination of accurate elemental profile, the most common method is atomic absorption spectrometry (AAS) though poor reproducibility and less sensitivity lead to the development of hydride-generated atomic absorption spectrometry (HG-AAS). Although, inductively coupled plasma (ICP), is a powerful and extremely sensitive method, yet, its high cost restricts its usage. Various variations of ICP are available, viz. ICP-optical emission spectrophotometry (ICP-OES), ICP-mass spectrometry (ICP-MS), and laser ablation-ICP-MS (LA-ICP-MS). The methods as mentioned earlier are destructive in natures so to overcome that challenge some non-destructive methods have also been introduced, viz. near-infrared reflectance spectrophotometry (NIRS), X-ray fluorescence spectrometry (XRF), secondary ion mass spectrometry (SIMS), synchrotron X-ray, fluorescence spectroscopy, and micro-X-ray fluorescence spectroscopy, but their approaches are costly and require much maintenance. The protocol for high-throughput screening for Fe, Zn, and Se in wheat grain was standardized using energy-dispersive-XRF (EDXRF) by (Paltridge et al. 2012). Similarly, for the rapid screening of provitamin A, Fe, and Zn, NIRS was utilized at Harvest Plus. Still, at last, the choice of the method would be dependent on the objective and accuracy required for estimation (Garcia-Oliveira et al. 2018).

## 11 Wheat Biofortification Programmes

Various global programmes have been launched to improve their nutritional status. At the international level, Consultative Group on International Agricultural Research (CGIAR), the International Maize and Wheat Improvement Center (CIMMYT), and the World Health Organization (WHO) are jointly working towards eliminating micronutrient malnutrition. Among them, 'Harvest Plus' is a popular programme by CGIAR, which uses conventional breeding to address the issue of micronutrient enhancement. It has aimed to diminish micronutrient malnutrition in countries of Africa, Asia, and Latin America. Among micronutrients, this programme has mainly focused on Fe, Zn, and provitamin A (Ortiz-Monasterio et al. 2007). In India, many centre and state government-funded programmes and schemes are operating. Some programmes are as follows, National Iron Plus Initiative (NIPI), National Iodine Deficiency Disorders Control Programme (NIDDCP), Integrated Child Development Services (ICDS) scheme, and Mid-Day Meal Programme which also helps to improve the health and nutrition status of the population. Moreover, in the year 2018, India's flagship programme – POSHAN Abhiyaan – was launched under the Ministry of Women and Child Development. This programme aimed to decrease stunting, undernutrition, and anaemia focusing mainly at children, adolescent girls, and women (NITI Aoyog 2019). Still, there is an urgent need to strengthen the innovative techniques related to agriculture systems, which are cost-effective and sustainable to enhance the micronutrient deficiencies to improve human health in the future (Gonmei and Toteja 2018).

## 12 Conclusion

In developing countries, populations are deficient in one or more essential vitamins and minerals, mainly due to low concentrations and reduced bioavailability of essential micronutrients present in commonly eaten foods. Many staple crops such as wheat are rich in protein and dietary fibres but low in micronutrients, so researchers are developing biofortified wheat varieties. Biofortified wheat has enormous potential to combat hidden hunger as the edible portions are denser in bioavailable, micronutrient, minerals, and vitamins. Wheat biofortification is a cost-effective, environment-friendly, and sustainable means to bridge the gap between consumer demand and supply in terms of high-quality produce. Effective utilization of naturally available genetic resources and employing modern technology such as CRISPR-Cas-based genetic modifications along with agronomic and conventional wheat breeding have displayed promising results. Still, it is crucial to maintain a balance between yield and micronutrient content to achieve worldwide acceptance by the farmers and consumers. In addition to this, the developed cultivars must be stable across different environments and climatic conditions. A better comprehension of the heritability, characterization, and introgression of genes, mechanism engaged in the transfer of micronutrients, is important in developing micronutrient-rich cultivars. Moreover, the improved understanding of micronutrient bioavailability and the effect of the environment will help in adopting the more efficient breeding methodology. Various aspects like cooking quality, palatability, and colour should be reviewed according to consumer's acceptance. The wide collaboration between agricultural research and government is a vital aspect for the success of biofortification programme.

## References


Abdel-Aal ESM, Young JC, Rabalski I (2006) Anthocyanin composition in black, blue, pink, purple, and red cereal grains. J Agric For 54:4696–4704. https://doi.org/10.1021/jf0606609

Abdel-Aal ESM, Abou-Arab AA, Gamel TH et al (2008) Fractionation of blue wheat anthocyanin compounds and their contribution to antioxidant properties. J Agric For 56:11171–11177. https://doi.org/10.1021/jf802168c

Abid N, Khatoon A, Maqbool A, Irfan M et al (2017) Transgenic expression of phytase in wheat endosperm increases bioavailability of iron and zinc in grains. Transgenic Res 26:109–122

Aciksoz SB, Ozturk L, Gokmen OO et al (2011a) Effect of nitrogen on root release of phytosiderophores and root uptake of Fe(III)-phytosiderophore in Fe-deficient wheat plants. Physiol Plant 142:287–296. https://doi.org/10.1111/j.1399-3054.2011.01460.x

Aciksoz SB, Yazici A, Ozturk L, Cakmak I (2011b) Biofortification of wheat with iron through soil and foliar application of nitrogen and iron fertilizers. Plant Soil 349:215–225. https://doi.org/10.1007/s11104-011-0863-2

Acquaah G (2016) Conventional plant breeding principles and techniques. In: Al-Khayri JM, Jain SM, Johnson DV (eds) Advances in plant breeding strategies: breeding, biotechnology and molecular tools. Springer, Heidelberg, p 116

Adisa IO, Pullagurala VLR, Peralta-Videa JR et al (2019) Recent advances in nano-enabled fertilizers and pesticides: a critical review of mechanisms of action. Environ Sci Nano 6:2002–2030. https://doi.org/10.1039/c9en00265k

Aggarwal S, Kumar A, Bhati KK et al (2018) RNAi-mediated downregulation of inositol pentakisphosphate kinase (IPK1) in wheat grains decreases phytic acid levels and increases Fe and Zn accumulation. Front Plant Sci 9:259. https://doi.org/10.3389/fpls.2018.00259

Ahmed T, Hossain M, Sanin KI (2012) Global burden of maternal and child undernutrition and micronutrient deficiencies. Ann Nutr Metab 61:8–17. https://doi.org/10.1159/000345165

Ali N, Paul S, Gayen D et al (2013) RNAi mediated down regulation of myo-inositol-3-phosphate synthase to generate low phytate rice. Rice 6:1–12. https://doi.org/10.1186/1939-8433-6-1

Alomari DZ, Eggert K, von Wirén N et al (2018) Identifying candidate genes for enhancing grain Zn concentration in wheat. Front Plant Sci 9:1–11. https://doi.org/10.3389/fpls.2018.01313

Alomari DZ, Eggert K, Von Wirén N et al (2019) Whole-genome association mapping and genomic prediction for iron concentration in wheat grains. Int J Mol Sci 20:1–14. https://doi.org/10.3390/ijms20010076

Alshaal T, El-Ramady H (2017) Foliar application: from plant nutrition to biofortification. Environ Biodivers Soil Secur 1:71–83. https://doi.org/10.21608/jenvbs.2017.1089.1006

Aluru M, Xu Y, Guo R et al (2008) Generation of transgenic maize with enhanced provitamin A content. J Exp Bot 59:3551–3562. https://doi.org/10.1093/jxb/ern212

Atkinson J, Harroun T, Wassall SR et al (2010) The location and behavior of α-tocopherol in membranes. Mol Nutr Food Res 54:641–651. https://doi.org/10.1002/mnfr.200900439

Aviram M, Kaplan M, Rosenblat M et al (2005) Dietary antioxidants and paraoxonases against LDL oxidation and atherosclerosis development. In: Atherosclerosis: diet and drugs 263–300. Springer, Heidelberg, pp 263–300. https://doi.org/10.1007/3-540-27661-0_9

Badakhshan H, Moradi N, Mohammadzadeh H, Zakeri MR (2013) Genetic variability analysis of grains Fe, Zn and Beta-carotene concentration of prevalent wheat varieties in Iran. Int J Agric Crop Sci 6:57–62

Balk J, Connorton JM, Wan Y et al (2019) Improving wheat as a source of iron and zinc for global nutrition. Nutr Bull 44:53–59. https://doi.org/10.1111/nbu.12361

Barone E, Calabrese V, Mancuso C (2009) Ferulic acid and its therapeutic potential as a hormetin for age-related diseases. Biogerontology 10:97–108. https://doi.org/10.1007/s10522-008-9160-8

Beal T, Massiot E, Arsenault JE et al (2017) Global trends in dietary micronutrient supplies and estimated prevalence of inadequate intakes. PLoS One 12:1–20. https://doi.org/10.1371/journal.pone.0175554

Beasley JT, Bonneau JP, Sánchez-Palacios JT et al (2019) Metabolic engineering of bread wheat improves grain iron concentration and bioavailability. Plant Biotechnol J 17:1514–1526. https://doi.org/10.1111/pbi.13074

Beede RH, Brown PH, Kallsen C et al (2005) Diagnosing and correcting nutrient deficiencies. 5. 4ª. Ferguson L (ed) University of California, Davis

Belderok B, Mesdag J, Mesdag H et al (2000) Bread-making quality of wheat: a century of breeding in Europe. Springer

Bhati KK, Alok A, Kumar A et al (2016) Silencing of ABCC13 transporter in wheat reveals its involvement in grain development, phytic acid accumulation and lateral root formation. J Exp Bot 67:4379–4438

Bhatta M, Stephen Baenziger P, Waters BM et al (2018) Genome-wide association study reveals novel genomic regions associated with 10 grain minerals in synthetic hexaploid wheat. Int J Mol Sci 19:1–18. https://doi.org/10.3390/ijms19103237

Birošová L, Mikulasova M, Vaverkova S (2005) Antimutagenic effect of phenolic acids. Biomed Pap 149:489–491. https://doi.org/10.5507/bp.2005.087

Boonyaves K, Gruissem W, Bhullar NK (2016) NOD promoter-controlled AtIRT1 expression functions synergistically with NAS and FERRITIN genes to increase iron in rice grains. Plant Mol Biol 90:207–215. https://doi.org/10.1007/s11103-015-0404-0

Borg S, Brinch-pedersen H, Tauris B et al (2012) Wheat ferritins: improving the iron content of the wheat grain. J Cereal Sci 56:204–213. https://doi.org/10.1016/j.jcs.2012.03.005

Borisjuk N, Kishchenko O, Eliby S et al (2019) Genetic modification for wheat improvement: from transgenesis to genome editing. Biomed Res Int 2019:1–18. https://doi.org/10.1155/2019/6216304

Bouis HE (2003) Micronutrient fortification of plants through plant breeding: can it improve nutrition in man at low cost? Proc Nutr Soc 62:403–411

Bouis HE, Welch RM (2010) Biofortification – a sustainable agricultural strategy for reducing micronutrient malnutrition in the global south. https://doi.org/10.2135/cropsci2009.09.0531

Brinch-Pedersen H, Hatzack F, Stöger E et al (2006) Heat-stable phytases in transgenic wheat (*Triticum aestivum* L.): deposition pattern, thermostability, and phytate hydrolysis. J Agric Food Chem 54:4624–4632. https://doi.org/10.1021/jf0600152

Cakmak I (2008) Enrichment of cereal grains with zinc: agronomic or genetic biofortification? Plant Soil 302:1–17. https://doi.org/10.1007/s11104-007-9466-3

Cakmak S, Gülüt KY, Marschner H et al (1994) Effect of zinc and iron deficiency on phytosiderophore release in wheat genotypes differing in zinc efficiency. J Plant Nutr 17:1–17

Cakmak I, Ozkan H, Braun HJ et al (2000) Zinc and iron concentrations in seeds of wild, primitive, and modern wheats. Food Nutr Bull 21:401–403

Cakmak I, Torun A, Özkan H et al (2004) *Triticum dicoccoides*: An important genetic resource for increasing zinc and iron concentration in modern cultivated wheat. Soil Sci Plant Nutr 50:1047–1054. https://doi.org/10.1080/00380768.2004.10408573

Calderini DF, Ortiz-monasterio I (2003) Are synthetic hexaploids a means of increasing grain element concentrations in wheat? Euphytica 134:169–178
Čermák T, Curtin SJ, Gil-Humanes J et al (2017) A multipurpose toolkit to enable advanced genome engineering in plants. Plant Cell 29:1196–1217
Chacko SA, Song Y, Nathan L et al (2009) Relations of dietary magnesium intake to biomarkers of inflammation and endothelial dysfunction in an ethnically diverse cohort of postmenopausal women. Diabetes Care 33:304–310. https://doi.org/10.2337/dc09-1402
Cheema SA, Rehman H, Kiran A et al (2018) Progress and prospects for micronutrient biofortification in rice/wheat. In: Plant micronutrient use efficiency. Academic, pp 261–278
Chen R, Xue G, Chen P et al (2008) Transgenic maize plants expressing a fungal phytase gene. Transgenic Res 17:633–643. https://doi.org/10.1007/s11248-007-9138-3
Chhuneja P, Dhaliwal HS, Bains NS, Singh K (2006) *Aegilops kotschyi* and *Aegilops tauschii* as sources for higher levels of grain Iron and Zinc. Plant Breed 125:529–531. https://doi.org/10.1111/j.1439-0523.2006.01223.x
Choi EY, Graham R, Stangoulis J (2007) Semi-quantitative analysis for selecting Fe- and Zn-dense genotypes of staple food crops. J Food Compos Anal 20:496–505. https://doi.org/10.1016/j.jfca.2007.01.004
CIMMYT (2019) Mexico City, Mexico. Retrieved December 02, 2019 from https://www.cimmyt.org/news/the-case-for-rushing-farmer-access-to-bari-gom-33/
Cong L, Wang C, Chen L et al (2009) Expression of phytoene synthase1 and carotene desaturase crtI genes result in an increase in the total carotenoids content in transgenic elite wheat (*Triticum aestivum* L.). J Agric Food Chem 57:8652–8660. https://doi.org/10.1021/jf9012218
Connorton JM, Jones ER, Rodríguez-Ramiro I et al (2017) Wheat vacuolar iron transporter TaVIT2 transports Fe and Mn and is effective for biofortification. Plant Physiol 174:2434–2444. https://doi.org/10.1104/pp.17.00672
Cornell H, Cauvain SP (2003) Bread making: improving quality. In: Dalmo RA, Bogwald J (eds) Fish and shellfish immunology, vol 384. Woodhead Publishing, Cambridge
Crespo-Herrera LA, Velu G, Singh RP (2016) Quantitative trait loci mapping reveals pleiotropic effect for grain iron and zinc concentrations in wheat. Ann Appl Biol 169:27–35. https://doi.org/10.1111/aab.12276
Crespo-Herrera LA, Govindan V, Stangoulis J et al (2017) QTL mapping of grain Zn and Fe concentrations in two hexaploid wheat RIL populations with ample transgressive segregation. Front Plant Sci 8:1–12. https://doi.org/10.3389/fpls.2017.01800
Deshpande P, Dapkekar A, Oak MD et al (2017) Zinc complexed chitosan/tpp nanoparticles: a promising micronutrient nanocarrier suited for foliar application. Carbohydr Polym 165:394–401
Distelfeld A, Cakmak I, Peleg Z et al (2007) Multiple QTL-effects of wheat Gpc-B1 locus on grain protein and micronutrient concentrations. Physiol Plant 129:635–643. https://doi.org/10.1111/j.1399-3054.2006.00841.x
Doria E, Galleschi L, Calucci L et al (2009) Phytic acid prevents oxidative stress in seeds: evidence from a maize (Zea mays L.) low phytic acid mutant. J Exp Bot 60:967–978. https://doi.org/10.1093/jxb/ern345
Duffy SJ, Vita JA (2003) Effects of phenolics on vascular endothelial function. Curr Opin Lipidol 14:21–27. https://doi.org/10.1097/00041433-200302000-00005
Eagling T, Neal AL, McGrath SP et al (2014) Distribution and speciation of iron and zinc in grain of two wheat genotypes. J Agric Food Chem 62:708–16
FAOSTAT Database (2019) Rome, Italy. Retrieved December 02, 2019 from http://www.fao.org/faostat/en/#data
Feregrino-Pérez A, Magaña-López E, Guzmán C et al (2018) A general overview of the benefits and possible negative effects of the nanotechnology in horticulture. Sci Hortic 238:126–137
Ficco DB, Mastrangelo AM, Trono D et al (2014) The colours of durum wheat: a review. Crop Pasture Sci 65:1. https://doi.org/10.1071/cp13293

Fung TT, Hu FB, Pereira MA et al (2002) Whole-grain intake and the risk of type 2 diabetes: a prospective study in men. Am J Clin Nutr:535–540. https://doi.org/10.1093/ajcn/76.3.535
Gama JJT, Sylos CM (2005) Major carotenoid composition of Brazilian Valencia orange juice: identification and quantification by HPLC. Food Res Int 38:899–903. https://doi.org/10.1016/j.foodres.2005.03.008
Garcia-Oliveira AL, Chander S, Ortiz R et al (2018) Genetic basis and breeding perspectives of grain iron and zinc enrichment in cereals. Front Plant Sci 9:1–13. https://doi.org/10.3389/fpls.2018.00937
Garg M, Chawla M, Chunduri V et al (2016) Transfer of grain colors to elite wheat cultivars and their characterization. J Cereal Sci 71:138–144. https://doi.org/10.1016/j.jcs.2016.08.004
Garvin DF, Welch RM, Finley JW (2006) Historical shifts in the seed mineral micronutrient concentration of US hard red winter wheat germplasm. J Sci Food Agric 86:2213–2220
Genc Y, Verbyla AP, Torun AA et al (2009) Quantitative trait loci analysis of zinc efficiency and grain zinc concentration in wheat using whole genome average interval mapping. Plant Soil 314:49–66. https://doi.org/10.1007/s11104-008-9704-3
Ghosh S, Watson A, Gonzalez-Navarro OE et al (2018) Speed breeding in growth chambers and glasshouses for crop breeding and model plant research. Nat Protoc 13:2944–2963. https://doi.org/10.1038/s41596-018-0072-z
Gomez-Becerra HF, Erdem H, Yazici A et al (2010) Grain concentrations of protein and mineral nutrients in a large collection of spelt wheat grown under different environments. J Cereal Sci 52:342–349. https://doi.org/10.1016/j.jcs.2010.05.003
Gonmei Z, Toteja GS (2018) Micronutrient status of Indian population. Ind J Med Res 148:511. https://doi.org/10.4103/ijmr.ijmr_1768_18
Goudia BD, Hash CT (2015) Breeding for high grain Fe and Zn levels in cereals. 12:342–354
Graf E (1992) Antioxidant potential of ferulic acid. Free Radic Biol Med 13:435–448. https://doi.org/10.1016/0891-5849(92)90184-i
Graham R, Senadhira D, Beebe S et al (1999) Breeding for micronutrient density in edible portions of staple food crops: conventional approaches. Food Crop Res 60:57–80. https://doi.org/10.1016/S0378-4290(98)00133-6
Guo Z, Zhang Z, Xu P et al (2012) Analysis of nutrient composition of purple wheat. Cereal Res Commun 41:293–303. https://doi.org/10.1556/crc.2012.0037
Guttieri M, Bowen D, Dorsch JA (2004) Identification and characterization of a low phytic acid wheat. Crop Sci 44:418–424
Guttieri MJ, Stephen Baenziger P, Frels K et al (2015) Prospects for selecting wheat with increased zinc and decreased cadmium concentration in grain. Crop Sci 55:1712–1728. https://doi.org/10.2135/cropsci2014.08.0559
Hao Y, Velu G, Pen RJ, Singh RP (2014) Genetic loci associated with high grain zinc concentration and pleiotropic effect on kernel weight in wheat (*Triticum aestivum* L.). 1893–1902. https://doi.org/10.1007/s11032-014-0147-7
Harvest Plus (2019) Washington, United States of America. https://www.harvestplus.org/sites/default/files/publications/Science%20Brief-Biofortification-1_Zinc%20Wheat_May2017.pdf
Havrlentová M, Pšenáková I, Žofajová A et al (2014) Anthocyanins in wheat seed – a mini review. Nova Biotechnol Chim 13:1–12. https://doi.org/10.2478/nbec-2014-0001
Heidari B, Padash S, Dadkhodaie A (2016) Variations in micronutrients, bread quality and agronomic traits of wheat landrace varieties and commercial cultivars. Aust J Crop Sci 10:377–384. https://doi.org/10.21475/ajcs.2016.10.03.p7231
Hodges SC, Constable G (2010) Plant responses to mineral deficiencies and toxicities. In: Physiology of cotton. Springer, Dordrecht, pp 142–161
Hong CY, Cheng KJ, Tseng TH et al (2004) Production of two highly active bacterial phytases with broad pH optima in germinated transgenic rice seeds. Transgenic Res 13:29–39. https://doi.org/10.1023/B:TRAG.0000017158.96765.67
IARI Database (2019) New Delhi, India. Retrieved December 02, 2019 from https://www.iari.res.in/index.php?option=com_content&view=article&id=1253&Itemid=951

Iqbal TH, Lewis KO, Cooper BT (1994) Phytase activity in the human and rat small intestine. Gut 35:1233–1236. https://doi.org/10.1136/gut.35.9.1233

Johnson AAT, Kyriacou B, Callahan DL et al (2011) Constitutive overexpression of the OsNAS gene family reveals single-gene strategies for effective iron- and zinc-biofortification of rice endosperm. PLoS One 6:e24476. https://doi.org/10.1371/journal.pone.0024476

Joppa LR, Du C, Hart GE, Hareland GA (1997) Mapping gene(s) for grain protein in tetraploid wheat (*Triticum turgidum* L.) using a population of recombinant inbred chromosome lines. Crop Sci 37:1586–1589

Kappara S, Neelamraju S, Ramanan R (2018) Down regulation of a heavy metal transporter gene influences several domestication traits and grain Fe-Zn content in rice. Plant Sci 276:208–219. https://doi.org/10.1016/j.plantsci.2018.09.003

Kibite S, Evans LE (1984) Causes of negative correlations between grain yield and grain protein concentration in common wheat. Euphytica 33:801–810. https://doi.org/10.1007/BF00021906

Kondić-Špika A, Mladenov N, Grahovac N et al (2019) Biometric analyses of yield, oil and protein contents of wheat (*Triticum aestivum* L.) genotypes in different environments. Agronomy 9:8–10. https://doi.org/10.3390/agronomy9060270

Kong L (2013) Maize residues, soil quality, and wheat growth in China. A review. Agron Sustain Dev 34:405–416. https://doi.org/10.1007/s13593-013-0182-5

Krishnappa G, Singh AM, Chaudhary S et al (2017) Molecular mapping of the grain iron and zinc concentration, protein content and thousand kernel weight in wheat (*Triticum aestivum* L.). PLoS One 12:e0174972. https://doi.org/10.1371/journal.pone.0174972

Kumar J, Saripalli G, Gahlaut V et al (2018) Genetics of Fe, Zn, β-carotene, GPC and yield traits in bread wheat (*Triticum aestivum* L.) using multi-locus and multi-traits GWAS. Euphytica 214:219. https://doi.org/10.1007/s10681-018-2284-2

Kumar A, Kapoor P, Chunduri V et al (2019) Potential of *Aegilops* sp. for improvement of grain processing and nutritional quality in wheat (*Triticum aestivum*). Front Plant Sci 10:1–19. https://doi.org/10.3389/fpls.2019.00308

Kurilich AC, Juvik JA (1999) Quantification of carotenoid and tocopherol antioxidants in *Zea mays*. J Agric Food Chem 475:1948–1955

Laidig F, Piepho HP, Rentel D et al (2017) Breeding progress, environmental variation and correlation of winter wheat yield and quality traits in German official variety trials and on-farm during 1983–2014. Theor Appl Genet 130:223–245. https://doi.org/10.1007/s00122-016-2810-3

Lee S, An G (2009) Over-expression of OsIRT1 leads to increased iron and zinc accumulations in rice. Plant Cell Environ 32:408–416. https://doi.org/10.1111/j.1365-3040.2009.01935.x

Leenhardt F, Lyan B, Rock E et al (2006) Genetic variability of carotenoid concentration, and lipoxygenase and peroxidase activities among cultivated wheat species and bread wheat varieties. Eur J Agron 25:170–176. https://doi.org/10.1016/j.eja.2006.04.010

Li WX, Zhao HJ, Pang WQ et al (2014) Seed-specific silencing of OsMRP5 reduces seed phytic acid and weight in rice. Transgenic Res 23:585–599. https://doi.org/10.1007/s11248-014-9792-1

Liu R, Lal R (2015) Potentials of engineered nanoparticles as fertilizers for increasing agronomic productions. Sci Total Environ 514:131–139

Lott JNA, Spitzer E (1980) X-ray analysis studies of elements stored in protein body globoid crystals of Triticum grains. Plant Physiol 66:494–499

Lucca P, Hurrell R, Potrykus I (2001) Genetic engineering approaches to improve the bioavailability and the level of iron in rice grains. Theor Appl Genet 102:392–397. https://doi.org/10.1007/s001220051659

Lucca P, Hurrell R, Potrykus I (2002) Fighting iron deficiency anemia with iron-rich rice. J Am Coll Nutr 21:184S–190S. https://doi.org/10.1080/07315724.2002.10719264

Ludwig Y, Slamet-Loedin IH (2019) Genetic biofortification to enrich rice and wheat grain iron: from genes to product. Front Plant Sci 10:1–10. https://doi.org/10.3389/fpls.2019.00833

Map Chart. https://mapchart.net/

Masuda H, Usuda K, Kobayashi T et al (2009) Overexpression of the barley nicotianamine synthase gene HvNAS1 increases iron and zinc concentrations in rice grains. Rice 2:155–166. https://doi.org/10.1007/s12284-009-9031-1
Masuda H, Ishimaru Y, Aung MS et al (2012) Iron biofortification in rice by the introduction of multiple genes involved in iron nutrition. Sci Rep 2:1–7. https://doi.org/10.1038/srep00543
Milman U, Blum S, Shapira C et al (2008) Vitamin E supplementation reduces cardiovascular events in a subgroup of middle-aged individuals with both type 2 diabetes mellitus and the haptoglobin 2-2 genotype. Arterioscler Thromb Vasc Biol 28(2):341–347. https://doi.org/10.1161/atvbaha.107.153965
Monasterio I, Graham RD (2000) Breeding for trace minerals in wheat. Food Nutr Bull 21:392–396. https://doi.org/10.1177/156482650002100409
Morgounov A, Gómez-Becerra HF, Abugalieva A et al (2007) Iron and zinc grain density in common wheat grown in Central Asia. Euphytica 155:193–203. https://doi.org/10.1007/s10681-006-9321-2
Muller O, Krawinkel M (2005) Malnutrition and health in developing countries. CMAJ 173:279–286. https://doi.org/10.1503/cmaj.050342
Murata Y, Ma JF, Yamaji N et al (2006) A specific transporter for iron(III)-phytosiderophore in barley roots. Plant J 46:563–572. https://doi.org/10.1111/j.1365-313X.2006.02714.x
Naqvi S, Zhu C, Farre G et al (2009) Transgenic multivitamin corn through biofortification of endosperm with three vitamins representing three distinct metabolic pathways. Proc Natl Acad Sci U S A 106:7762–7767. https://doi.org/10.1073/pnas.0901412106
Neelam K, Rawat N, Tiwari VK, Prasad R (2012) Evaluation and identification of wheat-Aegilops addition lines controlling high grain iron and zinc concentration and mugineic acid production. Cereal Res Commun 40:53–61. https://doi.org/10.1556/CRC.40.2012.1.7
NHS (2019) United Kingdom. Retrieved December 11, 2019 from https://www.nhs.uk/conditions/vitamins-and-minerals
NITI Aoyog (2019) India. Retrieved December 11, 2019 from https://niti.gov.in/poshan-abhiyaan
Niyigaba E, Twizerimana A, Mugenzi I et al (2019) Winter wheat grain quality, zinc and iron concentration affected by a combined foliar spray of zinc and iron fertilizers. Agronomy 9:250. https://doi.org/10.3390/agronomy9050250
Nozoye T, Nagasaka S, Kobayashi T et al (2011) Phytosiderophore efflux transporters are crucial for iron acquisition in graminaceous plants. J Biol Chem 286:5446–5454. https://doi.org/10.1074/jbc.M110.180026
Ogo Y, Itai RN, Kobayashi T et al (2011) OsIRO2 is responsible for iron utilization in rice and improves growth and yield in calcareous soil. Plant Mol Biol 75:593–605. https://doi.org/10.1007/s11103-011-9752-6
Ortiz-Monasterio JI, Palacios-Rojas N, Meng E et al (2007) Enhancing the mineral and vitamin content of wheat and maize through plant breeding. J Cereal Sci 46:293–307. https://doi.org/10.1016/j.jcs.2007.06.005
Ozaki Y (1992) Antiinflammatory effects of tetramethylpyrazine and ferulic acid. Chem Pharm Bull 40:954–956. https://doi.org/10.1248/cpb.40.954
Ozturk L, Yazici MA, Yucel C et al (2006) Concentration and localization of zinc during seed development and germination in wheat. Physiol Plant 128:144–152. https://doi.org/10.1111/j.1399-3054.2006.00737.x
Paine JA, Shipton CA, Chaggar S et al (2005) Improving the nutritional value of Golden Rice through increased pro-vitamin A content. Nat Biotechnol 23:482–487. https://doi.org/10.1038/nbt1082
Paltridge NG, Milham PJ, Ortiz-monasterio JI et al (2012) Energy-dispersive X-ray fluorescence spectrometry as a tool for zinc, iron and selenium analysis in whole grain wheat. Plant Soil 361:261–269. https://doi.org/10.1007/s11104-012-1423-0
Peleg Z, Cakmak I, Ozturk L et al (2009) Quantitative trait loci conferring grain mineral nutrient concentrations in durum wheat × wild emmer wheat RIL population. Theor Appl Genet 119:353–369. https://doi.org/10.1007/s00122-009-1044-z

Piergiovanni AR, Rizzi R, Pannacciulli E, Della Gatta C (1997) Mineral composition in hulled wheat grains: a comparison between emmer (*Triticum dicoccon* Schrank) and spelt (*T. spelta* L.) accessions. Int J Food Sci Nutr 48:381–386. https://doi.org/10.3109/09637489709028586

Rong-li SHI, Yi-ping T, Rui-lian J et al (2013) Characterization of quantitative trait loci for grain minerals in hexaploid wheat (*Triticum aestivum* L.). J Integr Agric 12:1512–1521. https://doi.org/10.1016/S2095-3119(13)60559-6

Schlemmer U, Frølich W, Prieto RM, Grases F (2009) Review Phytate in foods and significance for humans: food sources, intake, processing, bioavailability, protective role and analysis. Mol Nutr Food Res 53:330–375. https://doi.org/10.1002/mnfr.200900099

Shewry PR, Hey SJ (2015) The contribution of wheat to human diet and health. Food Energy Secur 4:178–202. https://doi.org/10.1002/fes3.64

Shewry PR, Charmet G, Branlard G et al (2012) Developing new types of wheat with enhanced health benefits. Trends Food Sci Tech 25:70–77. https://doi.org/10.1016/j.tifs.2012.01.003

Shi R, Li H, Tong Y (2008) Identification of quantitative trait locus of zinc and phosphorus density in wheat (*Triticum aestivum* L.) grain. Plant Soil 306:95–104. https://doi.org/10.1007/s11104-007-9483-2

Shukla V, Doyon Y, Miller J et al (2009) Precise genome modification in the crop species Zea mays using zinc-finger nucleases. Nature 459:437–441. https://doi.org/10.1038/nature07992

Shukla AK, Behera SK, Pakhre A et al (2018) Micronutrients in soils, plants, animals and humans. Indian J Fertil 14:30–54

Singh BD, Singh AK (2015) Marker-assisted plant breeding: principles and practices. Springer, Heidelberg, pp 125–146

Singh R, Velu G (2017) Zinc-Biofortified wheat: harnessing genetic diversity for improved nutritional quality. Science Brief: Biofortification No. 1 (No. 2187-2019-666). CIMMYT, HarvestPlus, and the Global Crop Diversity Trust. Bonn, Germany

Singh R, Govindan V, Andersson MS (2017a) Zinc-biofortified wheat: harnessing genetic diversity for improved nutritional quality (No. 2187-2019-666)

Singh SP, Keller B, Gruissem W, Bhullar NK (2017b) Rice NICOTIANAMINE SYNTHASE 2 expression improves dietary iron and zinc levels in wheat. Theor Appl Genet 130:283–292. https://doi.org/10.1007/s00122-016-2808-x

Slavin J, Savarino V, Paredes-Diaz A et al (2009) A review of the role of soluble fiber in health with specific reference to wheat dextrin. Int J Med Res 37:1–17. https://doi.org/10.1177/147323000903700101

Solanki P, Bhargava A, Chhipa H et al (2016) Nano-fertilizers and their smart delivery system, agriculture. In: Nanoscience in food and agriculture, pp 81–101

Sparvoli F, Cominelli E (2015) Seed biofortification and phytic acid reduction: a conflict of interest for the plant? Plan Theory 4:728–755. https://doi.org/10.3390/plants4040728

Šramková Z, Gregova E, Slikova S et al (2010) Wheat varieties released in Slovakia and their bread-making quality. Cereal Res Commun 38:386–394. https://doi.org/10.1556/crc.38.2010.3.9

Srinivasa J, Arun B, Kumar V (2014) Zinc and iron concentration QTL mapped in a *Triticum spelta* × *T. aestivum* cross. 1643–1651. https://doi.org/10.1007/s00122-014-2327-6

Stevenson L, Phillips F, O'sullivan K et al (2012) Wheat bran: its composition and benefits to health, a European perspective. Int J Food Sci Nutr 63:1001–1013. https://doi.org/10.3109/09637486.2012.687366

Suzuki M, Takahashi M, Tsukamoto T et al (2006) Biosynthesis and secretion of mugineic acid family phytosiderophores in zinc-deficient barley. Plant J 48:85–97. https://doi.org/10.1111/j.1365-313X.2006.02853.x

Tiwari VK, Rawat N, Chhuneja P et al (2009) Mapping of quantitative trait loci for grain iron and zinc concentration in diploid A genome wheat. J Hered 100:771–776

Tiwari C, Wallwork H, Arun B et al (2016) Molecular mapping of quantitative trait loci for zinc, iron and protein content in the grains of hexaploid wheat. Euphytica 207:563–570

Traber MG, Frei B, Beckman JS (2008) Vitamin E revisited: do new data validate benefits for chronic disease prevention? Curr Opin Inter Med 7:155–163. https://doi.org/10.1097/mol.0b013e3282f2dab6

Trethowan RM (2007) Breeding wheat for high iron and zinc at CIMMYT: state of the art, challenges and future prospects. In: Proceeding of the 7th international wheat conference. Mar del Plata, Argentina

Trethowan RM, Reynolds M, Sayre K, Ortiz-Monasterio I (2005) Adapting wheat cultivars to resource conserving farming practices and human nutritional needs. Ann Appl Biol 146:405–413. https://doi.org/10.1111/j.1744-7348.2005.040137.x

Trijatmiko KR, Duenãs C, Tsakirpaloglou N et al (2016) Biofortified indica rice attains iron and zinc nutrition dietary targets in the field. Sci Rep 6:1–13. https://doi.org/10.1038/srep19792

Uauy C, Distelfeld A, Fahima T et al (2006) A NAC gene regulating senescence improves grain protein, zinc, and iron content in wheat. Science 314:1298–1301

Velu G, Kulkarni VN, Muralidharan V et al (2006) A rapid screening method for grain iron content in pearl millet 158–161

Velu G, Bhattacharjee R, Rai KN et al (2008) A simple and rapid screening method for grain zinc content in pearl millet. 6:2006–2009

Velu G, Singh R, Huerta-Espino J et al (2011) Breeding for enhanced zinc and iron concentration in CIMMYT spring wheat germplasm. Czech J Genet Plant Breed 47:174–177

Velu G, Singh RP, Huerta-Espino J et al (2012) Performance of biofortified spring wheat genotypes in target environments for grain zinc and iron concentrations. Field Crop Res 137:261–267. https://doi.org/10.1016/j.fcr.2012.07.018

Velu G, Ortiz-Monasterio I, Cakmak I et al (2014) Biofortification strategies to increase grain zinc and iron concentrations in wheat. J Cereal Sci 59:365–372. https://doi.org/10.1016/j.jcs.2013.09.001

Velu G, Singh RG, Balasubramaniam A et al (2015) Reaching out to farmers with high zinc wheat varieties through public-private partnerships: an experience from Eastern-Gangetic plains of India. Adv Food Tech Nutr Sci 1:73–75

Velu G, Guzman C, Mondal S et al (2016) Effect of drought and elevated temperature on grain zinc and iron concentrations in CIMMYT spring wheat. J Cereal Sci 69:182–186

Velu G, Singh RP, Huerta J, Guzmán C (2017) Genetic impact of Rht dwarfing genes on grain micronutrients concentration in wheat. Field Crop Res 214:373–377. https://doi.org/10.1016/j.fcr.2017.09.030

Velu G, Singh RP, Crespo-Herrera L et al (2018) Genetic dissection of grain zinc concentration in spring wheat for mainstreaming biofortification in CIMMYT wheat breeding. Sci Rep 8:1–10. https://doi.org/10.1038/s41598-018-31951-z

Wang C, Zeng J, Li Y et al (2014a) Enrichment of provitamin A content in wheat (*Triticum aestivum* L.) by introduction of the bacterial carotenoid biosynthetic genes *CrtB* and *CrtI*. J Exp Bot 65:2545–2556

Wang S, Wong D, Forrest K et al (2014b) Characterization of polyploid wheat genomic diversity using a high-density 90000 single nucleotide polymorphism array. Plant Biotechnol J 12:787–796. https://doi.org/10.1111/pbi.12183

Watson A, Ghosh S, Williams MJ et al (2018) Speed breeding is a powerful tool to accelerate crop research and breeding. Nat Plants 4:23–29. https://doi.org/10.1038/s41477-017-0083-8

Weber EJ (1987) Carotenoids and tocols of corn grain determined by HPLC. J Am Oil Chem Soc 64:1129–1134

Wei Y, Shohag MJI, Yang X (2012) Biofortification and bioavailability of rice grain zinc as affected by different forms of foliar zinc fertilization. PLoS One 7:e45428. https://doi.org/10.1371/journal.pone.0045428

Welch RM, House WA, Ortiz-Monasterio I, Cheng Z (2005) Potential for improving bioavailable zinc in wheat grain (*Triticum* species) through plant breeding. J Agric Food Chem 53:2176–2180. https://doi.org/10.1021/jf040238x

WHO (2002) Rome, Italy. Retrieved December 02, 2019 from Mortality and Burden of Disease Estimates for WHO Member States in 2002.xls

Wirth J, Poletti S, Aeschlimann B et al (2009) Rice endosperm iron biofortification by targeted and synergistic action of nicotianamine synthase and ferritin. Plant Biotechnol J 7:631–644. https://doi.org/10.1111/j.1467-7652.2009.00430.x

Wu TY, Gruissem W, Bhullar NK (2018) Facilitated citrate-dependent iron translocation increases rice endosperm iron and zinc concentrations. Plant Sci 270:13–22. https://doi.org/10.1016/j.plantsci.2018.02.002

Wu TY, Gruissem W, Bhullar NK (2019) Targeting intracellular transport combined with efficient uptake and storage significantly increases grain iron and zinc levels in rice. Plant Biotechnol J 17:9–20. https://doi.org/10.1111/pbi.12943

Xu Y, An D, Li H, Xu H (2011) Review: breeding wheat for enhanced micronutrients. Can J Plant Sci 91:231–237. https://doi.org/10.4141/CJPS10117

Xu Y, An D, Liu D et al (2012) Molecular mapping of QTLs for grain zinc, iron and protein concentration of wheat across two environments. Field Crop Res 138:57–62. https://doi.org/10.1016/j.fcr.2012.09.017

Yadava DK, Choudhury PR, Hossain F, Kumar D (2017) Biofortified varieties: sustainable way to alleviate malnutrition. Indian Council of Agricultural Research, New Delhi

Ye X, Al-Babili S, Klöti A et al (2000) Engineering the provitamin A (β-carotene) biosynthetic pathway into (carotenoid-free) rice endosperm. Science 287:303–305. https://doi.org/10.1126/science.287.5451.303

Yeum K-J, Russell RM (2002) Carotenoid bioavailability and bioconversion. Annu Rev Nutr 22:483–504. https://doi.org/10.1146/annurev.nutr.22.010402.102834

Zhang Y, Song Q, Yan J et al (2010) Mineral element concentrations in grains of Chinese wheat cultivars. Euphytica 174:303–313. https://doi.org/10.1007/s10681-009-0082-6

Zhang Y, Xu YH, Yi HY, Gong JM (2012a) Vacuolar membrane transporters OsVIT1 and OsVIT2 modulate iron translocation between flag leaves and seeds in rice. Plant J 72:400–410. https://doi.org/10.1111/j.1365-313X.2012.05088.x

Zhang YQ, Sun YX, Ye YL et al (2012b) Zinc biofortification of wheat through fertilizer applications in different locations of China. Field Crop Res 125:1–7. https://doi.org/10.1016/j.fcr.2011.08.003

Zhang C, Shinwari KI, Luo L, Zheng L (2018) OsYSL13 is involved in iron distribution in rice. Int J Mol Sci 19:1–14. https://doi.org/10.3390/ijms19113537

Zhi-en P, Ma Y, Qiu-yi H et al (2014) Quantitative trait loci associated with micronutrient concentrations in two recombinant inbred wheat lines. J Integr 13:2322–2329. https://doi.org/10.1016/S2095-3119(13)60640-1

Zulfiqar F, Navarro M, Ashraf M et al (2019) Nanofertilizer use for sustainable agriculture: advantages and limitations. Plant Sci 289:110270. https://doi.org/10.1016/j.plantsci.2019.110270

Zuverza-Mena N, Martinez-Fernandez D, Du W et al (2017) Exposure of engineered nanomaterials to plants: insights into the physiological and biochemical responses – a review. Plant Physiol Biochem 110:236–264

# Changing Nutrition Scenario: Colored Wheat – A New Perspective

**Saloni Sharma, Payal Kapoor, Satveer Kaur, Anita Kumari, Natasha Sharma, Aman Kumar, Venkatesh Chunduri, and Monika Garg**

**Abstract** In the world of rapid economic growth, food security in terms of nutritional profile began to receive greater interest, especially in underdeveloped or developing countries. The development of biofortified bread wheat emerged with an idea of ensuring nutritional security. The anthocyanin-rich wheat developed through conventional breeding contains anthocyanins which are antioxidants capable of neutralizing the detrimental effects caused by destructive free radicals induced by various physiological processes going on in our body. The anthocyanin present in colored wheat has a broad spectrum of health implications such as protection against various metabolic syndromes like obesity, diabetes, hypertension, and dyslipidemia. The idea of developing anthocyanin-biofortified wheat is believed to shape the lifestyle of human beings as it is a staple food crop in many parts of the world. In this book chapter, we have covered various aspects of colored wheat such as its origin, biochemistry, agronomy, and health implications. This book chapter summarized the application of anthocyanin-rich colored wheat in ameliorating various clinical manifestations caused due to free radicals in both in vivo and in vitro environments.

**Keywords** Colored wheat · Black wheat · Blue wheat · Purple wheat Antioxidants

## 1 Introduction

Bread wheat is well known to provide elementary source of dietary carbohydrates throughout the world. Owing to its excellent processing quality, it is used globally in the form of bread, cookies, pasta, pizza, and other food items. In addition to being

S. Sharma · P. Kapoor · S. Kaur · A. Kumari · N. Sharma · A. Kumar
V. Chunduri · M. Garg (✉)
Agri-Biotechnology Division, National Agri-Food Biotechnology Institute (NABI), S.A.S. Nagar (Mohali), Punjab, India

© The Editor(s) (if applicable) and The Author(s), under exclusive license to Springer Nature Switzerland AG 2021

S. H. Wani et al. (eds.), *Physiological, Molecular, and Genetic Perspectives of Wheat Improvement*, https://doi.org/10.1007/978-3-030-59577-7_4

a carbohydrate source, whole wheat is also verified to be a rich source of dietary fibers, oligosaccharides, polyphenols, carotenoids, phytosterols, alkylresorcinols, and micronutrients (Gibson et al. 1995; Marlett et al. 2002; Slavin 2003; Fardet 2010). Hence, whole wheat is a package of inclusive range of bioactive components providing protection against many chronic diseases (Slavin 2003; Fardet 2010). On the other hand, existence of colored wheat, rich in anthocyanins, added up a new perspective to whole wheat health benefits and attracted several researchers, industrialists, and consumers toward colored wheat.

Anthocyanins are a type of phytochemicals, belonging to a flavonoid class of phenolic compounds imparting intense color to various flowers, vegetables, fruits, and plant parts. It is a well-characterized antioxidant, and in comparison to other bioactive components, it's also proved to be a more potent antioxidant (Guo and Ling 2015; Khoo et al. 2019). On the basis of studies in cell lines from different organs of the gastrointestinal tract, such as the intestines; esophageal, stomach, colorectal, liver, cervical, breast, and prostate cancers (Bowen-Forbes et al. 2010; Rugina et al. 2012; Hafidh et al. 2013; Jin et al. 2013; Li et al. 2014; Bishayee and Sethi 2016); and human clinical trials (https://clinicaltrials.gov), anthocyanins have antioxidant (He and Giusti 2010), anti-inflammation, bacteriostatic, anticancer, and antiaging functions (Bagchi et al. 2004; Cui and Li 2014; Chen et al. 2016). They can be exploited for the prevention of cardiovascular disease (Alvarez-Suarez et al. 2014; Cerletti et al. 2016), the alleviation of diabetes (Li et al. 2015b), cancer therapy (Bobe et al. 2006), and obesity control (Wu et al. 2013). No doubt dark-colored fruits and vegetables are a rich source of anthocyanins, but their feasibility to common men of underdeveloped and developing countries is limited, and regardless of the season, one can consume cereals effortlessly.

Considering its bioactive efficiency and easy availability, colored wheat received the attention as an alternative source of nutritional and functional food. Therefore, it's very useful and important to understand colored wheat in every aspect like, genetically, its antioxidant properties, agronomical qualities, and biological activity.

## 2 Origin and Genetics of Colored Wheat

Colored wheat exists in three forms, purple, blue, and black color, depending upon the types and position of the anthocyanins in wheat layers (Ficco et al. 2014; Abdel-Aal et al. 2006, 2008, Garg et al. 2016). None of the pigment originated naturally in wheat (Knievel et al. 2009; Havrlentova et al. 2014). In purple wheat, anthocyanins are present in the pericarp layer, while the blue wheat has anthocyanins in the aleurone layer, and the black wheat is a combination of both, i.e., anthocyanins are present in the pericarp and in the aleurone layer (Fig. 1; Abdel-Aal et al. 2006, 2008; Garg et al. 2016).

The purple grain trait is a mutant that was introgressed from Ethiopian (Abyssinia) origin *Triticum dicoccum* var. Arraseita Perc. into hexaploid wheat (Zeven 1991; Copp 1965). Previously its inheritance was thought to be controlled by a single

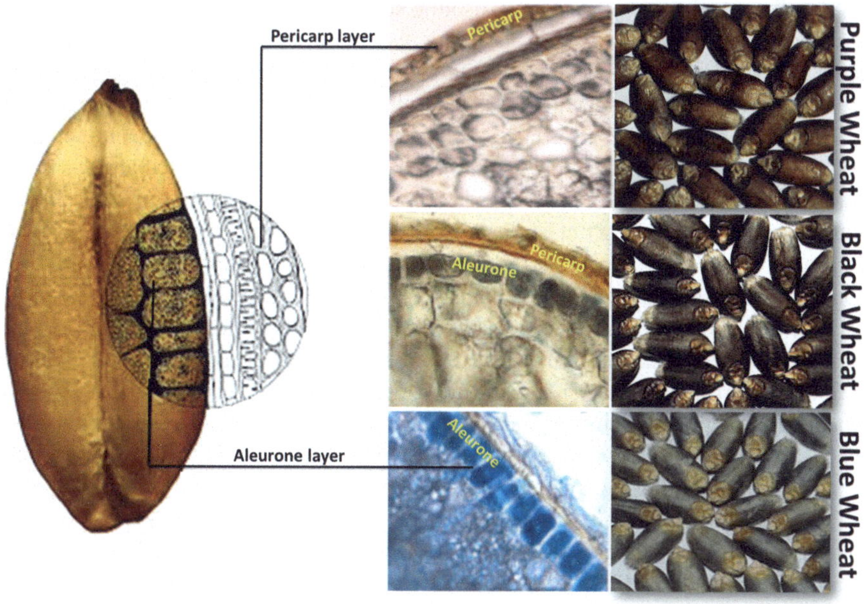


**Fig. 1** Illustrating different types of colored wheat with anthocyanin location. (Reproduced from Garg et al. 2016)

dominant gene in tetraploid wheat (Sharman 1958), while in hexaploid wheat, it was controlled by two incompletely dominant genes (Piech and Evans 1979; Griffin 1987). But later on researchers concluded that the purple pericarp trait is controlled by three dominant alleles in a complicated way as one allele is *Pp-1* designated: *Pp-B1* is located at 7BL (7B of *T. durum*, 7S of *Ae. speltoides*), and *Pp-D1* is located at 7D of *T. aestivum*, whereas the third allele *Pp3* is located at 2A of *T. aestivum* (Khlestkina et al. 2010; Tereshchenko et al. 2012). On the contrary, two complementary alleles are essential to impart purple pericarp color, which are located either in A or B and A or D subgenome of wheat (Tereshchenko et al. 2012). Because of this, the pigmented pericarp trait was seen to be expressed only in allopolyploid wheat not in diploid wheat relatives. However, the exact mechanism by which the purple pericarp trait is regulated in wheat is obscure, and till now it was believed that there were two transcription factors, viz., *TaPpm1* (purple pericarp-MYB 1) and *TaPpb1* (purple pericarp-bHLH 1), that co-regulate the anthocyanin synthesis by interacting with each other (Jiang et al. 2018).

On the other hand, the blue aleurone trait was inherited by introgressing chromosome from wheat wild relatives like *Agropyron trichophorum*, *Agropyron glaucum*, *Agropyron elongatum*, and *Triticum boeoticum* by means of an addition, substitution, or translocation line of wheat (Knott 1958; Zeven 1991; Singh et al. 2007; Garg et al. 2016). Its translocation line with replacement of 4BL of wheat with the homologous chromosome of *Th. ponticum* has been reported (Garg et al. 2016). In another line, the blue color is due to disomic substitution of 4A (4AmL) transferred to wheat from *T. boeoticum* (syn. *Triticum monococcum* ssp. aegilopoides) (Singh et al. 2007). Possibility of occurrence of the third gene on chromosome 4D has been proposed that has been substituted by homologous chromosomes from *Th. ponticum* (Lachman et al. 2017). Certain studies have mentioned about two complementary genes (Zeven 1991; Lan et al. 2008), while others have indicated about single dominant gene controlling the blue grain color (Knievel et al. 2009; Lan et al. 2008; Kuspira et al. 1989; Singh et al. 2007). There are three independent genes, (i) a dominant gene *Ba1* {syn. Ba (b)} located at 4AgL of *Th. ponticum*; (ii) a partial dominant gene Ba2 {syn. Ba (a)} located at 4Am and 4Abo on long arm of *T. monococcum* and *T. boeoticum*, respectively; and (iii) a dominant gene *BaThb* located at 4J of *Th. bessarabicum* (Zheng et al. 2006; Dubcovsky et al. 1996; Singh et al. 2007; William and Mujeeb-Kazi 1993), which controlled the blue color trait in wheat. Its molecular mechanism is also not clear yet, but a bHLH transcription factor *ThMYC4E* from *Th. ponticum* has been reported to control blue aleurone trait in an addition line of chromosome 4E of *Th. ponticum* in wheat (Li et al. 2017).

The black wheat was first of all produced in China by crossing earlier known blue and purple lines with the continued efforts of breeders for the last 20 years (Sun et al. 1996; Sun et al. 1999).

The inheritances of purple and blue color are complicated. The pericarp is a maternal tissue derived from the carpel, so it is not possible to observe segregation among the kernels inside the spike. In the case of the triploid aleurone layer, double dose of genetic information comes from the maternal parent and single dose from the paternal parent. Therefore, the expression of blue aleurone color depends upon the gene dose and therefore may result in the segregation of these colors among the kernels inside one spike or one plant.

## 3 Biochemical Composition of Colored Wheat

Researchers reported that major constituents (starch, proteins, and dietary fibers) of colored wheat vary with the background of genotypes and environment. Chinese black-grained wheat had shown high polysaccharide and protein content (Li et al. 2006; Sun et al. 2011), while Sun et al. (1999) observed high dietary fiber and low carbohydrate content in black-grained wheat as compared to non-pigmented wheat. Liu et al. (2018) observed that black wheat diet as compared to control showed higher intake of protein ($P = 0.012$) and dietary fiber and lower intake of carbohydrates.

However among minor elements, it contains diverse combinations of phytochemicals, and these are the key components which impart health protective and beneficiary effects of whole wheat.

## *Phytochemicals in Colored Wheat*

There are various types of phytochemicals composed by wheat grain such as phenolic compounds, phytosterols, lignins, betaine, and folate (Fardet 2010). Among them, phenolic compounds which are further divided into flavonoids and phenolic acids are the most abundant and assorted group of phytochemicals present in the whole colored wheat grain.

### Anthocyanins

Anthocyanins are well-known glycosides of anthocyanidins which belonged to subclass of flavonoids with basic structure as shown in Fig. 2 (Brouillard 1982). Structurally it consists of hydroxyl or methoxyl group on B-ring of 2-phenylbenzopyrylium or flavylium ion. This B-ring and positive charge at oxygen atom of C-ring (oxonium ion) are so reactive which make anthocyanin molecule a strong antioxidant (Fig. 2) (de Gaulejac et al. 1999; Huang et al. 2005; Kongpichitchoke et al. 2015).

There were several reports which documented the total anthocyanin content (TAC) in various pigmented wheat. Abdel-Aal et al. (2006) reported 7.1–211 ppm TAC in pigmented wheat; Varga et al. (2013) observed 5.3–17.4 ppm TAC in blue wheat; and Ficco et al. (2014) have seen 8–50 ppm and 83–174 ppm of TAC in purple and blue wheat, respectively, whereas Garg et al. (2016) and Sharma et al. (2018) observed 16–122 ppm TAC in purple wheat, 68–137 ppm in blue wheat, and 128–198 ppm in black-grained wheat. This shows that purple wheat has lower anthocyanin content than blue wheat, the highest being in black-grained wheat. It has been experiential that anthocyanin content varied with environment and background of colored wheat lines as different researchers are from different countries.

There are various forms of anthocyanins that prevailed in different colored wheat lines, and their composition is also seen to be variable (Table 1). However, the characterization of anthocyanin composition in pigmented wheat is still going on with

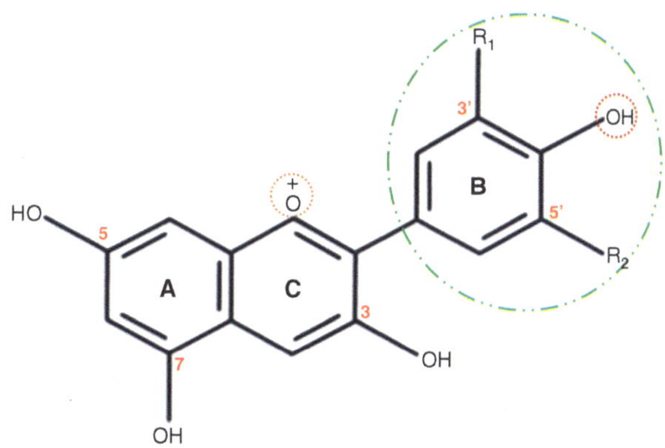


**Fig. 2** Anthocyanin moiety structure. Free radical scavenging capacity of anthocyanins depends upon (i) green color ring, number of hydroxyl groups; (ii) red color ring, catechol moiety in the B-ring; and (iii) orange color ring, oxonium ion in the C-ring

**Table 1** Anthocyanin composition in whole colored wheat grain

| Anthocyanidin name | Anthocyanins | References |
|---|---|---|
| Cyanidin 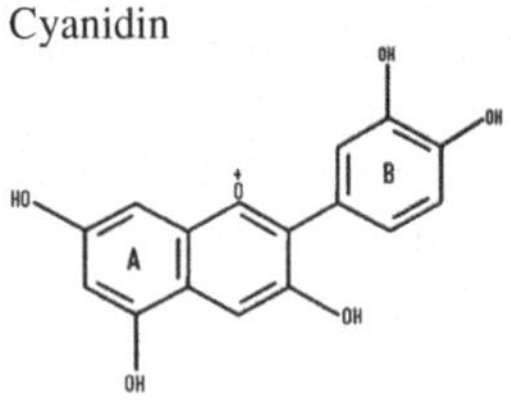 | Cyanidin 3-glucoside | Abdel-Aal et al. (2008) |
| | Cyanidin 3-rutinoside | |
| | Cyanidin 3,5-diglucoside | |
| | Cyanidin 3-arabidoside | |
| | Cyanidin 3-(6″-p-coumaryl)glucoside | |
| | Cyanidin 3-(6″-malonylglucoside) | |
| | Cyanidin 3-(6″-succinyl)glucoside | |
| | Cyanidin 3-galactoside | |
| | Cyanidin 3-(3″,6″-dimalonylglucoside) | |
| | Cyanidin 3-(3″,6″-malonylsuccinylglucoside) | |
| | Cyanidin 3-(6″-succinylglucoside) | |
| | Cyanidin 3-disuccinylglucoside | |
| | Cyanidin 3-(2G-xylosylrutinoside) | |
| | Cyanidin 3-(3″,6″-dimalonylglucoside) | |
| | Cyanidin 3-(6″-feruloylglucoside)-5-glucoside | |
| | Cyanidin 3-(6″-succinylglucoside) | |
| | Cyanidin 3-rutinoside-3′-glucoside | |
| | Catechin-(4,8)-cyanidin 3,5-diglucoside | |
| Delphinidin 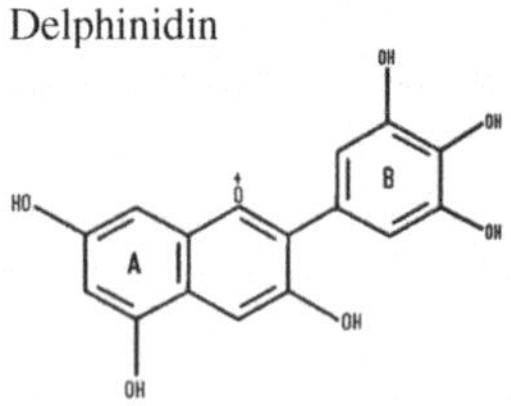 | Delphinidin 3-glucoside | Tyl and Bunzel (2012) |
| | Delphinidin 3-malonylglucoside | |
| | Delphinidin 3-rutinoside | |
| | Delphinidin with hexose/coumaric acid | |
| | Delphinidin 3-(6″-malonylglucoside) | |
| | Delphinidin 3-arabinoside | |
| | Delphinidin 3-caffeoylglucoside | |
| | Delphinidin 3-galactoside | |
| | Delphinidin 3-sambubioside | |
| Malvidin 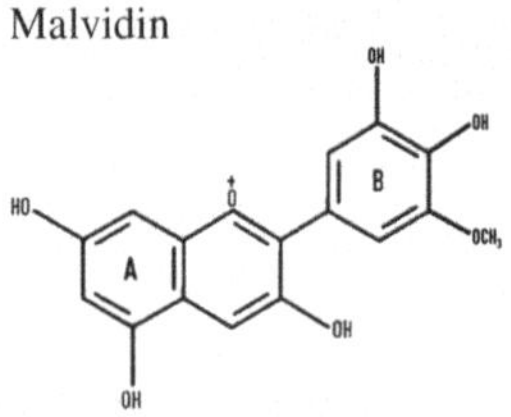 | Malvidin 3-glucoside | Chen et al. (2013) |
| | Malvidin 3-galactoside | |
| | Malvidin 3-rutinoside | |
| | Malvidin 3-rutinoside-5-glucoside | |
| | Malvidin 3-(6″-p-caffeoylglucoside) | |

(continued)

**Table 1** (continued)

| Anthocyanidin name | Anthocyanins | References |
|---|---|---|
| Pelargonidin 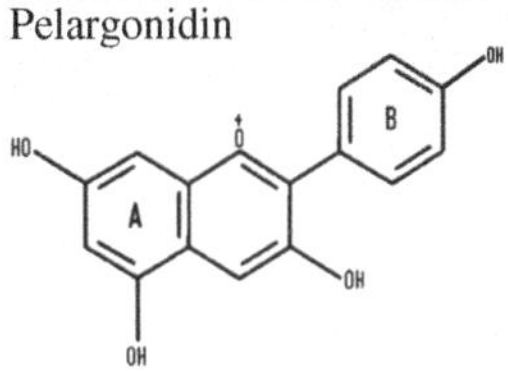 | Pelargonidin 3-glucoside | Ficco et al. (2014) |
| | Pelargonidin 3-(6″-malonylglucoside) | |
| | Pelargonidin 3-arabinoside | |
| | Pelargonidin 3-galactoside | |
| | Pelargonidin 3-rutinoside | |
| | Pelargonidin with hexose and acetic/malonic acid | |
| | Pelargonidin 3-rutinoside | |
| | Pelargonidin 3,5-diglucoside | |
| Peonidin 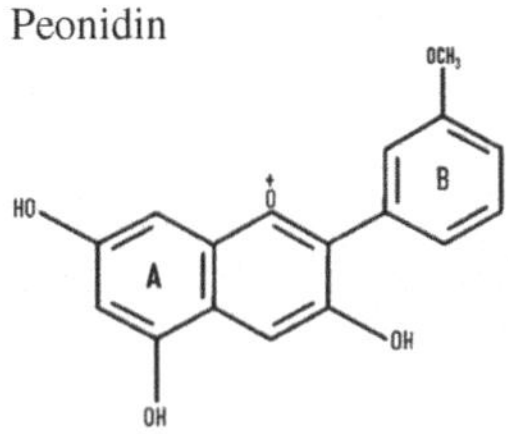 | Peonidin 3-(6″-p-coumaryl)glucoside | Garg et al. (2016) |
| | Peonidin 3-galactoside | |
| | Peonidin 3,5-diglucoside | |
| | Peonidin 3-arabinoside | |
| | Peonidin 3-rutinoside | |
| | Peonidin 3-rutinoside-5-glucoside | |
| | Peonidin 3-(6″-ethylmalonylglucoside) | |
| Petunidin 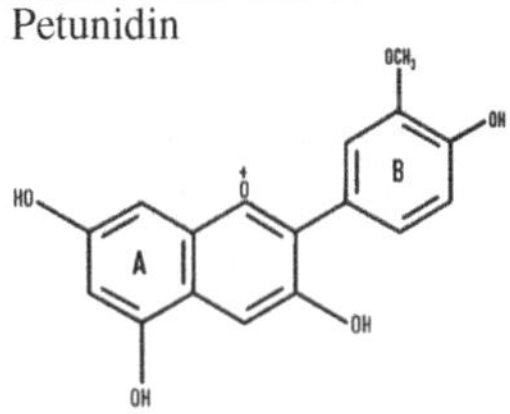 | Petunidin 3-glucoside | Lachman et al. (2017) |
| | Petunidin with hexose and rhamnose | |
| | Petunidin 3-rutinoside-5-glucoside | |

the advanced instrumentation. Recently Garg et al. (2016) reported 22, 23, and 26 types of anthocyanins in blue, purple, and black wheat lines, respectively. They all are sugar or phenolic group derivatives of six major anthocyanidins, i.e., cyanidin, delphinidin, malvidin, pelargonidin, peonidin, and petunidin (Table 1). It has been known that cyanidin 3-glucoside was a dominant anthocyanin in purple wheat (Abdel-Aal and Hucl 2003; Ficco et al. 2014). Contrary to that, in blue wheat, some found cyanidin 3-glucoside as the dominant anthocyanin (Abdel-Aal and Hucl 2003; Hu et al. 2007), while Abdel-Aal et al. (2008) found delphinidin 3-glucoside as the dominant one. This type of variations is also thought to be attributed by genotype and environmental conditions or could be because of variation in extraction and quantification methods (Escribano-Bailon et al. 2004).

Hence, anthocyanin content is affected by several factors likes genotype, climate, environmental conditions, and even the position of grain in spike (Abdel-Aal et al. 2008; Knievel et al. 2009; Lachman et al. 2017). Bustos et al. (2012) also reported that magnesium fertilization and early harvesting also increase the anthocyanin content in purple wheat.

### Phenolics

In addition to anthocyanin, colored wheat also contains other phenolic compounds specifically phenolic acids which were also characterized for antioxidant activity in non-pigmented wheat. Phenolic acids in wheat grains are known to be the main contributors of total antioxidant capacity of wheat (Siebenhandl et al. 2007; Liu et al. 2010; Zhang et al. 2018). Accordingly they are enhancing the antioxidant capacity of anthocyanin wheat in synergistic mode (Wang and Zhu 2017). Total soluble phenolic acid content was also observed to be high in colored wheat as compared to non-pigmented wheat (Sharma et al. 2018). Moreover, Zhang et al. (2018) found that colored wheat has higher content of bound phenolic acids as compared to free phenolic acids and hence has shown significant enhancement in antioxidant activity.

There were various forms of phenolic acids like gallic, protocatechuic, p-hydroxybenzoic, chlorogenic, vanillic, caffeic, syringic, p-coumaric, ferulic, and salicylic acids known to exist in non-pigmented whole wheat (Siebenhandl et al. 2007; Fares et al. 2010), whereas colored whole wheat was reported to be rich in gallic, protocatechuic, p-hydroxybenzoic, vanillic, syringic, p-coumaric, ferulic, isoferulic, and salicylic acids (Zhang et al. 2018). Phenolic acid composition and content are also affected by growing conditions (Vaher et al. 2010) and by agronomic practices, for example, spray of selenium into the soil can enhance it (Chu et al. 2010).

## *Minerals*

To overcome the micronutrient deficiencies, researchers have added the micronutrients to food in order to meet the human needs either by increasing mineral supplementation or by biofortification of staple crops. Wheat already contains various micronutrients like iron, zinc, copper, manganese, calcium, magnesium, and potassium which are very beneficial to human health (Li and Beta 2011). However, it has been observed by several researchers that pigmented grains have high mineral content. Sharma et al. (2018) observed that iron, zinc, copper, and manganese contents were higher in colored as compared to common wheat. He and Ning (2003) also reported that black wheat "Qinhei No. 1" had higher contents of iron, zinc, manganese, copper, selenium, magnesium, potassium, and phosphorus. Organic chromium (trivalent chromium) content in black wheat "03Z4–439" has been reported to be about four times higher than that of common wheat which is a characteristic that can be used to treat diabetes (He and Ning 2003). Nutrient composition analysis of purple wheat showed higher amount of different types of nutrients including iron, zinc, and magnesium than the common wheat (Guo et al. 2012).

## 4 Agronomic Traits of Colored Wheat

### *Yield*

The major challenge for a new variety of colored wheat is the improvement of yield along with biofortification of anthocyanins. Grain yield is the major hurdle in the popularization of colored wheat lines because of the linkage drag associated with the blue aleurone trait that is contributed by the wild wheats in the form of addition, substitution, or translocation lines (Martinek et al. 2014; Garg et al. 2016). Rigorous breeding is required to disrupt these linkages for creating lines with a high anthocyanin content and satisfactory yield level. In Austria, Skorpion is a commercial blue grain cultivar with about 25% lower yield in comparison to check cultivars (Martinek et al. 2014).

The other important trait is adaptability to the environment. In general, native colored wheat varieties are winter wheat in nature and need prolonged vegetative phase for biomass production, but when they are transferred to subtropical region, they experience multiple stresses because of poor adaptability. Consequently, colored wheat varieties can be adapted to various environments by using breeding strategies like crossing of exotic colored wheat lines to locally adapted cultivars, for example, colored wheat lines generated for Indian environment with better yield performance and high anthocyanin content (Garg et al. 2016).

### *Processing Quality of Colored Wheat*

In recent years, many food processing researchers throughout the world have come forward to exploit colored wheat, for example, purple wheat bran muffin (Li et al. 2007a) and antho-beer made from purple-grained wheat (Li et al. 2007b); soy sauce (Li et al. 2004), vinegar, breakfast cereal, and instant noodles produced from black-grained wheat; and fine dried noodles made from blue-grained wheat (Pei et al. 2002). It has been observed that anthocyanin content substantially influences the quality of wheat products such as bread, pasta, and noodles (Abdel-Aal and Hucl 2003). But because of health benefits of anthocyanins, it is very useful to understand the processing quality of these colored wheat grains in order to increase their production and use. We can accomplish the colored wheat varieties with desired characteristics by using breeding strategies.

Sharma et al. (2018) studied the nutritional profile of the colored wheat lines and showed that their lines have all the features required for commercial product development. The processing quality of wheat is largely determined by the quantity and quality of the storage proteins specifically glutenins and gliadins (Wall 1979), and thus, it is very important to understand the protein properties of blue, purple, and black colored wheats. SDS sedimentation provides the gluten strength of the flour, and the higher the SDSS value, the higher the gluten strength as they are positively correlated (Dick and Quick 1983). Li et al. (2006) have observed a reduction in SDSS value, gluten index, and mid peak time (MPT) in black-grained wheat when compared to the three check cultivars. The gluten index value of black-grained wheat (69.74%) lies in the optimum range (60–90%) for good bread making quality (Li et al. 2006). They also observed that black-grained wheat has low stickiness value and also has the HMW-GS pattern of 2* and 5 + 10 which means it has better baking properties (Li et al. 2006).

Usage of colored wheat in different bread products has been well studied. Janeckova et al. (2014) observed that addition of purple wheat bran (10–30%) affected the loaf volume, crust color and integrity, crumb structure, and taste of the resulting bread. The loaf volume was decreased when 20% of the baking flour was replaced with 10% purple wheat bran and 10% semolina, but crumb porosity is similar to that of control bread. Replacement of flour with 20% finely milled bran and 10% semolina resulted in further reduction in loaf volume, but the bread has the most preferred crust integrity and crumb porosity. The sensory parameters of the bread worsen when the flour is replaced with 30% unmilled bran and 10% semolina.

## 5 Stability of Anthocyanins and Phenolics on Processing of Colored Wheat

Anthocyanins are stable at lower temperature, and their stability decreases with increase in temperature and the duration of heat. Modern food processing technologies require high temperatures (160–180 °C), and studies showed that anthocyanin stability is decreased in foods produced after processing (Mercadante and Bobbio 2008). Initial studies on the effect of heat on anthocyanin wheat flour were performed by Abdel-Aal and Hucl (2003), and they found degradation of anthocyanin content in blue wheat. But the degradation is less when compared to anthocyanin extracts; this might be due to the protective effect of food matrix in whole wheat flour. Other reports such as Yu and Beta (2015) and Pasqualone et al. (2015) have also found anthocyanin reduction of 55% in bread and 57% in biscuit made from purple wheat flour, respectively. Bartl et al. (2015) have also reported the reduction in anthocyanins in bread made from purple and blue wheat. Similar results have been reported in other anthocyanin-rich foods.

Other than anthocyanins, the total phenolic content (TPC) of colored wheat is comparably more than the normal amber wheat (Sharma et al. 2018; Li et al. 2015a). Previous studies have shown the effect of total phenolic content on processed foods. Li et al. (2015a) have observed a decrease in TPC after processing the colored wheat flour into steam bread and noodles. In contrast, Yu and Beta (2015) have observed an increase in TPC after bread making in white and purple wheat. Similarly, Leenhardt et al. (2006) have reported that the degradation of carotenoids occurred during bread making from whole wheat flour.

Measurement of antioxidant activity after product making is important to assess the functionality of products and their commercial exploitation. Yu and Beta (2015) have observed more than 30% higher antioxidant activity of purple wheat bread compared to purple flour. Li and Beta (2011) have observed higher antioxidant activity of purple wheat bread. Pasqualone et al. (2015) have observed more than 15% increases in antioxidant activity of biscuits prepared from purple wheat. Li et al. (2015a) have reported a decrease in antioxidant activity of purple and black wheat noodles and steamed bread as compared to flour. Therefore, it depends on the method used for product making and extraction of anthocyanins from the flour and finished products. Alavi et al. (2014) have reported an increase in antioxidant activity of extruded products with apple and tomato pomace despite a decrease in antioxidants. Thus, most of these studies have indicated that although there is a decrease in anthocyanin content during heating and product making, still there is an increase in antioxidant activity. The reason might be increase in total phenolic content and other hypothesis may be that breakdown products of anthocyanins after heating might be having higher antioxidant activity than their colored and glycosylated forms or the synergistic effect of different phytochemicals.

## 6 Applications of Colored Wheat in Health

Anthocyanins are considered as biologically active compounds and known to play a vital role in the prevention of several metabolic diseases; thus, it is hailed as a nutraceutical agent in recent years. Owing to its strong antioxidant properties, anthocyanin acts as a panorama of biomedical functions (Zafra-Stone et al. 2007; Prior 2003; Wang and Stoner 2008). Numerous epidemiological studies have already established the anti-proliferative, antioxidant, antiaging, and anti-inflammatory properties of anthocyanins from diverse sources (Lin et al. 2017).

Anthocyanins from cereals like black rice, black sorghum, barley, and purple/black/blue wheat have been well characterized by pioneer researchers (Garg et al. 2016; Awika et al. 2005; Ryu et al. 1998). Some studies have very well documented the inhibitory effect of anthocyanins from black rice against cancer cells (Chen et al. 2006).

But very few studies are available supporting the role of anthocyanins from wheat in the aspect of health. It has been well documented in various researches that anthocyanins are known to possess antioxidant activity (Hu et al. 2007; Liu et al. 2010; Abdel-Aal et al. 2018; Sharma et al. 2018). In addition, blue-grained wheat has been found to obstruct the LDL cholesterol oxidation which might contribute to the development of various heart diseases (Abdel-Aal et al. 2008).

## In Vitro *Studies/Reports Supporting the Role of Anthocyanins from Colored Wheat*

Report published by Sharma et al. in 2018 used the murine-based raw macrophage cell lines to study the effect of the three types of anthocyanin-biofortified wheat. The cell lines induced with lipopolysaccharides produced the nitric oxide and pro-inflammatory cytokines. This effect was attenuated efficiently upon the treatment of cell lines with the anthocyanin extracts from the colored wheat. It has been suggested that despite the low anthocyanin content in purple wheat and relatively low antioxidant activity as compared to blue and black lines, it impedes the production of pro-inflammatory cytokines in cell line-based assays in a more effective manner.

## In Vivo *Studies*

### Animal Model Studies Supporting the Role of Anthocyanins from Colored Wheat

The work led by J Prokop (2018) appraised the effect of anthocyanins from two blue wheat varieties, UC66049 and Skorpion, on the drug-metabolizing cytochrome P450 enzyme which is implicated in the course of drug metabolism and also in the metabolism of steroid and cancer-causing substances (Rendic and Guengerich 2015; Guengerich 2015; Guengerich et al. 2016). Groups of rat fed with two blue-grained wheat showed an increase in cytochrome P450 enzyme, i.e., aspartate aminotransaminase activity. They also found that there was lower weight gain in the rats fed with anthocyanin-rich UC66049 blue-grained wheat compared with control and Skorpion blue wheat-fed rats. Upon analyzing the antioxidant status of the rat plasma, an elevated level of total -SH groups has been observed in blue-grained-fed rats, which is correlated with the ability to survive the oxidants produced in the body. Additionally, FRAP method has been also used to study the overall antioxidant capacity of plasma, and they found a positive result. Mechanistically they observed a moderate increase in the activity of almost all CYP (CYP1A2, CYP2C, CYP2E1, and CYP3A) enzymes which are also positively correlated with the corresponding mRNA expression in the case of rats fed with UC66049 blue wheat as compared to control.

Purple wheat (named as Karkula) has been assessed in terms of improvement in oxidative status and behavior of rats (Jansakova et al. 2016). They suggested a significant increase of total antioxidant capacity in serum ($P = 0.039$), whereas in the kidney a decreased level of advanced oxidation protein products. On the other hand, they found an increase in the levels of thiobarbituric acid reactive substances in the treatment group as compared to the control group (Jansakova et al. 2016).

Anthocyanin-rich wheat is a staple food crop for humans and is also used as animal feed. Mrkvicova and his colleagues (2017) studied the influence of feeding rats, chickens, and fish with purple Konini wheat. The activities of various enzymes like gamma-glutamyltransferase (GGT), aspartate aminotransferase (AST), alanine aminotransferase (ALT), and alkaline phosphatase (ALP) were measured in the collected tissues, and it has been found that in the animals fed with purple Konini wheat, the enzyme activities of the liver were greatly lowered, while the significant differences were only seen in the gamma-glutamyltransferase activity in the chicken's blood. In the case of rats fed with purple Konini wheat, notably higher antioxidant values have been observed as per determined by the DPPH and FR methods. In the case of the chickens fed with Konini wheat, a significant rise in the antioxidant values ($P < 0.05$) has been observed by DPPH and ABTS methods. On the other hand, FR method anticipated lower antioxidant values in chickens. Notably various antioxidant assays depicted no significant differences in the case of the hepatopancreas of fish. On a concluding remark, the overall results suggested that the purple Konini wheat with a higher content of anthocyanins may positively alter the antioxidant activity and functioning of the liver of an organism as suggested by selected enzyme activity.

Apart from causing metabolic dysfunction due to oxidative damage by free radicals, aging is another consequence possibly originating as a result of oxidative stress (Cutler 1985). Thus, Chen and his colleagues used *Caenorhabditis elegans,* a model organism, to study stress resistance and antiaging mechanisms (Olsen et al. 2006; Rothman et al. 2012). They carried out life span assays with wild-type and longevity mutant strains of *C. elegans mev-1(hn1)* to examine the antiaging and antioxidant potential of purple wheat (Kenyon et al. 1993; Kimura et al. 1997). The treatment of the respective cultures of *C. elegans* with purple wheat extracts revealed an extended or prolonged life span of wild type and *mev-1(hn1)* at the rate of 10.5% and 9.2%. Thus, it can be concluded that anthocyanins present in the purple wheat could augment the usual life span of an organism. The results suggested the antiaging effect exerted by the anthocyanins from purple wheat had prolonged the life span of *C. elegans* by inhibiting the insulin/IGF-1-like signaling pathway regulated by insulin receptor DAF-2.

## Human Intervention Studies Supporting the Role of Anthocyanins from Colored Wheat

The effect of black wheat on diabetes has been found in a clinical trial on 120 individuals affected with type 2 diabetes mellitus (T2DM) (Liu et al. 2018). Diet contained black wheat noodles, steamed bun, or kernels. The impact of the dietary intake of functional foods of black wheat was significant lowering of serum level of glycated albumin (GA) compared to the control diet. This finding supported the fact that the anthocyanins present in the black wheat could actually improve the hyperglycemia in T2DM patients. Another observation supporting the role of black wheat in alleviating the T2DM has been made in the same study. Black wheat was found to be more potent against the elevation in the levels of IL-6 (interleukin-6), TNF-α (tumor necrosis factor), and hypersensitive-C reactive protein (hs-CRP) in the circulating bloodstream as compared to the control diet. Besides, plasma glucose and HbA1c levels did not show any significant differences between the black wheat and control groups (*P*.0.05). All the findings concluded that black wheat could actually ameliorate the glycemic index and inflammatory profiles in the T2DM-affected population.

## References

Abdel-Aal ES, Hucl P (2003) Composition and stability of anthocyanins in blue-grained wheat. J Agric Food Chem 51(8):2174–2180

Abdel-Aal ES, Young JC, Rabalski I (2006) Anthocyanin composition in black, blue, pink, purple, and red cereal grains. J Agr Food Chem 54(13):4696–4704

Abdel-Aal ES, Abou-Arab AA, Gamel TH et al (2008) Fractionation of blue wheat anthocyanin compounds and their contribution to antioxidant properties. J Agric Food Chem 56(23):11171–11177

Abdel-Aal ESM, Hucl P, Rabalski I (2018) Compositional and antioxidant properties of anthocyanin-rich products prepared from purple wheat. Food Chem 254:13–19

Alavi S, Giannetta F, Nanjundaswamy A et al (2014) Delivery of antioxidants through fruits and vegetables in extruded foods. Cereal Foods World 59(4):179–185

Alvarez-Suarez JM, Giampieri F, Tulipani S et al (2014) One-month strawberry-rich anthocyanin supplementation ameliorates cardiovascular risk, oxidative stress markers and platelet activation in humans. J Nutr Biochem 25:289–294

Awika JM, Rooney LW, Waniska RD (2005) Anthocyanins from black sorghum and their antioxidant properties. Food Chem 90(1–2):293–301

Bagchi D, Sen CK, Bagchi M et al (2004) Anti-angiogenic, antioxidant and anti-carcinogenic properties of a novel anthocyanin-rich berry extract formula. Biochem Mosc 69(1):75–80

Bartl P, Albreht A, Skrt M et al (2015) Anthocyanins in purple and blue wheat grains and in resulting bread: quantity, composition, and thermal stability. Int J Food Sci Nutr 66(5):514–519

Bishayee A, Sethi G (2016) Bioactive natural products in cancer prevention and therapy: progress and promise. In: Seminars in cancer biology, vol 40. Academic, pp 1–3

Bobe G, Wang B, Seeram NP et al (2006) Dietary anthocyanin-rich tart cherry extract inhibits intestinal tumorigenesis in APC (Min) mice fed suboptimal levels of sulindac. J Agric Food Chem 54:9322–9328

Bowen-Forbes CS, Zhang Y, Nair MG (2010) Anthocyanin content, antioxidant, anti-inflammatory and anticancer properties of blackberry and raspberry fruits. J Food Compos Anal 23(6):554–560

Brouillard R (1982) Chemical structure of anthocyanins. In: Markakis P (ed) Anthocyanins as food color. Academic, New York, pp 1–40

Bustos DV, Riegel R, Calderini DF (2012) Anthocyanin content of grains in purple wheat is affected by grain position, assimilate availability and agronomic management. J Cereal Sci 55(3):257–264

Cerletti C, de Curtis A, Bracone F et al (2016) Dietary anthocyanins and health: data from FLORA and ATHENA EU projects. Br J Clin Pharmacol 83(1):103–106. https://doi.org/10.1111/bcp.12943

Chen PN, Kuo WH, Chiang CL et al (2006) Black rice anthocyanins inhibit cancer cells invasion via repressions of MMPs and u-PA expression. Chem Biol Interact 163(3):218–229

Chen W, Mulle D, Richling E et al (2013) Anthocyanin-rich purple wheat prolongs the life span of *Caenorhabditis elegans* probably by activating the DAF-16/FOXO transcription factor. J Agric Food Chem 61(12):3047–3053

Chen WC, Liu HM, Liu JS (2016) Progress in the research on anticarcinogenic activities of anthocyanins (Chinese). Food Res Dev 37:211–215

Chu J, Yao X, Zhang Z (2010) Responses of wheat seedlings to exogenous selenium supply under cold stress. Biol Trace Elem Res 136(3):355–363

Copp LGL (1965) Purple grains in hexaploid wheat. Wheat Inf Serv 18:19–20

Cui J, Li XY (2014) Progress on anti-tumor mechanisms of anthocyanins (Chinese). Food Sci 35:310–315

Cutler RG (1985) Antioxidants and longevity of mammalian species. In: Molecular biology of aging. Springer, Boston, pp 15–73

de Gaulejac NS, Glories Y, Vivas N (1999) Free radical scavenging effect of anthocyanins in red wines. Food Res Int 32(5):327–333

Dick JW, Quick JS (1983) A modified screening test for rapid estimation of gluten strength in early generation durum wheat breeding lines. Cereal Chem 60:315–318

Dubcovsky J, Luo M-C, Zhong G-Y et al (1996) Genetic map of diploid wheat, *Triticum monococcum* L., and its comparison with maps of *Hordeum vulgare* L. Genetics 143:983–999

Escribano-Bailon MT, Santos-Buelga C, Rivas-Gonzalo JC (2004) Anthocyanins in cereals. J Chromatogr A 1054:129–141

Fardet A (2010) New hypotheses for the health-protective mechanisms of whole-grain cereals: what is beyond fibre? Nutr Res Rev 23(1):65–134. https://doi.org/10.1017/S0954422410000041

Fares C, Platani C, Baiano A et al (2010) Effect of processing and cooking on phenolic acid profile and antioxidant capacity of durum wheat pasta enriched with debranning fractions of wheat. Food Chem 119(3):1023–1029

Ficco DB, De Simone V, Colecchia SA et al (2014) Genetic variability in anthocyanin composition and nutritional properties of blue, purple, and red bread (*Triticum aestivum L.*) and durum (*Triticum turgidum L.* ssp. *turgidum convar.* durum) wheats. J Agric Food Chem 62(34):8686–8695

Garg M, Chawla M, Chunduri V et al (2016) Transfer of grain colors to elite wheat cultivars and their characterization. J Cereal Sci 71:138–144

Gibson GR, Beatty ER, Wang XI et al (1995) Selective stimulation of bifidobacteria in the human colon by oligofructose and inulin. Gastroenterology 108(4):975–982

Griffin WB (1987) Out-crossing in New Zealand wheat measured by occurrence of purple grain. N Z J Agric Res 30:287–290

Guengerich F (2015) Human cytochrome P450 enzymes. In: Ortiz de Montellano P (ed) Cytochrome P450. Springer, Cham, pp 523–785

Guengerich FP, Waterman MR, Egli M (2016) Recent structural insights into cytochrome P450 function. Trends Pharmacol Sci 37(8):625–640

Guo H, Ling W (2015) The update of anthocyanins on obesity and type 2 diabetes: experimental evidence and clinical perspectives. Rev Endocr Metab Disord 16(1):1–13. https://doi.org/10.1007/s11154-014-9302-z

Guo Z, Zhang Z, Xu P (2012) Analysis of nutrient composition of purple wheat. Cereal Res Commun 41(2):293–303

Hafidh RR, Abdulamir AS, Bakar FA et al (2013) Novel anticancer activity and anticancer mechanisms of *Brassica oleracea* L. var. capitata f. rubra. Eur J Integr Med 5(5):450–464

Havrlentova M, Psenakova I, Zofajova A et al (2014) Anthocyanins in wheat seed – a mini review. Nova Biotechnol Chim 13(1):1–2

He J, Giusti MM (2010) Anthocyanins: natural colorants with health-promoting properties. Annu Rev Food Sci Technol 10(1):163–187

He YZ, Ning JF (2003) Analysis of nutrition composition in the special purple grain wheat "Qinhei 1" containing rich Fe and Zn. J Northwest A&F Univ Nat Sci Ed 31:87–90. (in Chinese)

Hu C, Cai YZ, Li W et al (2007) Anthocyanin characterization and bioactivity assessment of a dark blue grained wheat (*Triticum aestivum* L. cv. Hedong Wumai) extract. Food Chem 104(3):955–961

Huang D, Ou B, Prior RL (2005) The chemistry behind antioxidant capacity assays. J Agric Food Chem 53(6):1841–1856

Janeckova M, Hrivna L, Juzl M et al (2014) Possibilities of using purple wheat in producing bakery products. Mendel Net
Jansakova K, Babickova J, Havrlentova M et al (2016) The effects of anthocyanin rich wheat diet on the oxidative status and behaviour of rats. Croat Med J 57(2):119–129
Jiang W, Liu T, Nan W et al (2018) Two transcription factors TaPpm1 and TaPpb1 co-regulate anthocyanin biosynthesis in purple pericarps of wheat. J Exp Bot 69(10):2555–2567
Jin HL, Lim JD, Choung MG (2013) Studies on the anthocyanin profile and biological properties from the fruits of Acanthopanax senticosus (Siberian ginseng). J Funct Foods 5:380–388
Kenyon C, Chang J, Gensch E et al (1993) A *C. elegans* mutant that lives twice as long as wild type. Nature 366(6454):461–464
Khlestkina EK, Roder MS, Borner A et al (2010) Mapping genes controlling anthocyanin pigmentation on the glume and pericarp in tetraploid wheat (*Triticum durum* L.). Euphytica 171(1):65–69
Khoo HE, Lim SM, Azlan A (2019) Evidence-based therapeutic effects of anthocyanins from foods. Pak J Nutr 18(1):1–11
Kimura KD, Tissenbaum HA, Liu YX et al (1997) Daf-2, an insulin receptor-like gene that regulates longevity and diapause in *Caenorhabditis elegans*. Science 277:942–946
Knievel DC, Abdel-Aal ESM, Rabalski I et al (2009) Grain color development and the inheritance of high anthocyanin blue aleurone and purple pericarp in spring wheat (*Triticum aestivum* L). J Cereal Sci 50:113–120
Knott DR (1958) The inheritance in wheat of a blue endosperm colour derived from *Agropyron elongatum*. Can J Plant Sci 36:571e574
Kongpichitchoke T, Hsu J, Huang T (2015) Number of hydroxyl groups on the B-ring of flavonoids affects their antioxidant activity and interaction with Phorbol Ester binding site of PKCδ C1B domain: *in vitro* and *in silico* studies. J Agric Food Chem 63(18):4580–4586. https://doi.org/10.1021/acs.jafc.5b00312
Kuspira J, Maclagan J, Bhambhani RN et al (1989) Genetic and cytogenetic analyses of the A genome of Triticum monococcum LV inheritance and linkage relationships of genes determining the expression of 12 qualitative characters. Genome 32(5):869–881
Lachman J, Martinek P, Kotikova Z et al (2017) Genetics and chemistry of pigments in wheat grain – a review. J Cereal Sci 74:145–154
Lan SQ, Li XP, Liu XP (2008) Genetic of seed pigment of blue kernel wheat. Acta Agric Bor Sin 23:12–14
Leenhardt F, Lyan B, Rock E et al (2006) Genetic variability of carotenoid concentration, and lipoxygenase and peroxidase activities among cultivated wheat species and bread wheat varieties. Eur J Agron 25(2):170–176
Li W, Beta T (2011) Flour and bread from black, purple and blue-colored wheats. In: Flour and breads and their fortification in health and disease prevention. Academic, London, pp 59–67
Li WD, Sun SC, Ren GX (2004) Study on characteristic of black wheat and its exploitation and utilization. China Condiment 1:9–11
Li W, Beta T, Sun S et al (2006) Protein characteristics of Chinese black-grained wheat. Food Chem 98(3):463–472
Li W, Pickard MD, Beta T (2007a) Effect of thermal processing on antioxidant properties of purple wheat bran. Food Chem 104(3):1080–1086
Li W, Pickard MD, Beta T (2007b) Evaluation of antioxidant activity and electronic taste and aroma properties of antho-beers from purple wheat grain. J Agric Food Chem 55(22):8958–8966
Li L, Wang L, Wu Z, et al (2014) Anthocyanin-rich fractions from red raspberries attenuate inflammation in both RAW264.7 macrophages and a mouse model of colitis. Scientific Reports, 4(1). https://doi.org/10.1038/srep06234
Li Y, Ma D, Sun D et al (2015a) Total phenolic, flavonoid content, and antioxidant activity of flour, noodles, and steamed bread made from different colored wheat grains by three milling methods. Crop J 3(4):328–334

Li D, Zhang Y, Liu Y et al (2015b) Purified anthocyanin supplementation reduces dyslipidemia, enhances antioxidant capacity, and prevents insulin resistance in diabetic patients. J Nutr 145:742–748

Li N, Li S, Zhang K et al (2017) ThMYC4E, candidate blue aleurone 1 gene controlling the associated trait in Triticum aestivum. PLoS One 12(7):e0181116

Lin BW, Gong CC, Song HF et al (2017) Effects of anthocyanins on the prevention and treatment of cancer. Br J Pharmacol 174(11):1226–1243

Liu Q, Qiu Y, Beta T (2010) Comparison of antioxidant activities of different colored wheat grains and analysis of phenolic compounds. J Agric Food Chem 58:9235–9241

Liu Y, Qiu J, Yue Y et al (2018) Dietary black-grained wheat intake improves glycemic control and inflammatory profile in patients with type 2 diabetes: a randomized controlled trial. Clin Risk Manag 14:247

Marlett JA, McBurney MI, Slavin JL (2002) Position of the American Dietetic Association. J Am Diet Assoc 102(7):993–1000. https://doi.org/10.1016/s0002-8223(02)90228-2

Martinek P, Jirsa O, Vaculova K et al (2014) Use of wheat gene resources with different grain colour in breeding. In: Proceedings of the Tagungsband der 64

Mercadante AZ, Bobbio FO (2008) Anthocyanins in foods: occurrence and physicochemical properties. In: Food colorants: chemical and functional properties, pp 241–276

Mrkvicova E, Pavlata L, Karasek F (2017) The influence of feeding purple wheat with higher content of anthocyanins on antioxidant status and selected enzyme activity of animals. Acta Vet Brno 85(4):371–376

Olsen A, Vantipalli MC, Lithgow GJ (2006) Using *Caenorhabditis elegans* as a model 490 for aging and age related diseases. Ann N Y Acad Sci 1067:120–128

Pasqualone A, Bianco AM, Paradiso VM et al (2015) Production and characterization of functional biscuits obtained from purple wheat. Food Chem 80:64–70

Pei C, Sun Y, Sun C et al (2002) Research and utilization of black-grained wheat. Seed (Chinese) 4:42–44

Piech J, Evans LE (1979) Monosomic analysis of purple grain colour in hexaploid wheat. Plant Breed 1979:212–217

Prior RL (2003) Fruits and vegetables in the prevention of cellular oxidative damage. Am J Clin Nutr 78(3):570S–578S

Prokop J, Anzenbacher P, Mrkvicova E et al (2018) In vivo evaluation of effect of anthocyanin-rich wheat on rat liver microsomal drug-metabolizing cytochromes P450 and on biochemical and antioxidant parameters in rats. Food Chem Toxicol 122:225–233

Rendic S, Guengerich FP (2015) Survey of human oxidoreductases and cytochrome P450 enzymes involved in the metabolism of xenobiotic and natural chemicals. Chem Res Toxicol 28:38–42

Rothman JH et al (2012) *Caenorhabditis elegans*: cell biology and physiology, vol 107. Academic

Rugina D, Sconta Z, Leopold L, Pintea A, Bunea A, Socaciu C (2012) Antioxidant activities of chokeberry extracts and the cytotoxic action of their anthocyanin fraction on HeLa human cervical tumor cells. J Med Food 15(8):700–706

Ryu SN, Park SZ, Ho CT (1998) High performance liquid chromatographic determination of anthocyanin pigments in some varieties of black rice. J Food Drug Anal 6(4):729–736

Sharma S, Chunduri V, Kumar A et al (2018) Anthocyanin bio-fortified colored wheat: nutritional and functional characterization. PLoS One 13(4):e0194367

Sharman BC (1958) 'Purple pericarp': a monofactorial dominant in tetraploid wheats. Nature 181(4613):929

Siebenhandl S, Grausgruber H, Pellegrini N et al (2007) Phytochemical profile of main antioxidants in different fractions of purple and blue wheat, and black barley. J Agric Food Chem 55(21):8541–8547

Singh K, Ghai M, Garg M et al (2007) An integrated molecular linkage map of diploid wheat based on a *Triticum boeoticum* x *T. monococcum* RIL population. Theor Appl Gene 115:301e312

Slavin J (2003) Why whole grains are protective: biological mechanisms. Proc Nutr Soc 62(1):129–134. https://doi.org/10.1079/pns2002221

Sun SC, Yuan WY, Liu SX et al (1996) A study on the production of tetrageneric hybrids crossed between Triticum, agropyron, haynaldia, secale, and their morphology and cytology. Chin J Genet 23:131–153

Sun SC, Sun Y, Yuan WY et al (1999) Breeding and qualitative analysis for black wheat 76 of superior quality. Acta Agron Sin 25(1):50–54

Sun Y, Cui SW, Gu X et al (2011) Isolation and structural characterization of water unextractable arabinoxylans from Chinese black-grained wheat bran. Carbohyd Polym 85(3):615–621

Tereshchenko OY, Gordeeva EI, Arbuzova VS et al (2012) The D genome carries a gene determining purple grain colour in wheat. Cereal Res Commun 40(3):334–341

Tyl CE, Bunzel M (2012) Antioxidant activity-guided fractionation of blue wheat (UC66049 Triticum aestivum L.). J Agric Food Chem 60(3):731–739

Vaher M, Matso K, Levandi T et al (2010) Phenolic compounds and the antioxidant activity of the bran, flour and whole grain of different wheat varieties. Procedia Chem 2(1):76–82

Varga M, Banhidy J, Cseuz L et al (2013) The anthocyanin content of blue and purple coloured wheat cultivars and their hybrid generations. Cereal Res Commun 41(2):284–292

Wall JS (1979) The role of wheat proteins in determining baking quality. In: Recent advances in the biochemistry of cereals

Wang LS, Stoner GD (2008) Anthocyanins and their role in cancer prevention. Cancer Lett 269(2):281–290

Wang S, Zhu F (2017) Dietary antioxidant synergy in chemical and biological systems. Crit Rev Food Sci Nutr 57:2343–2357

William MD, Mujeeb-Kazi A (1993) Thinopyrum bessarabicum: biochemical and cytological markers for the detection of genetic introgression in its hybrid derivatives with Triticum aestivum L. Theor Appl Genet 86(2–3):365–367

Wu T, Yu Z, Tang Q et al (2013) Honeysuckle anthocyanin supplementation prevents diet-induced obesity in C57BL/6 mice. Food Funct 4:1654–1661

Yu L, Beta T (2015) Identification and antioxidant properties of phenolic compounds during production of bread from purple wheat grains. Molecules (Basel, Switzerland) 20(9):5525–15549. https://doi.org/10.3390/molecules200915525

Zafra-Stone S, Yasmin T, Bagchi M et al (2007) Berry anthocyanins as novel antioxidants in human health and disease prevention. Mol Nutr Food Res 51(6):675–683

Zeven AC (1991) Wheats with purple and blue grains: a review. Euphytica 56:243e258

Zhang J, Ding Y, Dong H (2018) Distribution of phenolic acids and antioxidant activities of different bran fractions from three pigmented wheat varieties. J Chem:2018

Zheng Q, Li B, Mu SM et al (2006) Physical mapping of the blue-grained gene(s) from Thinopyrum ponticum by GISH and FISH in a set of translocation lines with different seed colors in wheat. Genome 49:1109–1114

# Genetics and Breeding of Fe and Zn Improvement in Wheat

Rahul Kumar, Sachin Kumar, Shailendra Sharma, and Rajeev Kumar

**Abstract** Hidden hunger arises when the food consumed by people does not provide adequate micronutrients like vitamins and minerals, iron (Fe) and zinc (Zn), etc. Bio-availability of Fe and Zn in soil have significant roles in the mineral micronutrient uptake and concentration in plants. The inherited low concentration and low bioavailability of Fe and Zn in cereal grains contributed a lot to Fe and Zn deficiency in people, which is widespread, mainly in areas where cereal-based foods are dominant in the diets. Genetic biofortification may enrich cereal grains with micronutrient especially iron and zinc. QTLs for grain iron and zinc have also been mapped in populations derived from crosses between diploid wheat, durum wheat, and wild Emmer wheat and also in synthetic hexaploid wheat and *T. spelta*. A number of wild species of *Triticum, Aegilops*, and other genera have been shown to have in their grains 2–3-fold higher Fe and Zn relative to modern hexaploid wheat cultivars. Synthetic hexaploid (SH) wheat (AABBD'D') has been developed and utilized to bridge gene transfer from *Ae. tauschii* and durum wheat to hexaploid bread wheat. A more recent outcome of the utilization of SH has been the development and release of high grain Zn varieties. Utilizing this variation, HarvestPlus has released several varieties of wheat with 4–10 ppm higher zinc content.

**Keywords** Iron · Zinc · Wheat · Micronutrients · QTL · GWAS

## 1 Introduction

Global biofortification research for a number of crops, including wheat, can be traced back to 1995 when Consultative Group for International Agricultural Research (CGIAR) launched its "CGIAR Micronutrients Program," which continued till 2002. When CGIAR approved its major "Biofortification Challenge

R. Kumar (✉) · S. Kumar · S. Sharma
Department of Genetics and Plant Breeding, Chaudhary Charan Singh University, Meerut, India

R. Kumar
Department of Agricultural Botany, C.S.S.S. (P.G.) College Machhra, Meerut, India

© The Editor(s) (if applicable) and The Author(s), under exclusive license to Springer Nature Switzerland AG 2021

S. H. Wani et al. (eds.), *Physiological, Molecular, and Genetic Perspectives of Wheat Improvement*, https://doi.org/10.1007/978-3-030-59577-7_5

Program" that was later renamed as "HarvestPlus," the program also covered South East Asia and South Asia including India. Thus, studies on genetics and plant breeding for producing biofortified crops, including wheat, have been underway in many countries during the last two decades. The program on biofortification of wheat was undertaken and coordinated by the CIMMYT, Mexico. Improvement of grain micronutrients did not receive the desired attention in the past, although it was recognized that there was a significant loss of genetic variation for Fe and Zn in contemporary cultivars and also that there was limited variability for grain Fe and grain Zn contents in wheat cultivars grown in India (Rawat et al. 2009b). Consequently, the study of genetics and its use for improvement of grain nutritional composition, especially for density of Fe and Zn without any yield penalty received the desired attention during the last ~15 years, although much more remains to be done.

## 2 Fe and Zn Malnutrition: A Global Health Problem

Micronutrient-deficient food seriously causes "hidden hunger" in people worldwide (Palmgren et al. 2008; Poletti et al. 2004; White and Broadley 2005). Hidden hunger is occurring when the food consumed by people does not provide adequate vitamins and minerals, particularly iron (Fe) and zinc (Zn), in their daily diet. In developing countries, dietary deficiency of micronutrients including iron, selenium, calcium, iodine, and vitamin A has serious health implications (Bhaskaram 2002; Demment et al. 2003; Hotz and Brown 2004; Ramakrishnan et al. 2009). Dietary deficiency of essential micronutrients such as iron (Fe) and Zinc (Zn) is a serious public health concern which affects more than three billion people worldwide. Such deficiency leads to malnutrition syndromes when the food consumed by people does not provide enough vitamins and minerals (Bouis 2007; Welch and Graham 2004; White and Broadley 2009).

The World Health Organization (WHO) estimates that approximately 25% of the world's population suffer from anemia (World Health Organization, 2008) and that Fe-deficiency anemia led to the loss of over 46,000 disability adjusted life years (DALYs) in 2010 alone (Murray and Lopez 2013). More than 30% of world population has been severely affected by iron deficiency mainly affecting children (47.4%), pregnant (41.8%), and nonpregnant women (30.2%) (McLean et al. 2009). Iron deficiency during pregnancy causes pregnancy complications, maternal death, birth defects, low birth weight (LBW), etc. (Ortiz-Monasterio et al. 2007; Pathak et al. 2004; Stoltzfus et al. 2004). Clinical or subclinical micronutrient deficiency may affect growth, cognition, and reproductive performance (Seshadri 2001). In India, 58% children, 53% women, and 23% adult men were found to be anemic due to iron deficiency (IIPS and ICF, 2017). Low intake of dietary iron results in higher morbidity and mortality rates, prenatal birth defects impairment in cognitive skills and physical strength, and adverse effects on neuropsychological functions (Brabin et al. 2001; Stein et al. 2008).

It has been estimated that 17.3% of world's population are at risk of inadequate Zn intake (Wessells and Brown 2012), and Zn deficiency leads to estimated annual deaths of 450,000 children (Black and Walker, 2012). Zinc deficiency causes serious health problems, including poor physical growth, incompetent immune system, reduced learning capacity, reproductive inability, and adverse effect on mother and child during the course of pregnancy. According to WHO, zinc deficiency accounts 11th in the world and 5th in the developing countries as a key risk factor causing disease burden in humans (Cakmak 2008). In the past, WHO declared recommended dietary allowance (RDA) of iron 10 mg/day and zinc 15 mg/day for men and 10 mg/day iron and 12 mg/day zinc for women in the age group of 25–50 years (FAO/WHO 2000). Nevertheless, micronutrient deficiency has a prolonged effect on the entire human life cycle. The non-diversified diet of developing countries contains mainly starch-rich cereals, roots, tubers, banana, and plantain food for calorie requirements and lags far behind the RDA in terms of micronutrients (Joint FAO/WHO Expert Committee on Food Additives 2004).

## 3 Causes of Fe and Zn Deficiency

In the developing countries, most of the dietary calories diversified carbohydrate-rich food including rice, wheat, potato, maize, and banana, Whereas, fruits, vegetables, milk, and dairy products have negligible proportion (Joint FAO/WHO Expert Committee on Food Additives 2004). According to IRRI report (2006), it has been found that the polished rice contains only 2 mg/kg of Fe and 12 mg/kg of Zn, whereas the minimum RDA for Fe and Zn is 10–15 mg and 12–15 mg, respectively. Hence, to achieve the RDA for better nutrition, cereal grains should contain around 40–60 mg/kg of Fe and Zn (Cakmak 2000). Most of the staple food crops have very low micronutrient content. Most of the micronutrients are present in the aleurone layer of the cereal grains, and various processing methods like milling, polishing, etc. remove the outermost micronutrient-rich layer of grain resulting in nutrient-poor diet. The majority of the agricultural land is Zn deficit (Cakmak 2002; Mori 1999). The Fe and Zn content in soil have significant roles in the mineral micronutrient uptake and concentration in the edible part of plants. This ultimately results in severe yield loss, stunted plant growth, poor grain quality, and poor nutrition content of grains (Brown et al. 1962; Cakmak 2008; Mori et al. 1990).

## 4 Importance of Fe and Zn in Human Nutrition

The inherited low concentration and low bioavailability of Fe and Zn in cereal grains contributed a lot to Fe and Zn deficiency in people, which is widespread, mainly in areas where cereal-based foods are dominant in the diets (Cakmak 2008). Cereal grains are inherently low both in concentration and bioavailability of Fe and

Zn, mainly when grown on potentially Fe- and Zn deficient soils (Cakmak et al. 2010; Welch and Graham 2004). Increasing cropping intensity and accompanying changes in the soil and fertilizer management practices have lowered the macronutrients as well as micronutrient like Fe and Zn status of soils and its availability. The processing of wheat grains substantially reduces the concentration of Zn and other essential elements, which further increases the Zn deficiency in humans (Cakmak 2008; Kutman et al. 2011; Zhang et al. 2010). Currently, Zn and Fe concentration of cereal grains represents an important challenge to be met by using agricultural tools such as breeding and fertilization.

To alleviate micronutrient malnutrition, a comprehensive strategy involves dietary diversification, supplementation, fortification, and biofortification adapted to conditions in different countries (Stein 2010; Zimmermann and Hurrell 2007). (i) Dietary diversification interventions are interventions that change food consumption at the household level. (ii) Supplementation involving the oral delivery of micronutrients in the form of pills and syrups has been used in chronic deficiencies. (iii) Fortification is the addition of the desired minerals to food stuffs like iodine in salts. Recurring expenditure and lack of a robust distribution system and careful implementation are some of the problems associated with these approaches. (iv) Biofortification is a strategy for producing staple food cultivars whose edible portions have a higher concentration of bioavailable minerals and vitamins. The most economical and feasible approach to alleviate hidden hunger is biofortificaton. The deficiency of micronutrients can be alleviated in human beings using supplements like liquid, tablets, etc. or micronutrient-biofortified food. The high-income group can avail the first option, but poor people cannot purchase the costly supplements for correction of iron and zinc deficiency. Therefore, It will be important to provide the micronutrient-enriched food grains, especially iron and zinc to the people to alleviate micronutrient deficiency.

## 5 Role of Iron and Zinc Metabolism

(i) In Humans

Iron is a very essential mineral required in several vital functions in all living organisms, including several metabolic processes, electron transport, oxygen transport, and DNA synthesis (Lieu et al. 2001). Iron plays a significant role in hematopoiesis (Red blood cell production), hemoglobin formation and conversion of blood sugar to energy, regulation of metabolic energy, immune system, normal brain development, muscle development, and proper growth and development of the body. With respect to absorption mechanism, the dietary iron is classified as two types of iron: heme iron and non-heme iron. The primary source of heme is hemoglobin and myoglobin from fish, poultry, and meat products, whereas cereals, legumes, fruits, and vegetables are the primary source of non-heme iron (Hallberg 1981). There are three mechanisms to control iron balance and regulation of iron absorption. (i) First mechanism is the continued re-utilization of Fe from catabolized erythrocytes. (ii) The second mechanism is through iron storage protein ferritin to store and release iron, especially under the iron demand conditions, e.g., menstrual cycle and pregnancy. (iii) The third mechanism involves regulation of iron absorption in the intestinal tract by establishing equilibrium between absorption and requirement (Hallberg 1981). Zinc is an essential component of several enzymes (>300) and plays a significant role in the synthesis and degradation of carbohydrates, lipids, nucleic acid, and protein. Zinc also plays a significant role in the maintenance of cell and organ integrity by stabilizing molecular structure of cellular components and membranes. Zinc plays a central role in several immune processes (Shankar and Prasad 1998). The zinc deficiency in humans cause the stunted growth, poor bone maturation and fertility, skin lesions, alopecia, diarrhea, impaired appetite, and defect in the immune system and wound healing (Hambidge et al. 1987). Zinc metabolism and absorption are concentration dependent and occur in small intestine (Sandstrom 1997).

(ii) In Plants

Iron is a very essential micronutrient and is involved in various plant metabolic reactions, including most of the redox processes of electron transport chain and photosynthesis (photosystem I and II), chlorophyll synthesis, and nitrogen fixation (Garrido et al. 2006; Kim and Guerinot 2007). Although iron is the fourth most abundant mineral in the earth's crust, it constitutes only 0.005% of plant mass (Graham et al. 2007; Meng et al. 2005). Most of the iron is present in the form of oxides, hydroxides, phosphates, and other complex forms in plants. Similarly, zinc is key mineral for the plant and is involved as a co-factor in nearly 300 different enzymes and plays a very crucial role in the structure of many proteins and gene regulatory elements (Hershfinkel 2005; Palmgren et al. 2008).

About one third of the world's cereal-growing area are iron deficient with high soil pH and half has zinc deficient soil (Cakmak 2002; Mori 1999). Inefficient mineral uptake due to calcareous or salt-stressed alkaline soil results in severe loss in yield and poor plant growth and nutritional quality of grains (Cakmak 2008; Mori et al. 1990). In these mineral-deficient soils with abiotic stress conditions, plants show high susceptibility to environmental stress, including drought stress, pathogenic infections, stunted growth, and leaf necrosis. The grain micronutrient content depends on the amount of mineral uptake by plant roots from soil during different developmental stages and their remobilization and distribution in grain from the vegetative tissues via phloem. The mobility of each mineral element differs significantly from each other in the phloem tissues. It has been found that Zn shows good mobility, Fe has intermediate mobility and copper (Cu), and manganese (Mn) has lower mobility in the phloem tissues (Kochian 1991; Pearson and Rengel 1994). It has been found that in wheat and rice only 4–5% of shoot iron is being translocated into grains at maturity (Hocking 1994; Marr et al. 1995). Enhancing mineral use efficiency (MUE), which involve uptake, translocation, and storage of essential minerals, are the high priority areas of research for biofortification of cereal crops.

## 6 Agronomic Biofortification of Wheat with Iron and Zinc

For improvement of cereal grains with iron and zinc, there are two alternatives: (i) development of the cereal varieties rich in micronutrients through conventional breeding (genetic biofortification) and (ii) enhancement of the accumulation of targeted micronutrients through fertilization (agronomic biofortification). The agronomic biofortification can be used to take care of micronutrient deficiency until the new biofortified varieties are being developed. Different crops can be enriched with iron and zinc through agronomic biofortification. A successful biofortification strategy should meet the following criteria: (i) grain yield capacity of the biofortified crop must be maintained or even improved to guarantee farmer acceptance, (ii) the resultant increase in micronutrient levels must have a significant impact on human health, and (iii) the micronutrient levels achieved must be relatively stable across various locations and climatic zones (Welch and Graham 2004). Due to lack of available Fe and Zn in soil and the limited uptake of Fe and Zn by roots at the time of grain filling in the dry season (Liu et al. 2010), the rate of foliar Fe and Zn application may be the major factor which could help to determine the size of the Fe and Zn pool in the vegetative parts of wheat and hence to increase concentration in grain through nutrient fertilization.

## 7 Genetics of Iron and Zinc Biofortification in Wheat

Under the biofortification program, a number of studies involving QTL analysis have been conducted (Distelfeld and Fahima 2007; Shi et al. 2008; Peleg et al. 2009; Tiwari et al. 2009a, b). In these studies, QTLs for grain zinc and iron have also been mapped in populations derived from crosses between diploid wheat (Tiwari et al. 2009a, b), durum wheat, and wild Emmer wheat (Peleg et al. 2009) and also in synthetic hexaploid wheat and *T. spelta* (Krishnappa et al. 2017; Crespo-Herrera et al. 2017). There are several reports on QTL mapping to identify the genetic basis of high grain Fe and Zn (for a review, see Garcia-Oliveira et al. 2018). These reports involved several mapping populations and led to the identification of a number of QTL (Table 1). In particular, QTL for grain Fe (GFe) were identified in diploid wheat in the interval Xwmc382-Xbarc124 on chromosomes 2A and interval Xgwm473-Xbarc29 on 7A. Similarly, QTL for grain Zn (GZn) were identified within the marker interval Xcfd31-Xcfa2049 on chromosome 7A (Tiwari et al. 2009a, b). Subsequently, one QTL located on wheat chromosome 2A (Xgwm501-Xgwm156.2) showed additive×additive epistatic interaction with the other QTL (Xwmc181-Xcfd267.1) located on the same chromosome 2A for GZn concentration, and one QTL on chromosome 2B (Xbarc1138.2-Xcfd238) showed same additive×additive epistatic interaction with the other QTL (Xgwm617-Xcfa2114) located on the chromosome 6A for GFe clearly indicated the role of epistasis in the expression of these traits in wheat grains (Xu et al. 2012). Shi et al. (2008) detected

**Table 1** A summary of QTL studies on grain Fe (GFe) and grain Zn 9GZn) content in wheat

| Mapping population | QTL-chromosome | | Range of PVE % | | References |
|---|---|---|---|---|---|
| | GFe | GZn | GFe | GZn | |
| DH (RAC875–2 × cascades) | **Fe concentration**<br>2B, 3D, 4B, 4B, 4B, 4B, 4D, 4D, 4D, 4D, 4D | **Zn deficiency**<br>2A, 4A, 7B, 7D<br>**SZn concentration**<br>2D, 4A, 4B, 4D, 4D, 4D, 4D, 5B, 5D, 5D, 6A, 6A | – | – | Genc et al. (2009) |
| DH (Hanxuan10 × Lumai 14) | – | **GZn concentration** 4A, 4D, 5A,7A<br>**GZn content**<br>1A, 2D, 3A, 4A, 4D, 5A, 7A | – | 5.3–11.9<br>4.6–14.6 | Shi et al. (2008) |
| RIL (durum wheat × wild emmer wheat) | 2A, 2A, 2B, 3A, 4B, 3B, 5A, 6A, 6B, 7A, 7B | 2A, 2A, 5A, 6B, 7A, 7B | 2–18 | 1–23 | Peleg et al. (2009) |
| *Triticum boeoticum* (pau5088) × *Triticum monococcum* (pau14087) | 2A, 7A | 7A | 11.7–12.6 | 18.8 | Tiwari et al. (2009a, b) |
| RIL wheat (Xiaoyan 54 × Jing 411) | 5A1, 5A2 | 4B,5A | 1.14–9.25 | 1.14–9.25 | Xu et al. (2012) |
| *Tabassi* × *Taifun* | 2A, 3D, 4D,7B,7D | 1A,4A | 8.94–47.0 | 6.28–7.11 | Roshanzamir et al. (2013) |
| DH (Hanxuan10 × Lumai 14) | 4D, 5A, 7A,7B | – | 6.1–14.6 | – | Shi et al. (2008) |
| *Triticum spelta* × *T. aestivum* | 1A.1, 1A.2, 3, 2A, 3B | 2A, 2B, 3D, 6A, 6B | 3.27–13.3 | 7.0–16.0 | Srinivasa et al. (2014) |
| RIL (PBW343 × Kenya Swara) | – | 1BS, 2Bcc, 3AL, 4AS, 5BL, 2Bc, 2Dd, 3AL, 6AL, 1BS, 2Bc, 3AL | – | 7–15 | Hao et al. (2014) |
| RILs {SHW-L1 /Chuanmai 32 (SC); Chuanmai 42/Chuannong 16 (CC)} | **SC population**<br>2B, 5B, 5D, 7D<br>**CC population**<br>4A, 4D, 5A, 5B | **SC population**<br>2D, 3D, 4D, 5B,<br>**CC population**<br>3D, 4D, 5B | 5.4–9.5<br>9.2–19.1 | 5.5–8.6<br>13.8–15.9 | Pu et al. (2014) |

(continued)

**Table 1** (continued)

| Mapping population | QTL-chromosome GFe | QTL-chromosome GZn | Range of PVE % GFe | Range of PVE % GZn | References |
|---|---|---|---|---|---|
| DH (Berkut' × 'Krichauff); | 2B | 1B, 2B | 22.2 | 23.1–35.9 | Tiwari et al. (2016) |
| SeriM82 × SHWCWI76364 | 6DS, 5BL,5BL,6BL, 5BS, 2BL, 2DS, 4BS, 6AL 7DS | 4BS, 4BS, 6BL, 4BS, 6AL,6BL | 7.2–14.5 | 8.3–19.6 | Crespo-Herrera et al. (2016) |
| RILs **Tetraploid** (Saricanak98 × MM5/4); **Hexaploid** (Adana99 × 70,711) | $2B_{tsk}$, 1B, $2B_{tkm}$,6B 2A,3A,7B, $1B_{TSK}$, $1B_{MCO}$, $5B_{TKM}$, $5B_{MCO}$, 3A/3BB | 1B,1D, 2B, 3A, 7A, 6A, 6B, 7A, 3D, 7B | 10–16.9 9–18 | 9.1 11.7 9–31 | Velu et al. (2017) |
| WH542 × synthetic wheat[a] | 2A, 5A, 7A, 7B, | 2A, 4A, 5A, 7A, 7B | 2.3–6.8 | 3.2–14.4 | Krishnappa et al. (2017) |
| RIL (synthetic hexaploid wheat ×*T. spelta* L.) | **Population 1** 3A_P, 14B_P1, 5B_P1 | 1B_P1,6A_P1, 7B_1P1, 7B_2P1 | 5.49–10.35 | 2.86–16.75 | Crespo-Herrera et al. (2017) |
| | **Population 2** 2A, 2B, 1B, 2B 4A, 4D, 5B | 1A, 1B, 1B, 2B, 3D 4A, 5B, 6A, 1B, 2B, 3B, 7D | 5.79–21.14 | 3.30–32.79 | |

[a]Synthetic wheat (*Triticum dicoccon* PI94624/*Aegilops squarrosa*[409]//BCN)

four QTLs for Zn concentration and seven QTLs for Zn content; they suggested a possibility to improve simultaneously both grain Zn concentration and Zn content because all the four QTLs for Zn concentration were co-located with the QTLs for Zn content. In another study, ten QTLs (five each for Zn and Fe accumulation) were detected on seven different chromosomes (Srinivasa et al. 2014). In a study involving a DH population derived from the cross Berkut x Krichauff, two QTL for Zn were identified on chromosomes 1B (flanked by wmc036–cfa2129) and 2B (flanked by gwm120–wpt2430) (Tiwari et al. 2016). In another study involving two populations derived from spelt × bread wheat crosses (H+ 35 × HUW 468 and H+ 15 × HUW 234), four genes were found to control inheritance of grain Zn concentration (Srinivasa et al. 2014). Recently in 2017, several significant QTLs identified on chromosome 7B explaining the largest proportion (32.7%) of the total phenotypic variance for GZn and one QTL on chromosome 4A, explaining the largest (21.14%) proportion of phenotypic variance of the GFe in two RIL populations derived from *T. spelta* L. and synthetic hexaploid wheat crosses (Crespo-Herrera et al. 2017). There were also regions containing QTL for more than one micronutrient. For instance, a common region in the interval *Xgwm359-Xwmc407* on chromosome 2A was associated with Fe, Zn, and GPC. Two more regions on 5A (*Xgwm126-Xgwm595*) and 7A (*Xbarc49-Xwmc525*) were found to be associated with both Fe and Zn (Krishnappa et al. 2017). Among these studies, some have reported a significant positive correlation between GZn and GFe across different environments indicating co-localization of QTL or pleiotropic effect regulating the concentrations of both GZn and GFe in wheat. Co-localization of QTLs for GZn and GFe on other chromosomes, such as 2A (Krishnappa et al. 2017), 2B (Tiwari et al. 2016), 4BS (Crespo-Herrera et al. 2016), 5A (Xu et al. 2012; Krishnappa et al. 2017), and 6B (Velu et al. 2016), has also been reported. This co-localization of QTLs provides the opportunity to take up only one MAS program to raise the concentrations of both GZn and GFe, simultaneously.

## 8 Genome-Wide Association Mapping for Iron and Zinc in Wheat

Genome-wide association studies (GWAS) were also conducted to identify MTAs for grain Fe and Zn concentration. One such study involved the HarvestPlus Association Mapping (HPAM) panel consisting of 330 bread wheat genotypes and the other involving a Spring Wheat Reference Set (SWRS) consisting of ~320 genotypes. The HPAM panel gave 39 Zn MTAs including two larger effect QTL regions, one each on chromosomes 2 and 7 (Velu et al. 2018). In the other study, nine most important MTAs were selected for three traits (GPC, Fe content and yield per plot) (Kumar et al. 2018). Using markers, the new wheat varieties developed by CIMMYT under HarvestPlus project are 20–40% superior in grain Zn concentration and are agronomically at par or superior to the popular wheat cultivars of South Asia (Velu et al. 2018). A GWAS for grain Zn concentrations using 369 European wheat genotypes, as many as 40 marker-trait associations (MTAs), were detected on chromosomes 2A, 3A, 3B, 4A, 4D, 5A, 5B, 5D, 6D, 7A, 7B, and 7D, whereas the most significant and consistent MTAs were located on chromosomes 3B (723,504,241–723,611,488 bp) and 5A (462,763,758–466,582,184 bp) having major effects. The number of MTAs in the subpanel increased to 161 MTAs. These genomic regions include newly identified putative candidate genes, which are related to Zn uptake and transport, or represent bZIP and mitogen-activated protein kinase genes (Alomari et al. 2018). In another study, wide variation for micronutrients was observed in a panel of 167 *Ae. tauschii* accessions. To identify potential new genetic regions for improving grain micronutrient concentration, a set of 114 non-redundant *Ae. tauschii* accessions were subjected to genotyping-by-sequencing (GBS) and therefore 5249 markers were identified. A total of 19 SNP MTAs were detected on all 7 chromosomes. Significant associations were detected five for grain Fe and four Zn concentrations. The associations were linked to the genes encoding transcription factor regulators, transporters, and phytosiderophore synthesis (Arora et al. 2019). Synthetic hexaploid wheat (SHW; *Triticum durum* L. × *Aegilops tauschii* Coss.) is a means of introducing novel genes/genomic regions into bread wheat (*T. aestivum* L.) and a potential genetic resource for improving grain mineral concentrations. A total of ten grain minerals (Ca, Cd, Cu, Co, Fe, Li, Mg, Mn, Ni, and Zn) were quantified using an inductively coupled mass spectrometer in 123 SHWs through GWAS. Another set of 92 MTAs were identified in a GWAS using 35,648 SNP; in this study 60 MTAs were novel and 40 were within genes, and the genes underlying 20 MTAs had annotations suggesting a potential role in grain mineral concentration. Moreover, superior SHW lines in comparison to checks, in terms of beneficial grain minerals (Cu, Fe, Mg, Mn, Ni, and Zn), were identified and recommended for utilization in the breeding program for the genetic biofortification (Bhatta et al. 2018).

## 9 Breeding for Iron and Zinc in Wheat

Biofortification is a strategy that uses plant breeding techniques to produce staple food crops with higher level of micronutrient level and reduced level of antinutrients. It should also bring about an increase in the level of substances that promote nutrient absorption. For this purpose, the first step was to screen the germplasm not only for higher level of Fe and Zn but also for genes that would allow an increase in micronutrient uptake, transport, and sequestration. Transfer of these traits from various sources to high yielding variety was required to develop biofortified superior varieties.

## *Synthetic Hybrid Wheat (SHW)*

Bread wheat (*Triticum aestivum L.*; hexaploid genome = AABBDD) naturally evolved via natural hybridization between wild goat grass *Aegilops tauschii* (DD) and a cultivated emmer plant *T. turgidum* L. ssp. *dicoccon* (Schrank) Thell. (2n = 28; AABB, a progenitor of modern durum wheat) around 8000 years ago. Thus, it consists of three diploid progenitor genomes, AA from *Triticum urartu*, BB from an unknown species, and DD from *Ae. tauschii* (Fig. 1). In current breeding programs, three major factors have contributed to the narrow genetic variability in wheat germplasm. These factors include (i) evolutionary bottleneck as a result of rare hybridization events between the domesticated emmer wheat (*T. turgidum* ssp. *dicoccum*; AABB genome) and the wild goat grass, i.e., *Ae. tauschii* (DD genome) during the course of the evolution of hexaploid wheat (AABBDD genome), (ii) restricted gene flow due to self-fertilizing nature of wheat and the progenitor species, and (iii) stringent natural and human-mediated selection during wheat domestication and breeding. The narrow genetic variability has hindered progress in breeding wheat varieties that are rich in grain Fe and Zn levels. However, the accessions of tetraploid wheats (*T. turgidum* ssp. *dicoccum*, *T. turgidum* ssp. *dicoccoides* and *T. turgidum* ssp. *durum*), Ae. *tauschii* and also the *T. aestivum* ssp. *spelta* (spelt wheat) have twice the levels of Fe (38 mg/kg) and Zn (up to 45 mg/kg) than the common wheat (Chhuneja et al. 2006; Ortiz-Monasterio et al. 2007; Guzman et al. 2014). Thus, to broaden the genetic variability for grain Fe and Zn (or some other traits), primary SHWs were developed by crossing the accessions of tetraploid wheat (*T. turgidum* ssp. *dicoccum*, *T. turgidum* ssp. *dicoccoides* and *T. turgidum* ssp. *durum*) with *Ae. tauschii* at a large-scale during mid-1980s at CYMMYT, Mexico. Although, in 1940s, the first attempts to reproduce the bread wheat original crosses were made in Japan (Kihara 1944) and the USA (McFadden and Sears 1944). These

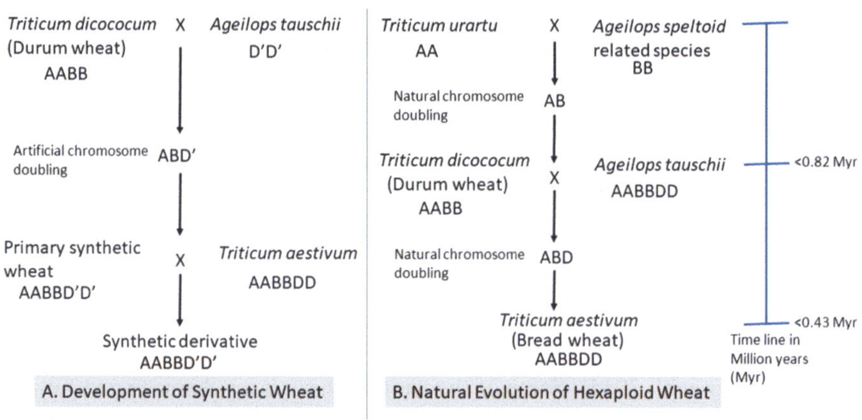


**Fig. 1** Development of synthetic wheat and natural evolution of hexaploid wheat. (Rosyara et al. 2019)

attempts led to the development of the first SH wheat (Gill et al. 1985). In 1980s, CIMMYT started to explore the value of wide crosses and the development of SH wheat to increase D genome diversity. Recently, CIMMYT has developed more than 1500SHWs (Rosyara et al. 2019). In addition, many SHW were developed using *T. turgidum* ssp. *dicoccum* and *T. turgidum* ssp. *dicoccoides*, utilizing the useful genetic diversity present among the SHWs. However, evidence of linkage drags, tall phenotype, and poor agronomic performance has limited their direct use in wheat breeding programs. Therefore, sooner or later a majority of SHW lines were developed by crossing durum wheat (*turgidum* ssp. *durum*, AABB) and *Ae. tauschii* (Li et al. 2018). SHW were crossed with desirable high yielding wheat varieties/germplasm and SHW derivatives with high yield potential, and other desirable attributes were selected and used in wheat breeding programs all over the world. More than 60 SHW derivatives have also been directly released as cultivars in several developing countries. Altogether, the availability of Fe- and Zn-rich SHW also verified superiority of tetraploid wheat and is used for developing SHW derivatives for higher Fe and Zn levels as rich sources (Gomez-Becerra et al. 2010; Guzman et al. 2014).

The ability of SHW have higher Fe and Zn levels in grain and their greater ability to uptake the micronutrients, CIMMYT, Mexico, in collaboration with the National Agriculture Research Systems (NARS) and Agricultural Universities of several developing countries (including India, Pakistan, and Bangladesh) used these genetic resources (SHW and derivatives) for developing Fe- and Zn-rich wheat varieties. In India, the breeding targets were aimed at improving the levels of Fe and Zn by 25 and 10 mg/kg above the mean of the popular varieties which was treated as a baseline (Joshi et al. 2010). Since the significant and positive correlation between the Fe and Zn concentrations (Gomez-Becerra et al. 2010; Morgounov et al. 2007; Zhao et al. 2009; Guzman et al. 2014) and also the lower bioavailability of Fe, focus was mainly placed on improvement of Zn alone, although adequate variability was also available for Fe concentration. CIMMMYT, Mexico, also mainly focused for the transfer of desirable genetic variability for increasing the level of Zn concentration in the high yielding elite wheat backgrounds particularly from *T. turgidum* ssp. *dicoccum*-based SHW and also other high Zn sources (Velu et al. 2014). As a result of the CIMMYT's effort under the auspices of HarvestPlus program, the first proof of concept 1 HarvestPlus Yield Trial (1HPYT) comprising 40 high-yielding, biofortified, wheat lines was conducted. A number of these lines exceeded the intermediate to full target levels of Zn in multilocation trials suggesting the possibility of developing biofortified wheat varieties using SHW; these lines also had consumer-preferred and end-use traits (Velu et al. 2012). Subsequently, during 2011–2012, 2HPYT trial of 50 biofortified lines conducted in target environments identified 6–7 lines containing grain Zn with 75–150% improvement above the checks; these lines also had high yield potential, had resistance to rusts, and preferred end-use quality traits (see Velu et al. 2014). Some of the varieties have been characterized for the sources of genes for high Zn levels contributed via particularly SHW. For example, "Zinc Shakti" has genes from *Ae. tauschii,* "Zincol2016" has genes for *T. aestivum* ssp. *spelta*, and "WB02" and "HPBW-01" have genes from *Ae. squarrosa* and

*T. turgidum* ssp. *dicoccum*, respectively (Saini et al. 2020). In general, there has been significant contribution of SHW to develop micronutrient rich wheat varieties. In future, this is expected to continue the importance of SHW in breeding programs.

## 10 Alien Chromosome Transfer for Fe and Zn in Wheat

Useful genes for biofortification were also transferred in a study involving 80 accessions belonging to nine alien species of wild *Triticum* and *Aegilops*; 15 semi-dwarf cultivars of bread wheat and durum wheat were also evaluated for grain Fe and Zn contents. It was observed that the related non-progenitor wild species of wheat with S, U, and M genomes had up to 3–4-fold higher Fe and Zn contents compared to bread and durum wheat. In particular, two accessions of *Ae. kotschyi* had > 75% higher Fe and 60% higher Zn relative to wheat (Rawat et al. 2009b). A number of wild species of *Triticum, Aegilops*, and other genera have been shown to have in their grains 2–3-fold higher Fe and Zn relative to modern hexaploid wheat cultivars. Synthetic amphiploids (AABBDDUU$S^lS^l$), with seeds as large as that of wheat cultivars, had higher grain, flag leaf, and grain ash Fe and Zn concentrations than *Ae. kotschyi* parent, thus also suggesting that *Ae. kotschyi* possesses a distinctive genetic system for the micronutrient uptake, translocation, and sequestration than the wheat cultivars (Rawat et al. 2009a). Particularly, three species, namely, *Ae. longissima, Ae. peregrina* and *Ae. kotschyi*, were found to be promising for biofortification involving Fe and Zn (Table 2; Chhuneja et al. 2006; Rawat et al. 2009b; Neelam et al. 2011). These species were used for developing amphiploids, which were found to have higher Fe and Zn content (Tiwari et al. 2008). From the three species used, major emphasis was laid on *Ae kotschyi*, which was later used for transfer of alien genes for biofortification. Three different approaches were used for transfer of alien segments from chromosomes of *Ae. kotschyi*:

(i) Interspecific $F_1$ hybrids from Chinese Spring (CS) × *Ae. kotschyi* crosses. The $F_1$ hybrids from these crosses were backcrossed, and $BC_1F_1$ and $BC_2F_1$ plants were selfed; plants with high grain Fe and Zn concentration were selected, which had 50–120% increased Fe and Zn contents relative to recipient wheat cultivars. It was also possible to use anchored wheat SSR markers, for transfer genes/QTL for high grain Fe and Zn from chromosomes of homoeologous groups 2 and 7 from *Ae. kotschyi* (Tiwari et al. 2009a, b, 2010; Rawat et al. 2011).

(ii) *Use of Ph1b /Mono5B for inducing homoeologous pairing*. The interspecific hybrid plants without 5B chromosome (developed through crossing with monosomic 5B) showed much higher chromosome pairing relative to plants with 5B (Fig. 2). This facilitated transfer of alien segments to wheat chromosomes, so that the $BC_2F_2$ plants showed up to 125% increase in Fe and 158% increase in Zn relative to recipient cv. PBW343 with Lr24 and Yr36 (Verma et al. 2016a).

**Table 2** A summary of grain Fe and Zn% increase and the transfer of alien genes/chromosome for these traits to wheat (Gupta et al. 2020)

| Alien species | Genomic constitution | Chromosome | Fe increase (%) | Zn increase (%) | References |
|---|---|---|---|---|---|
| *Ae. kotschyi* | $U^kU^kS^KS^k$ | $2S^k$, $7U^k$ | 75, 89 | 75, 93 | Tiwari et al. (2010), Verma et al. (2016a) |
| *Ae. kotschyi* | $U^kU^kS^KS^k$ | Hybrid line[a] | 47 | 54 | Prażak and Krzepiłko (2018) |
| *Ae. longissima* | $S^lS^l$ | $2S^l$ | 124 | 132 | Tiwari et al. (2008), Sharma et al. (2018) |
| *Ae. longissima* | $S^lS^l$ | $1S^l$, $2S^l$ | 55, 38 | 124, 74 | Wang et al. (2011) |
| *Ae. peregrina* | $U^PU^PS^PS^P$ | $4S^P$, $7S^P$, $7U^P$ | 46, 133, 92 | 125, 107, 251 | Neelam et al. (2011) |
| *Ae. peregrina* | $U^PU^PS^PS^P$ | $4S^p$ | 36 | 69 | Wang et al. (2011) |
| *Ae. searsii* | SS | $1S^S$, $2S^S$ | 84, 61 | 143, 129 | Wang et al. (2011) |
| *Ae. umbellulata* | UU | 2 U, 6 U | 47, 70 | 79, 32 | Wang et al. (2011) |
| *Ae. caudata* | CC | B | 41 | 161 | Wang et al. (2011) |
| *Ae. Geniculata* | $M^gM^gU^gU^g$ | $5M^g$ | 14 | 47 | Wang et al. (2011) |
| *Ae. variabilis* | $UUS^vS^v$ | Hybrid line[b] | 59 | 71 | Prażak and Krzepiłko (2018) |
| *Secale cereale* | RR | 1R | – | 18 | Velu et al. (2019) |

[a]*Ae. kotschyi* × *T. aestivum*
[b]*Ae. variabilis* × *T. aestivum*

(iii) *Irradiation of amphiploids*. Wheat–*Aegilops kotschyi* substitution lines were also developed and evaluated. Pollen from wheat-*Ae. kotschyi* $2A/2S^k$ and $7A/7S^k$ substitution lines with high Fe and Zn were irradiated with gamma rays using a dose of 40 krad (Fig. 3; Verma et al. 2016b; Tiwari et al. 2010).

Some of the derivatives had 65% higher Fe and 54% higher Zn contents coupled with better harvest index than the elite wheat cultivars WL711 and PBW343 indicating effective and compensating translocations of fragments into wheat genome (Verma et al. 2016b; Sharma et al. 2018). In these improved lines, the grain Fe content was highly positively correlated with Fe content in the plant tissues. Most of the lines had much higher Fe/Zn content in all tissues during grain-filling period indicating higher Fe/Zn uptake from soil during this stage. Although Fe/Zn contents are nearly similar in grains, there was much less Zn content in the plant tissues of all the lines suggesting that the uptake of Zn in *Triticeae* species was low but is mobilized to grains more effectively than Fe (Sharma et al. 2017). Similarly, 47 wheat-*Aegilops* disomic addition lines derived from 6different *Aegilops* species evaluated and identified the chromosomes $1S^l$ and $2S^l$ of *Ae. longissima*, 1SS and 2SS of *Ae. searsii*, 2 U and 6 U of *Ae. umbellulata*, B of *Ae. caudata*, $4S^v$ of *Ae.*

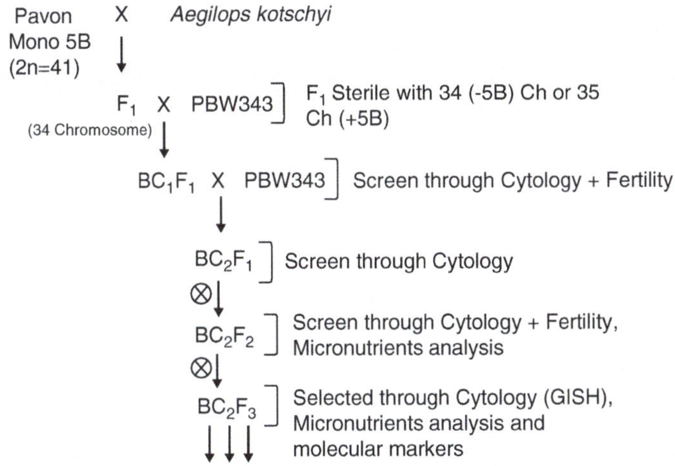


**Fig. 2** Schematic representation of alien gene transfer for high GFe and GZn in hexploid wheat. (Verma et al. 2016b)

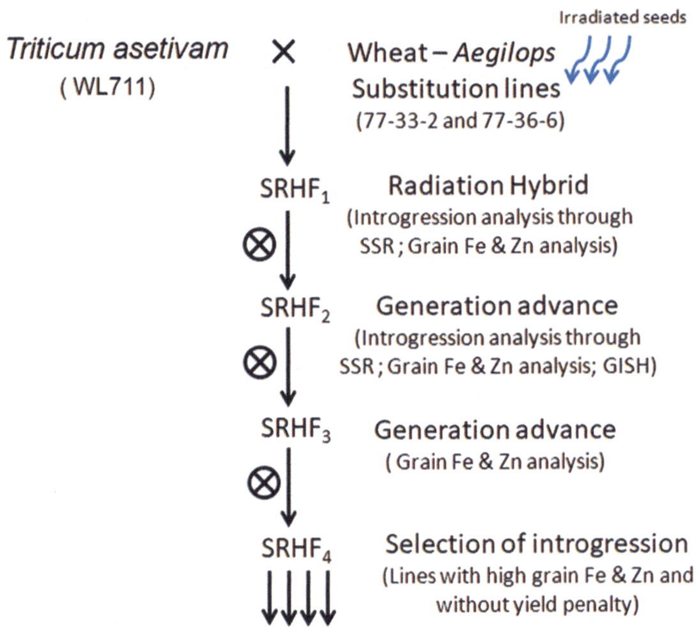


**Fig. 3** Schematic representation of precise transfer of alien gene using radiation hybrid breeding method. (Verma et al. 2016b)

*peregrina*, and 5 $M^g$ of *Ae. geniculata* carrying genes for high grain Fe and Zn concentrations ranging from 50% to 248% compared with the Chinese Spring recipient cultivar (Wang et al. 2011). Prażak and Krzepiłko (2018) detected the presence of DNA fragments specific to *Ae. kotschyi* Boiss (2n = 4x = 28, UUSS) using two ISSR markers – ISSR23690 and ISSR33650 – to characterize the hybrid lines derived from *Ae. kotschyi* Boiss x *T. aestivum* L. Sixty-two translocation lines at CIMMYT from rye and Aegilops species in a "Pavon-76" wheat genetic background were evaluated, and identified four translocation lines (disomic line) with 1R translocation had significantly Zn advantage over the trial mean of 62 lines. The results of this study demonstrate that large genetic variation is available in translocation lines for improving the nutritional quality of wheat and could be used in wheat breeding program (Velu et al. 2019).

## 11 Marker-Assisted Selection (MAS)

A large number of studies involving GWAS and interval mapping for Fe and Zn contents have been carried out in wheat. These studies have identified hundreds of markers, which can be used for marker-assisted breeding. However, only few examples are available, where these markers have been utilized in MAS for biofortification. The grain protein content gene (*Gpc-B1*) or *NAM-B1* has pleiotropic effects on whole plant senescence, grain protein, zinc, and iron content. The grain protein content gene (*Gpc-B1*) or *NAM-B1* is an important example where MAS has been used for improvement in Fe and Zn content (Distelfeld and Fahima 2007).The *Gpc-B1* or *NAM-B1* locus for high grain protein content (GPC) was originally discovered in wild emmer wheats (*Triticum turgidum* ssp. *dicoccoides*), which is ancestor/progenitor of cultivated pasta wheat (*T. turgidum* ssp. *durum*). Israel accession, FA15-3 of cultivated pasta wheat is a widely studied source of high protein, and it was referred as DIC by Aviviin 1978. This accession, FA15-3, has been widely used to introduce the high GPC trait into durum and bread wheats. Joppa and Cantrell (1990) developed substitution lines of the DIC (FA15-3) chromosomes in the cultivar 'Langdon' (LDN) and showed that a QTL (*QGpc.ndsu-6Bb*) for high GPC is present on chromosome 6B and was mapped to the short arm of chromosome 6B (6BS) using recombinant substitution lines derived from a cross of the substitution line Langdon (DIC 6B) × Langdon. The *QGpc.ndsu-6Bb* explained 66% of the variation for GPC (Joppa et al. 1997).

Uauy et al. (2006b) discovered an ancestral wild wheat allele encodes a NAC transcription factor (*Gpc-B1* or *NAM-B1*) for NAC genes which play important roles in developmental processes like auxin signaling, defense and abiotic stress responses, and leaf senescence in vegetative parts of the plant, resulting in increased transfer of nitrogen and mineral remobilization to the developing grains. The wild *Gpc-B1* allele accelerates senescence in flag leaves producing and its pleiotropic effect are reduction in grain size and yield due to acceleration in monocarpic senescence (Uauy et al. 2006a). RNA interference technique, used to knock down the

*Gpc-B1* gene, has allowed to conclude that this, *Gpc-B1* allele regulates two processes, viz., senescence and nutrient remobilization of Fe, Zn, and N from vegetative tissues to grain (Uauyet al. 2006b; Waters et al. 2009). Generally, cultivated varieties of wheat carry a non-functional *NAM-B1* allele (Asplund et al. 2013), which was produced due to frameshift mutation in the original/wild *NAM-B1* allele, and this non-functional allele was selected during domestication of wheat from wild to cultivated. This non-functional *NAM-B1* allele increases the grain size and ultimately higher yield as it delays senescence and thereby more time is available for developing grains. This non-functional *NAM-B1* allele is located on chromosome 6BS in wheat (Brevis and Dubcovsky 2010). In a study conducted on a worldwide core collection of 367 bread wheat genotypes and found that out of these 367 genotypes only five cultivated hexaploid Fennoscandian wheat cultivars were carrying wild-type/functional *Gpc-B1* or *NAM-B1* allele and were adapted to very short growing seasons in northern Europe. The wild-type/functional *Gpc-B1* allele conserved during domestication (Hagenblad et al. 2012). Eagles et al. (2014) introgressed the wild-type *Gpc-B1* gene from the Canadian cultivars into the Australian cultivars using "Xuhw89" marker and found reduced grain weight, with no effect on grain yield. Uauy et al. (2006b) silenced *Gpc-B1* allele using RNA interference (RNAi) technique and demonstrated reductions in GPC, Fe, and Zn levels (>30%) in the wheat grain as well as senescence were delayed by 3 weeks in comparison to control lines. Similarly, loss-of-function was reported by generated mutations for *GPC1* and *GPC2* (Avni et al. 2014; Pearce et al. 2014). This *Gpc-B1*allele introgression in cultivated wheat from wild relatives resulted in the increase of Zn and Fe grain concentrations by 12 and 18%, respectively, in addition to grain protein content. In fact, this *Gpc-B1*QTL has multiple effects and contain gene(s) possibly encoding for (either one or more) transporters, chelators, chelator biosynthesis enzymes, regulatory factors such as protein kinases, membrane receptors, or transcription factors (Distelfeld and Fahima 2007). However, *GPC-B1* or *NAM-B1* allele studied from *Triticum turgidum* L. var. *dicoccoides* was responsible for increase in GPC and also have pleiotropic effect on Fe and Zn concentrations in wheat grain. In general, GPC is negatively correlated with grain yield and strongly affected by the genetic background (Brevis and Dubcovsky 2010). The functional copy of the *GPC-B1* allele is associated with higher protein, iron and zinc content with only marginally negative impacts on yield (Tabbita et al. 2017) and discussed the possibilities for its application in wheat breeding. Venegas et al. (2018) demonstrated that LPA (low phytic acid)-GPC, both should be introgressed into a well-adapted cultivar which may simultaneously increases total grain Fe and Zn concentrations and grain protein without any pleiotropic effects on grain yield.

## 12 Biofortified Wheat Cultivars

In the 1990s, CIMMYT produced synthetic wheats (using *T. durum* or *T. dicoccum* and diverse sources of *Ae. tauschii*) to create new genetic variation in wheat and then crossing these with elite breeding lines to improve traits, including stress tolerance and agronomic and nutritional quality traits. Wide variation in grain iron and zinc concentrations in wheat and its closely related wild species has been observed that it can be exploited for improvement of modern elite cultivars (Cakmak et al. 2004). In recent years, utilizing this variation, breeding efforts and subsequent testing broadly for adaptation and stability in target locations, 11 Zn biofortified varieties, 1 durum, and 10 from common wheat have been released for cultivation in different countries (Table 3; Velu et al. 2012, 2016; Baloch et al. 2015). In India, ICAR-IIWBR, Karnal, PAU, Ludhiana and IARI, New Delhi have released eight Zn biofortified varieties: WB 02, HPBW 01, Pusa Tejas (HI 8759), Pusa Ujala (HI 1605), MACS 4028 (durum wheat). PBW1Zn, However, two varieties Zinc Shakti (Chitra) and Ankur Shiva were also developed by private sector (private seed companies like Ankur Seeds). Similarly, a Zn-biofortified variety 'Zincol2016' was released in Pakistan. These varieties have up to 42 mg/kg Zn and up to 46.1 mg/kg Fe; the improved Zn level in these varieties was 20–40% higher than the level of Zn in local varieties (Singh et al. 2017; Saini et al. 2020 for review). Other recently released Zn biofortified wheat varieties include 'Nohely-F2018' released in Mexico for the Mexicali valley of northern Sonora region and 'BARI Gom 33' released in Bangladesh which showed 7–8 mg/kg Zn advantage; this latter variety also has resistance to wheat blast caused by *Magnaporthe oryzae*. Some of these varieties have also been characterized for the sources of genes for high Zn levels contributed via particularly SHW. Interestingly, for example, "Zinc Shakti" has genes from *Ae. tauschii,* 'Zincol2016' has genes for *T. aestivum* ssp. *spelta*, and 'WB02' and 'HPBW-01' has genes from *Ae. squarrosa* and *T. turgidum* ssp. *dicoccum*, respectively (see Singh et al. 2017; Saini et al. 2020). These materials represent a significant steppingstone to achieve the ultimate goal of micronutrient-enriched wheat varieties, and this is likely to continue in the future, suggesting the importance of SHW breeding program.

## 13 Conclusions and Future Directions

The above account on biofortification suggests that biofortified wheat can be developed using the available genetic variability in the gene pool. It has also been shown that there are significant positive correlations among Zn, Fe, and protein contents and a negative correlation between the contents of micronutrients and important agronomic characteristics like plant height, grain yield, and thousand-grain weight. Absence of these negative correlations between grain yield and the concentrations of Fe and Zn has also been reported. It is worth noting that the strength of these relationships is influenced greatly by the environment. This makes it difficult to breed wheat with high Zn concentration and high grain yield. In some studies, the concentration of Fe was shown to be positively correlated with grain weight, indicating that possibility of simultaneous improvement by traditional breeding strategies. The levels of bioavailability have been shown to be low for grain Fe and Zn in staple food crops. The grains of cereal crops contain various anti-nutrient factors,

**Table 3** Characteristic feature of biofortified wheat cultivars released for commercial cultivation

| S. No. | Wheat cultivars | Release by | Remarks |
|---|---|---|---|
| 1. | WB 02 | India | It is rich in zinc (42.0 ppm) and iron (40.0 ppm) and average grain yield is 51.6 q/ha. It matures in 142 days and is suitable for irrigated timely sown conditions |
| 2. | HPBW 01 | India | It contains high iron (40.0 ppm) and zinc (40.6 ppm). Its average grain yield is 51.7 q/ha and matures in 141 days and is suitable for irrigated timely sown conditions |
| 3. | PusaTejas (HI 8759) | India | It is a pure line variety with high protein (12%), iron (42.1 ppm) and zinc (42.8 ppm). It is a durum wheat variety suitable for making chapatti (Indian bread), pasta, and other traditional food products. The average yield of this variety is 50.0 q/ha under timely sown irrigated conditions |
| 4. | PusaUjala (HI 1605) | India | It is a pure line variety with high protein (13%), iron (43 ppm) and zinc (35 ppm) and having excellent chapatti making quality. Its average yield is 30.0 q/ha under timely sown, restricted irrigation conditions |
| 5. | MACS 4028 | India | It is a pure line durum wheat variety with high protein (14.7%), iron (46.1 ppm) and zinc (40.3 ppm). Its average grain yield is 19.3 q/ha under rainfed low fertility, timely sown conditions in peninsular zone. It matures in 102 days |
| 6. | PBW1Zn | India | PBW 1 Zn recorded 7–9 ppm more grain zinc concentration than check varieties and yields at par with best check PBW725 |
| 7. | Zinc Shakti (Chitra) | India | Zinc Shakti (Chitra) was developed through participatory variety selection (registered by private seed companies and growers). It has profitable yield potential and matures nearly 2 weeks earlier than common wheat |
| 8. | Ankur Shiva | India | Wheat varieties released in India by public and private partners (Ankur seeds). It is same as WB02 or its sister line |
| 9. | NR- 421 (Zincol-16) | Pakistan | It has more than 6 ppm Zn compared to best local check and released from Pakistan in 2015 |
| 10. | BARI Gom 33 | Bangladesh | BARI Gom 33″ (="Kachu"/"Solala") released in Bangladesh during 2017 showed 7–8 mg/kg Zn advantage, and also resistance to wheat blast disease |
| 11. | Nohely-F2018 | Mexico | Nohely-F2018 released in Mexico for the Mexicali valley of northern Sonora region |

such as phytic acid and tannins, which reduces the bioavailability of micronutrients. Thus, bioavailability of micronutrients and the micronutrient concentration should also be considered in breeding for biofortification.

**Acknowledgments** Thanks are due to the Department of Biotechnology (DBT), Government of India, for providing funds in the form research projects and to the Head, Department of Genetics and Plant Breeding, CCS University, Meerut, India, for providing the necessary facilities.

**Author's Contributions** RK wrote the first draft of the manuscript, and each author contributed different sections jointly. The manuscript was read and revised by all the authors and finalized by RK.

## References

Alomari DZ, Eggert K, von Wirén N, Alqudah AM, Polley A, Plieske J, Ganal MW, Pillen K and Röder MS (2018) Identifying Candidate Genes for Enhancing Grain Zn Concentration in Wheat. Front Plant Sci 9:1313. https://doi.org/10.3389/fpls.2018.01313

Arora S, Cheema J, Poland J, Uauy C, Chhuneja P (2019) Genome-wide association mapping of grain micronutrients concentration in *Aegilops tauschii* Front Plant Sci 10:54

Asplund L, Bergkvist G, Leino MW, Westerbergh A, Weih M (2013) Swedish spring wheat varieties with the rare high grain protein allele of *NAM-B1* differ in leaf senescence and grain mineral content. PLoS One 8(3):1–11. https://doi.org/10.1371/journal.pone.0059704

Avivi LYDIA (1978) High protein content in wild tetraploid *Triticum dicoccoides* Korn. In Proceedings of the 5th International wheat genetics symposium. Indian Society of Genetics and Plant Breeding (ISGPB), New Delhi, pp 372–380

Avni R, Zhao R, Pearce S, Jun Y, Uauy C, Tabbita F, Fahima T, Slade A, Dubcovsky J, Distelfeld A (2014) Functional characterization of *GPC-1* genes in hexaploid wheat. Planta 239(2):313–324

Baloch M, Baloch A, Shaikh D, Asad M, Gandahi N, Baloch G, Ali M (2015) Inheritance studies for leaf rust resistance and different economic traits in F2 generation developed through introgression of functional genes in bread wheat genotypes. Sindh Univ Res J 47(4)

Bhaskaram P (2002) Micronutrient malnutrition, infection, and immunity: an overview. Nutr Rev 60(suppl 5):S40–S45

Bhatta M, Baenziger PS, Waters BM, Poudel R, Belamkar V, Poland J, Morgounov A (2018) Genome-wide association study reveals novel genomic regions associated with 10 grain minerals in synthetic hexaploid wheat. Int J Mol Sci 19(10):3237

Black RE, Walker CF (2012) Role of zinc in child health and survival. In Meeting Micronutrient Requirements for Health and Development (Vol. 70, pp. 37–42). Karger Publishers.

Bouis HE (2007) The potential of genetically modified food crops to improve human nutrition in developing countries. J Dev Stud 43(1):79–96

Brabin BJ, Hakimi M, Pelletier D (2001) An analysis of anemia and pregnancy-related maternal mortality. J Nutr 131(2):604S–615S

Brevis JC, Dubcovsky J (2010) Effects of the chromosome region including the *Gpc-B1* locus on wheat grain and protein yield. Crop Sci 50(1):93–104. https://doi.org/10.2135/cropsci2009.02.0057

Brown G (1972) The X-ray identification and crystal structures of clay minerals. Pp. 580. London: Mineralogical Society, Clay Minerals Group; 2nd Edition .

Brown G Britai MSOG, Ireland (1962) The X-ray identification and crystal structures of clay minerals

Cakmak I (2000) Possible roles of zinc in protecting plant cells from damage by reactive oxygen species. New Phytol 146:185–205

Cakmak I (2002) Plant nutrition research: Priorities to meet human needs for food in sustainable ways. Plant Soil 247(1):3–24

Cakmak I (2008) Enrichment of cereal grains with zinc: Agronomic or genetic biofortification? Plant Soil 302(1–2):1–17

Çakmak İ, Torun A, Millet E, Feldman M, Fahima T, Korol A, Özkan H (2004) *Triticum dicoccoides*: an important genetic resource for increasing zinc and iron concentration in modern cultivated wheat. Soil Sci Plant Nutr 50(7):1047–1054

Cakmak I, Pfeiffer WH, McClafferty B (2010) Biofortification of durum wheat with zinc and iron. Cereal Chem 87(1):10–20

Chhuneja P, Dhaliwal HS, Bains NS, Singh K (2006) *Aegilops kotschyi* and *Aegilops tauschii* as sources for higher levels of grain iron and zinc. Plant Breed 125(5):529–531

Crespo-Herrera LA, Velu G, Singh RP (2016) Quantitative trait loci mapping reveals pleiotropic effect for grain iron and zinc concentrations in wheat. Ann Appl Biol 169(1):27–35

Crespo-Herrera LA, Govindan V, Stangoulis J, Hao Y, Singh RP (2017) QTL mapping of grain Zn and Fe concentrations in two hexaploid wheat RIL populations with ample transgressive segregation. Front Plant Sci 8:1800

Demment MW, Young MM, Sensenig RL (2003) Providing micronutrients through food-based solutions: a key to human and national development. J Nutr 133(11):3879S–3885S

Distelfeld A, Fahima T (2007) Wild emmer wheat as a source for high-grain-protein genes: map-based cloning of *Gpc-B1*. Israel J Plant Sci 55(3–4):297–306

Eagles HA, Robyn ML, Eastwood RF, Appelbee MJ, Karen C, Martin PJ, Wallwork H (2014) High-yielding lines of wheat carrying *Gpc-B1* adapted to Mediterranean-type environments of the south and west of Australia. Crop Pasture Sci. https://doi.org/10.1071/CP14106

Garcia-Oliveira AL, Chander S, Ortiz R, Menkir A, Gedil M (2018) Genetic basis and breeding perspectives of grain iron and zinc enrichment in cereals. Front Plant Sci 9:937

Garrido C, Galluzzi L, Brunet M, Puig PE, Didelot C, Kroemer G (2006) Mechanisms of cytochrome c release from mitochondria. Cell Death Differ 13(9):1423–1433

Genc Y, Verbyla AP, Torun AA, Cakmak I, Willsmore K, Wallwork H, McDonald GK (2009) Quantitative trait loci analysis of zinc efficiency and grain zinc concentration in wheat using whole genome average interval mapping. Plant Soil 314(1–2):49

Gill BS et al (1985) Evaluation of *Aegilops* species for resistance to wheat powdery mildew, wheat leaf rust, Hessian fy, and Greenbug. Plant Dis 69:314–316

Gomez-Becerra HF, Yazici A, Ozturk L, Budak H, Peleg Z, Morgounov A et al (2010) Genetic variation and environmental stability of grain mineral nutrient concentrations in *Triticum dicoccoides* under five environments. Euphytica 171(1):39–52

Graham RD, Welch RM, Saunders DA, Ortiz-Monasterio I, Bouis HE, Bonierbale M et al (2007) Nutritious subsistence food systems. Adv Agron 92:1–74

Gupta PK, Balyan HS, Sharma S, Kumar R (2020) Genetics of yield, abiotic stress tolerance and biofortification in wheat (*Triticum aestivum* L.). Theor Appl Genet:1–34

Guzmán C, Medina-Larqué AS, Velu G, González-Santoyo H, Singh RP, Huerta-Espino J et al (2014) Use of wheat genetic resources to develop biofortified wheat with enhanced grain zinc and iron concentrations and desirable processing quality. J Cereal Sci 60(3):617–622

Hagenblad J, Asplund L, Balfourier F, Ravel C, Leino MW (2012) Strong presence of the high grain protein content allele of *NAM-B1* in Fennoscandian wheat. Theor Appl Genet 125(8):1677–1686. https://doi.org/10.1007/s00122-012-1943-2

Hallberg L (1981) Bioavailability of dietary iron in man. Annu Rev Nutr 1(1):123–147

Hambidge KM, Krebs NF, Sibley L, English J (1987) Acute effects of iron therapy on zinc status during pregnancy. Obstet Gynecol 70(4):593–596

Hao Y, Velu G, Peña RJ, Singh S, Singh RP (2014) Genetic loci associated with high grain zinc concentration and pleiotropic effect on kernel weight in wheat (*Triticum aestivum* L.). Mol Breed 34(4):1893–1902

Hershfinkel M (2005) Zn 2+, a dynamic signaling molecule. In: Molecular biology of metal homeostasis and detoxification. Springer, Berlin/Heidelberg, pp 131–153

Hocking PJ (1994) Dry-matter production, mineral nutrient concentrations, and nutrient distribution and redistribution in irrigated spring wheat. J Plant Nutr 17(8):1289–1308

Hotz C, Brown KH (2004) Assessment of the risk of zinc deficiency in populations and options for its control

International Institute for Population Sciences (IIPS) and ICF (2017) National Family Health Survey (NFHS-4), 2015–16: India. Mumbai: IIPS

Joint, F. A. O., WHO Expert Committee on Food Additives (2000) Evaluation of certain food additives and contaminants. World Health Organ Tech Rep Ser 896:1
Joint, F. A. O., WHO Expert Committee on Food Additives, & World Health Organization (2004) Evaluation of certain food additives and contaminants: sixty-first report of the Joint FAO/WHO Expert Committee on Food Additives
Joppa LR, Cantrell RG (1990) Chromosomal location of genes for grain protein content of wild tetraploid wheat. Crop Sci 30(5):1059. https://doi.org/10.2135/cropsci1990.0011183X003000050021x
Joppa LR, Du C, Hart GE, Hareland GA (1997) Mapping gene(s) for grain protein in tetraploid wheat (*Triticum turgidum* L.) using a population of recombinant inbred chromosome lines. Crop Sci 37(5):1586. https://doi.org/10.2135/cropsci1997.0011183X003700050030x
Joshi AK, Crossa I, Arun B, Chand R, Trethowan R, Vargas M, Ortiz-Monasterio I (2010). Genotype × environment interaction for zinc and iron concentration of wheat grain in eastern Gangetic plains of India. Field Crop Res. 116:268– 277.
Kihara H (1944) Discovery of the DD analyser, one of the ancestors of *Triticum vulgare*. Agric Hortic 19:889–890
Kim SA, Guerinot ML (2007) Mining iron: iron uptake and transport in plants. FEBS Lett 581(12):2273–2280
Kochian LV (1991) Mechanisms of micronutrient uptake and translocation in plants. In: Micronutrients in agriculture (micronutrientsi 2), pp 229–296
Krishnappa G, Singh AM, Chaudhary S, Ahlawat AK, Singh SK, Shukla RB et al (2017) Molecular mapping of the grain iron and zinc concentration, protein content and thousand kernel weight in wheat (*Triticum aestivum* L.). PLoS One 12(4): e0174972.
Kumar J, Saripalli G, Gahlaut V, Goel N, Meher PK, Mishra KK et al (2018) Genetics of Fe, Zn, β-carotene, GPC and yield traits in bread wheat (*Triticum aestivum* L.) using multi-locus and multi-traits GWAS. Euphytica *214*(11):219
Kutman UB, Yildiz B, Cakmak I (2011) Improved nitrogen status enhances zinc and iron concentrations both in the whole grain and the endosperm fraction of wheat. J Cereal Sci 53(1):118–125
Li A, Liu D, Yang W, Kishii M, Mao L (2018) Synthetic hexaploid wheat: yesterday, today, and tomorrow. Engineering 4(4):552–558
Lieu PT, Heiskala M, Peterson PA, Yang Y (2001) The roles of iron in health and disease. Mol Asp Med 22(1–2):1–87
Liu F, Tang Y, Du R, Yang H, Wu Q, Qiu R (2010) Root foraging for zinc and cadmium requirement in the Zn/Cd hyperaccumulator plant Sedum alfredii. Plant Soil 327(1–2):365–375
Marr KM, Batten GD, Blakeney AB (1995) Relationships between minerals in Australian brown rice. J Sci Food Agric 68(3):285–291
McFadden ES, Sears ER (1944) The artificial synthesis of *Triticum* spelta. Rec Genet Soc Am 13:26–27
McLean E, Cogswell M, Egli I, Wojdyla D, De Benoist B (2009) Worldwide prevalence of anaemia, WHO vitamin and mineral nutrition information system, 1993–2005. Public Health Nutr 12(4):444–454
Meng F, Wei Y, Yang X (2005) Iron content and bioavailability in rice. J Trace Elem Med Biol 18(4):333–338
Morgounov A, Gómez-Becerra HF, Abugalieva A, Dzhunusova M, Yessimbekova M, Muminjanov H et al (2007) Iron and zinc grain density in common wheat grown in Central Asia. Euphytica 155(1–2):193–203
Mori S (1999) Iron acquisition by plants. Curr Opin Plant Biol 2(3):250–253
Mori S, Kishi-Nishizawa N, FUJUGAKI J (1990) Identification of rye chromosome 5R as a carrier of the genes for mugineic acid synthetase and 3-hydroxymugineic acid synthetase using wheat-rye addition lines. Jpn J Genet 65(5):343–352
Murray CJ, Lopez AD (2013) Measuring the global burden of disease. N Engl J Med 369(5):448–457

Neelam K, Rawat N, Tiwari VK, Kumar S, Chhuneja P, Singh K et al (2011) Introgression of group 4 and 7 chromosomes of *Ae. Peregrina* in wheat enhances grain iron and zinc density. Mol Breed 28(4):623–634

Ortiz-Monasterio JI, Palacios-Rojas N, Meng E, Pixley K, Trethowan R, Pena RJ (2007) Enhancing the mineral and vitamin content of wheat and maize through plant breeding. J Cereal Sci 46(3):293–307

Palmgren MG, Clemens S, Williams LE, Krämer U, Borg S, Schjørring JK, Sanders D (2008) Zinc biofortification of cereals: problems and solutions. Trends Plant Sci 13(9):464–473

Pathak P, Kapil U, Kapoor SK, Saxena R, Kumar A, Gupta N et al (2004) Prevalence of multiple micronutrient deficiencies amongst pregnant women in a rural area of Haryana. Indian J Pediatr 71(11):1007–1014

Pearce S, Tabbita F, Cantu D, Buffalo V, Avni R, Vazquez-Gross H, Zhao R, Conley CJ, Distelfeld A, Dubcovksy J (2014) Regulation of Zn and Fe transporters by the *GPC1* gene during early wheat monocarpic senescence. BMC Plant Biol 14(1):368

Pearson JN, Rengel Z (1994) Distribution and remobilization of Zn and Mn during grain development in wheat. J Exp Bot 45(12):1829–1835

Peleg Z, Cakmak I, Ozturk L, Yazici A, Jun Y, Budak H et al (2009) Quantitative trait loci conferring grain mineral nutrient concentrations in durum wheat× wild emmer wheat RIL population. Theor Appl Genet 119(2):353–369

Poletti S, Gruissem W, Sautter C (2004) The nutritional fortification of cereals. Curr Opin Biotechnol 15(2):162–165

Prażak R, Krzepiłko A (2018) Evaluation of iron and zinc content in grain of *Aegilops* l. × *Triticum Aestivum* l. hybrid lines. J Elem 23(2):545–557. https://doi.org/10.5601/jelem.2017.22.3.1486

Pu ZE, Ma YU, He QY, Chen GY, Wang JR, Liu YX et al (2014) Quantitative trait loci associated with micronutrient concentrations in two recombinant inbred wheat lines. J Integr Agric 13(11):2322–2329

Ramakrishnan U, Nguyen P, Martorell R (2009) Effects of micronutrients on growth of children under 5 y of age: meta-analyses of single and multiple nutrient interventions. Am J Clin Nutr 89(1):191–203

Rawat N, Tiwari VK, Neelam K, Randhawa GS, Chhuneja P, Singh K, Dhaliwal HS (2009a) Development and characterization of *Triticum aestivum–Aegilops kotschyi* amphiploids with high grain iron and zinc contents. Plant Genet Resour 7(3):271–280

Rawat N, Tiwari VK, Singh N, Randhawa GS, Singh K, Chhuneja P, Dhaliwal HS (2009b) Evaluation and utilization of *Aegilops* and wild *Triticum* species for enhancing iron and zinc content in wheat. Genet Resour Crop Evol 56(1):53

Rawat N, Neelam K, Tiwari VK, Randhawa GS, Friebe B, Gill BS, Dhaliwal HS (2011) Development and molecular characterization of wheat-*Aegilops kotschyi* addition and substitution lines with high grain protein, iron, and zinc. Genome 54(11):943–953

Roshanzamir H, Kordenaeej A, Bostani A (2013) Mapping QTLs related to Zn and Fe concentrations in bread wheat (*Triticum aestivum*) grain using microsatellite markers. Iran J Genet Plant Breed 2(1):10–17

Rosyara U, Kishii M, Payne T, Sansaloni CP, Singh RP, Braun HJ, Dreisigacker S (2019) Genetic contribution of synthetic Hexaploid wheat to CIMMYT's spring bread wheat breeding germplasm. Sci Rep 9(1):1–11

Saini DK, Devi P, Kaushik P (2020) Advances in genomic interventions for wheat biofortification: a review. Agronomy 10(1):62

Sandström B (1997) Bioavailability of zinc. Eur J Clin Nutr Suppl 51(1):S17–S19

Seshadri S (2001) Prevalence of micronutrient deficiency particularly of iron, zinc and folic acid in pregnant women in South East Asia. Br J Nutr 85(S2):S87–S92

Shankar AH, Prasad AS (1998) Zinc and immune function: the biological basis of altered resistance to infection. Am J Clin Nutr 68(2):447S–463S

Sharma P, Sheikh I, Singh D, Kumar S, Verma SK, Kumar R et al (2017) Uptake, distribution, and remobilization of iron and zinc among various tissues of wheat–*Aegilops* substitution lines at different growth stages. Acta Physiol Plant 39(8):185

Sharma P, Sheikh I, Kumar S, Verma SK, Kumar R, Vyas P, Dhaliwal HS (2018) Precise transfers of genes for high grain iron and zinc from wheat-*Aegilops* substitution lines into wheat through pollen irradiation. Mol Breed 38(6):81

Shi R, Li H, Tong Y, Jing R, Zhang F, Zou C (2008) Identification of quantitative trait locus of zinc and phosphorus density in wheat (*Triticum aestivum* L.) grain. Plant Soil *306*(1–2):95–104

Singh R, Govindan V, Andersson MS (2017) Zinc-biofortified wheat: harnessing genetic diversity for improved nutritional quality (No. 2187-2019-666)

Srinivasa J, Arun B, Mishra VK, Singh GP, Velu G, Babu R, Vasistha NK, Joshi AK (2014) Zinc and iron concentration QTL mapped in a *Triticum spelta* × *T. aestivum* cross. Theor Appl Genet 127(7):1643–1651. https://doi.org/10.1007/s00122-014-2327-6

Stein AJ (2010) Global impacts of human mineral malnutrition. Plant Soil 335(1–2):133–154

Stein AJ, Meenakshi JV, Qaim M, Nestel P, Sachdev, HPS, Bhutta ZA (2008) Potential impacts of iron biofortification in India. Social Science & Medicine, 66(8):1797–1808

Stoltzfus RJ, Mullany L, Black RE (2004) Iron deficiency anaemia. In: Comparative quantification of health risks: global and regional burden of disease attributable to selected major risk factors, vol 1, pp 163–209

Tabbita F, Pearce S, Barneix AJ (2017) Breeding for increased grain protein and micronutrient content in wheat: ten years of the *GPC-B1* gene. J Cereal Sci 73:183–191. https://doi.org/10.1016/j.jcs.2017.01.003

Tiwari VK, Rawat N, Neelam K, Randhawa GS, Singh K, Chhuneja P, Dhaliwal HS (2008) Development of *Triticum turgidum* subsp. *durum–Aegilops longissima* amphiploids with high iron and zinc content through unreduced gamete formation in F1 hybrids. Genome 51(9):757–766

Tiwari VK, Rawat N, Chhuneja P, Neelam K, Aggarwal R, Randhawa GS et al (2009a) Mapping of quantitative trait loci for grain iron and zinc concentration in diploid A genome wheat. J Hered 100(6):771–776

Tiwari VK, Rawat N, Chhuneja P, Neelam K, Aggarwal R, Randhawa GS, Dhaliwal HS, Keller B, Singh K (2009b) Mapping of quantitative trait loci for grain iron and zinc concentration in diploid A genome wheat. J Hered 100:771–776

Tiwari VK, Rawat N, Neelam K, Kumar S, Randhawa GS, Dhaliwal HS (2010) Substitutions of 2S and 7U chromosomes of *Aegilops kotschyi* in wheat enhance grain iron and zinc concentration. Theor Appl Genet 121(2):259–269

Tiwari C, Wallwork H, Arun B, Mishra VK, Velu G, Stangoulis J et al (2016) Molecular mapping of quantitative trait loci for zinc, iron and protein content in the grains of hexaploid wheat. Euphytica 207(3):563–570

Uauy C, Brevis JC, Dubcovsky J (2006a) The high grain protein content gene Gpc-B1 accelerates senescence and has pleiotropic effects on protein content in wheat. J Exp Bot 57(11):2785–2794. https://doi.org/10.1093/jxb/erl047

Uauy C, Distelfeld A, Fahima T, Blechl A, Dubcovsky J (2006b) A NAC gene regulating senescence improves grain protein, zinc, and iron content in wheat. Science 314(5803):1298–1301. https://doi.org/10.1126/science.1133649

Velu G, Singh RP, Huerta-Espino J, Pena RJ, Arun B, Mahendru-Singh A et al (2012) Performance of biofortified spring wheat genotypes in target environments for grain zinc and iron concentrations. Field Crop Res 137:261–267

Velu G, Ortiz-Monasterio I, Cakmak I, Hao Y, Singh RP (2014) Biofortification strategies to increase grain zinc and iron concentrations in wheat. J Cereal Sci 59(3):365–372

Velu G, Crossa J, Singh RP, Hao Y, Dreisigacker S, Perez-Rodriguez P et al (2016) Genomic prediction for grain zinc and iron concentrations in spring wheat. Theor Appl Genet 129(8):1595–1605

Velu G, Tutus Y, Gomez-Becerra HF, Hao Y, Demir L, Kara R, Cakmak I (2017) QTL mapping for grain zinc and iron concentrations and zinc efficiency in a tetraploid and hexaploid wheat mapping populations. Plant and soil, 411(1–2):81–99

Velu G, Singh RP, Crespo-Herrera L, Juliana P, Dreisigacker S, Valluru R et al (2018) Genetic dissection of grain zinc concentration in spring wheat for mainstreaming biofortification in CIMMYT wheat breeding. Sci Rep 8(1):1–10

Velu G, Crespo Herrera L, Guzman C, Huerta J, Payne T, Singh RP (2019) Assessing genetic diversity to breed competitive biofortified wheat with enhanced grain Zn and Fe concentrations. Front Plant Sci 9:1971

Venegas JP, Graybosch RA, Wienhold B, Rose DJ, Waters BM, Stephen Baenziger P, Eskridge K, Bai G, St. Amand P (2018) Biofortification of hard red winter wheat by genes conditioning low Phytate and high grain protein concentration. Crop Sci 58(5):1942–1953. https://doi.org/10.2135/cropsci2018.03.0175

Verma SK, Kumar S et al (2016a) Induced homoeologous pairing for transfer of useful variability for high grain Fe and Zn from *Aegilops kotschyi* into wheat. Plant Mol Biol Rep 34:1083–1094

Verma SK, Kumar S, Sheikh I, Malik S, Mathpal P, Chugh V, Dhaliwal HS (2016b) Transfer of useful variability of high grain iron and zinc from *Aegilops kotschyi* into wheat through seed irradiation approach. Int J Radiat Biol 92(3):132–139

Wang YX, Specht A, Horst WJ (2011) Stable isotope labelling and zinc distribution in grains studied by laser ablation ICP-MS in an ear culture system reveals zinc transport barriers during grain filling in wheat. New Phytol 189(2):428–437

Waters BM, Uauy C, Dubcovsky J, Grusak MA (2009) Wheat (*Triticum aestivum*) NAM proteins regulate the translocation of Iron, zinc, and nitrogen compounds from vegetative tissues to grain. J Exp Bot 60(15):4263–4274. https://doi.org/10.1093/jxb/erp257

Welch RM, Graham RD (2004) Breeding for micronutrients in staple food crops from a human nutrition perspective. J Exp Bot 55(396):353–364

Wessells KR, Brown KH (2012) Estimating the global prevalence of zinc deficiency: results based on zinc availability in national food supplies and the prevalence of stunting. PLoS One 7(11)

World Health Organization (2008) Worldwide prevalence of anaemia 1993–2005: WHO global database on anaemia

White PJ, Broadley MR (2005) Biofortifying crops with essential mineral elements. Trends Plant Sci 10(12):586–593

White PJ, Broadley MR (2009) Biofortification of crops with seven mineral elements often lacking in human diets–iron, zinc, copper, calcium, magnesium, selenium and iodine. New Phytol 182(1):49–84

Xu Y, An D, Liu D, Zhang A, Xu H, Li B (2012) Molecular mapping of QTLs for grain zinc, iron and protein concentration of wheat across two environments. Field Crop Res 138:57–62

Zhang Y, Song Q, Yan J, Tang J, Zhao R, Zhang Y et al (2010) Mineral element concentrations in grains of Chinese wheat cultivars. Euphytica 174(3):303–313

Zhao FJ, Su YH, Dunham SJ, Rakszegi M, Bedo Z, McGrath SP, Shewry PR (2009) Variation in mineral micronutrient concentrations in grain of wheat lines of diverse origin. J Cereal Sci 49(2):290–295

Zimmermann MB, Hurrell RF (2007) Nutritional iron deficiency. Lancet 370(9586):511–520

# Membrane Fluidity and Compositional Changes in Response to High Temperature Stress in Wheat

**Sruthi Narayanan**

**Abstract** Membranes are prime targets of high temperature stress in plants. Thus, cell membrane stability has been used as a measure of heat tolerance in wheat. Under optimal temperature conditions, membranes are lipid bilayers that are largely in fluid phase. High temperatures or dehydration can cause phase transitions of membranes to non-bilayer phases. In order to maintain optimal fluidity and stability of membranes under high temperature conditions, wheat plants alter lipid compositions and reduce unsaturation levels in the fatty acid chains. Besides altering the fatty acid chains synthesized, the composition of chloroplast and thylakoid membranes may be adjusted by adjusting the diacylglycerol species channeled from the endoplasmic reticulum to chloroplasts under heat stress conditions.

**Keywords** Wheat · High temperature stress · Membranes · Lipids · Fatty acids Lipid unsaturation

Membranes are prime targets of high temperature stress in plants. For example, high temperatures cause damage to thylakoid membranes, which are the sites of the photosynthetic reaction centers and electron transport chains in photosynthesis, resulting in reduction or interruption of photosynthesis (Ristic et al. 2007; Prasad et al. 2008; Narayanan et al. 2016; Djanaguiraman et al. 2018). High temperatures also damage plasma membranes, which results in cell content leakage, leading to cell death and loss of physiological function and yield (Narayanan et al. 2014; Djanaguiraman et al. 2018). This chapter presents an overview of membrane structure and mechanisms leading to changes in membrane fluidity under high temperature stress in wheat.

S. Narayanan (✉)
Department of Plant and Environmental Sciences, Clemson University, Clemson, SC, USA
e-mail: skutty@clemson.edu

© The Editor(s) (if applicable) and The Author(s), under exclusive license to Springer Nature Switzerland AG 2021

S. H. Wani et al. (eds.), *Physiological, Molecular, and Genetic Perspectives of Wheat Improvement*, https://doi.org/10.1007/978-3-030-59577-7_6

# 1 Membrane Structure and Phases

Membranes consist of both lipids and proteins and form a barrier between two aqueous environments. In the case of plasma membrane, these environments are the inside and the outside of the cell. Proteins diffuse within the lipid matrix, while lipids undergo multiple types of motion/transverse diffusion (rarely), rotation, and lateral diffusion (more often).The fundamental structure of a membrane is described by the fluid mosaic model proposed by Singer and Nicolson(1972). In this model, proteins are described as "icebergs" floating in a "sea" of lipids in a random or "mosaic" fashion. Singer and Nicolson's model is inadequate to completely describe membrane structure, as it understates variations in motional freedom and local order within bilayers, neglects the possibility that lipids can be distributed non-randomly within the membrane, and disregards the potential for non-bilayer-phase lipids(Jouhet 2013). Therefore, this model was updated with the introduction of the membrane domain concept, where domains are defined as patches of lipids with composition and physical state that differ from the average molecular composition and properties [see reviews on membrane model history in Edidin (2003) and Sonnino and Prinetti (2013)].

Biological membranes maintain primarily a fluid bilayer phase, with patches in the membrane that may transition to gel and non-bilayer phases (e.g., hexagonal I and II and cubic phases) under some circumstances. In gel and fluid bilayer phases, the polar head groups of the lipids face the aqueous phase on both sides of the bilayer, and the nonpolar hydrocarbon tails (fatty acid chains) oppose each other in the bilayer (i.e., head groups outside and hydrocarbon tails inside). The presence of *cis* double bonds, which are common in most plant cell membrane fatty acyl chains, introduces bends in the chains, reducing tight packing of adjacent lipid molecules, thus, contributing to membrane fluidity (Huang 2006). The hydrocarbon tails of fluid phase lipids have the ability to flex, while gel phase lipids are more closely packed and have more extended and ordered lipid chains (Garvey et al. 2013; Voet et al. 2008). In addition to small amounts of gel phase (Welti et al. 1981), functional membranes may contain small amounts of hexagonal II phase, which may occur during the membrane fusion events required in vesicular trafficking (Jouhet 2013). Hexagonal II phase is characterized by a reverse cylinder morphology with polar head groups inside and the hydrophobic, hydrocarbon tails outside (Garvey et al. 2013; Cullis and DeKruijff 1979). Gel and hexagonal phases are minor membrane components when membranes are functioning normally. However, when cells are subjected to stress, such as stress caused by temperatures change or dehydration, the amounts of these phases can change, and increases can cause membranes to lose functionality and stability. While the temperatures at which the membranes undergo phase transitions are not properties of single lipid molecules and are properties of groups of lipids in bilayers, the lipid composition of membranes does affect the transition temperatures.

Polar glycerolipids are the primary constituents of membranes in plant cells and include two large groups of membrane lipids: phospholipids and glycolipids. Polar glycerolipids are amphiphilic molecules with a 3-carbon glycerol scaffold in which the carbons are numbered as sn-1, sn-2, and sn-3. One or two hydrophobic acyl chains are esterified at sn-1 and sn-2, and a hydrophilic polar head at sn-3. Major membrane phospholipids found in wheat include phosphatidylcholine (PC), phosphatidylethanolamine (PE), phosphatidylglycerol (PG), phosphatidylinositol (PI), phosphatidylserine (PS), and phosphatidic acid (PA). Major glycolipids in wheat include monogalactosyldiacylglycerol (MGDG), digalactosyldiacylglycerol (DGDG), and sulfoquinovosyldiacylglycerol (SQDG). Glycerolipids also include neutral lipids such as triacylglycerol (TAG) and diacylglycerol (DAG), which are not major components of membranes. The size of the hydrophilic polar head group in comparison with the hydrophobic acyl-glycerol group affects lipid behavior in aqueous dispersions. For example, lipids with large negative curvature, due to small head group vs acyl chain diameter, such as MGDG, PE, PS, and PA, tend to form hexagonal II phase (or cubic phase), while lipids with small curvature (similar head group and acyl chain diameter), such as DGDG, SQDG, PC, PG, and PI, form bilayer phases (Shipley et al. 1973; Seddon 1990; Hansbro et al. 1992; Vikstrom et al. 2000). Apart from the characteristics of the head groups, fatty acyl chain composition also influences membrane phases. Decreases in the length and increases in the unsaturation levels of fatty acid chains lead to a decrease in van der Waals' interactions, increase the disorder (fluidity) of lipid molecules within membranes, and, thus, increase the propensity of a bilayer to non-bilayer phase transition (Seddon 1990). Thus, lipid molecules within the same head group class may be different in terms of whether they favor a bilayer or non-bilayer phase. Phase transitions are generally spontaneous and reversible (Siegel and Tenchov 2008).

## 2 Impact of Temperature on Membrane Fluidity and Stability

Compared to warm-blooded animals, plant cells are more likely to encounter temperature changes that induce phase transitions of membranes (Horvath et al. 1998; Orvar et al. 2000), affecting cell functions and, thus, plant growth and development. Increases in temperature cause transitions from gel phase to fluid phase and then to hexagonal phase (Quinn 1985). Cell membrane stability, measured through ion leakage, has been historically used as a measure of heat tolerance in plants including wheat (Sullivan 1972; Martineau et al. 1979; Blum and Ebercon 1981; Sairam et al. 1997; Ibrahim and Quick 2001; Djanaguiraman et al. 2010). Heat-tolerant genotypes maintain lower ion leakage as a result of stable membranes. Early research attempted to explain the mechanism of ion leakage (Simon 1974; Crowe et al. 1989; Hoekstra et al. 1992). The prevailing hypothesis is that membranes become transiently leaky in the event of phase changes (Simon 1974; Hammoudah et al. 1981; Hoekstra et al. 1992). Defects at the boundary between these phases lead to increased permeability. For example, transition of a membrane from a bilayer to hexagonal II phase would form hydrophilic pores in the membrane through which leakage may occur (Simmon 1974; Simons and Sampaio 2011; Ortiz et al. 1999). The leakage may stop if the bilayer conformation has been restored.

Optimal membrane fluidity is the major determinant of membrane stability during temperature stress (i.e., a fluid phase relates to stable membranes) and influences plant adaptation to stress (Harwood 1991; Murata and Los 1997; Iba 2002). The rest of this chapter describes three interconnected lipid-related mechanisms that affect membrane fluidity and, thus, membrane stability and function under high temperature stress conditions in wheat.

## *Alterations in Membrane Lipid Composition*

High temperatures lead to lipid remodeling in wheat leaf and pollen (Narayanan et al. 2016, 2018). Lipid remodeling refers to decreases in the amounts of certain lipids and increases in others (Zheng et al. 2011). For example, Narayanan et al. (2016) found that the amounts of plastidic glycolipids (DGDG,MGDG, and SQDG), plastidic phospholipids (PG) and extraplastidic phospholipids (PC and PE) decrease under high temperature stress, while the amounts of sterol lipids[sterol glycoside (SG) and acylated sterol glycoside (ASG)], 18:3-acyl-containing TAGs, and oxidized lipids increase in wheat leaves. Narayanan et al. (2018) evaluated membrane lipid changes in wheat pollen and found that most heat-responsive lipids were extraplastidic phospholipids, including PC, PE, PI, PA, and PS. Lipid remodeling is likely to prevent the phase transition of membranes from bilayer phase to excessive non-bilayer phase at high temperatures. As explained above, lipids such as MGDG and PE tend to form non-bilayer phases, whereas DGDG, SQDG, PC, and PG form bilayers (Shipley et al. 1973; Seddon 1990; Hansbro et al. 1992; Vikstrom et al. 2000). Higher ratios of DGDG to MGDG and PC to PE reduce the propensity of membranes to form non-bilayer phases (Suss and Yordanov 1986; Webb and Green 1991; Williams 1998; de Vries et al. 2004). Increasing DGDG:MGDG and PC:PE ratios is another strategy employed by wheat plants to adapt to high temperatures by maintaining membrane fluidity, presumably avoiding high-temperature-induced non-bilayer phase formation (Narayanan et al. 2016, 2018).

Narayanan et al. (2016) have identified phospholipids containing 15:0, 17:0, 17:1 or 17:2 acyl chains in wheat plants, the levels of which increased under high temperature stress. Such lipids containing fatty acid chains with an odd number of carbons have been reported in humans, animals, and microorganisms and very rarely in plants (Sperl et al. 2000; Řezanka and Sigler 2009). The odd-chain fatty acids were reported as biomarkers for risks of specific human diseases (Jenkins et al. 2015). In fungi, they were produced under alcoholic and hypoxic stress conditions (Jeennor et al. 2006). Narayanan et al. (2016) hypothesized that the increased formation of phospholipids with odd-chain fatty acyl components in wheat under high temperature stress might be an indication of the extent of stress damage. Propionyl-CoA acts as a primer in the biosynthesis of odd-chain fatty acids, and excess propionyl-CoA leads to enhanced synthesis of 15- and 17-carbon fatty acids in humans (Wendel 1989). Thus, it may be possible that the incorporation of propionic acid into fatty acids could be regulated by altered enzymatic specificity in wheat as a function of temperature.

Certain changes in the membrane lipid profile might be associated with heat tolerance or susceptibility. Narayanan et al. (2016) found that a heat-tolerant winter wheat genotype had greater ability to increase the amounts of sterol derivatives(SGs and saturated species of ASGs) at high temperatures, compared with a susceptible genotype. Sterol glycosides and ASGs are ubiquitous constituents of cells in vascular plants and function as membrane components, storage forms of sterols, transporters, and signaling molecules (Grille et al. 2010). Sterol glycosides have a modulatory effect on membranes that helps to eliminate phase transitions to non-bilayer phases at high temperatures (Muramatsu et al. 2000). Thus, increasing the levels of SGs and ASGs under high temperature stress could be part of a lipid remodeling mechanism that helps maintain membrane bilayer structure and improve heat tolerance in wheat.

High temperatures cause oxidation of membrane lipids in wheat, the extent of which may be an indication of heat susceptibility. Trienoic species of MGDG and DGDG have been found to be highly vulnerable to peroxidation, non-enzymatically by reactive oxygen species (ROS) and enzymatically by lipoxygenase in wheat (Mene-Saffrane et al. 2009; Farmer and Mueller 2013). Since, MGDG and DGDG are the major chloroplast lipids found in wheat, the peroxidation of their acyl chains may significantly damage the photosynthetic apparatus. Narayanan et al. (2016) found that the trienoic species of PC and PE were oxidized and the levels of the oxidized products were significantly increased under high temperature stress conditions in the heat susceptible wheat genotype, whereas the tolerant genotype maintained a basal level of most oxidized lipids(lipids with oxidized acyl chains). It has been found that trienoic fatty acids act as sinks for ROS (Mene-Saffrane et al. 2009). The non-enzymatic oxidation of trienoic fatty acids by ROS is a mechanism for immediately consuming ROS produced under stress conditions, without activating genes encoding ROS catabolizing enzymes (Mene-Saffrane et al. 2009). Thus, the amount of oxidized lipids could reflect the degree of oxidative stress a plant is experiencing.

## *Changes in Membrane Lipid Unsaturation Levels*

Decreasing the level of unsaturation of membrane lipids in order to maintain optimal fluidity and stability of membranes is another adaptation mechanism in wheat to high temperatures (Larkindale and Huang 2004; Narayanan et al. 2016, 2018). The decrease in unsaturation level is mainly because of the decrease in the polyunsaturated fatty acid and linolenic acid (18:3) and the increase in the less unsaturated fatty acids such as oleic acid (18:1) and saturated fatty acids such as palmitic acid (16:0). Since the *cis* double bonds in the membrane lipids reduce close interchain packing, increased unsaturation in the fatty acid chains may increase membrane fluidity above the optimal level and cause membranes to transition to non-bilayer phases. Recent reports suggest that the 18:3 acyl chains sequestered from the membrane lipids under high temperature conditions are recycled in TAGs, which likely occur in lipid droplets in the cytosol or plastid (Narayanan et al. 2016; Djanaguiraman et al. 2018).

### *Differential Channeling of DAG Moieties from the Endoplasmic Reticulum to Chloroplasts Under High Temperatures*

The plastidic and extraplastidic compartments possess their own unique pathways for glycerolipid assembly in plant cells (Browse and Somerville 1991; Ohlrogge and Browse 1995; Holzl et al. 2009). The lipid pathway located in the plastid (chloroplast) traces its origin to symbiogenesis and is called the prokaryotic pathway. On the other hand, the extraplastidic lipid pathway, which takes place in the endoplasmic reticulum (ER), is called the eukaryotic pathway. The two pathways are separated by membrane barriers but coordinate closely for biogenesis, maintenance, and proper functioning of membranes (Kunst et al. 1988; Ohlrogge and Browse 1995). However, the relative contributions of the two pathways are different among plant species (Heinz and Roughan 1983; Mongrand et al. 1998). In wheat, the biosynthesis of chloroplast glycerolipids is almost entirely dependent on the eukaryotic pathway, while the contribution of prokaryotic pathway is limited to the production of PG only (Arunga and Morrison 1971; Heinz and Roughan 1983). The eukaryotic pathway produces glycerolipid molecules that have only C18 fatty acid at the sn-2 position, which undergoes a series of desaturations, leading to the formation of 18:3 fatty acids. Therefore, wheat is called as an "18:3 plant," in contrast to plants that use both pathways, which are known as "16:3 plants" (Heinz and Roughan 1983; Browse et al. 1986; Ohlrogge and Browse 1995). Thus, glycerolipid channeling from the ER to chloroplasts occurs in wheat for proper functioning of cells (Li et al. 2015). Rebalancing of the two pathways influences the degree of fatty acid unsaturation in glycerolipid molecules.

Diacylglycerols are used to synthesize various glycerolipids. The C18 fatty acid at the sn-2 position derived through the eukaryotic pathway produces C36 (C18/C18) and C34(C16:0/C18) DAGs in the ER. The 16:0 fatty acyl moiety at the sn-1 position usually does not undergo any further desaturation. Therefore, the eukaryotic 34:3 DAGs (16:0/18:3, sn-1/sn-2) have a lower level of unsaturation than that of the prokaryotic 34:6 DAGs (18:3/16:3, sn-1/sn-2). Li et al. (2015) proposed that wheat plants maintain preferential channeling of C34 DAGs(16:0/C18, sn-1/sn-2) from the ER to the chloroplast, compared to C36 DAGs (C18/C18, sn-1/sn-2) under high temperatures in order to reduce the level of fatty acid unsaturation in the chloroplast under high temperature stress. This metabolic alteration could improve high temperature adaptations of wheat by decreasing the fluidity of chloroplast membranes, in particular the thylakoid membranes, where the light-dependent reactions of photosynthesis are carried out.

## 3 Conclusions

Under optimal temperature conditions, membranes are lipid bilayers that are largely in fluid phase. High temperatures alter motional freedom of lipids within the membranes and affect membrane fluidity, which will lead to phase transitions of membranes to non-bilayer phases. Recent reports suggest that wheat plants can alter lipid compositions and reduce unsaturation levels in the fatty acid chains in order to maintain optimal fluidity and stability of membranes under high temperature conditions. Differential channeling of DAG moieties from the ER to chloroplasts is another cellular mechanism that can aid in membrane fluidity adjustments under sub-optimal temperature conditions in wheat.

**Acknowledgments** The author thanks Dr. Ruth Welti (Professor at Kansas State University and Director of Kansas Lipidomics Research Center) for critical review and editing of the chapter.

## References

Arunga RO, Morrison WR (1971) The structural analysis of wheat flour glycerolipids. Lipids 6:768–776

Blum A, Ebercon A (1981) Cell membrane stability as a measure of drought and heat tolerance in wheat. Crop Sci 21:43–47

Browse J, Somerville C (1991) Glycerolipids synthesis: biochemistry and regulation. Annu Rev Plant Physiol Plant Mol Biol 42:467–506

Browse J, Warwick N, Somerville CR, Slack CR (1986) Fluxes through the prokaryotic and eukaryotic pathways of lipid synthesis in the '16:3' plant Arabidopsis thaliana. Biochem J 235:25–31

Crowe JH, Hoekstra FA, Crowe LM (1989) Membrane phase transitions are responsible for imbibitional damage in dry pollen. Proc Natl Acad Sci U S A 86:520–523

Cullis PR, DeKruijff B (1979) Lipid polymorphism and the functional roles of lipids in biological membranes. Biochim Biophys Acta 559:399–420

de Vries AH, Mark AE, Marrink SJ (2004) The binary mixing behavior of phospholipids in a bilayer: amolecular dynamics study. J Phys Chem B 108:2454–2463

Djanaguiraman M, Prasad PVV, Seppanen M (2010) Selenium protects sorghum leaves from oxidative damage under high temperature stress by enhancing antioxidant defense system. Plant Physiol Biochem 48:999–1007

Djanaguiraman M, Boyle DL, Welti R, Jagadish SVK, Prasad PVV (2018) Decreased photosynthetic rate under high temperature in wheat is due to lipid desaturation, oxidation, acylation, and damage of organelles. BMC Plant Biol 18:55

Edidin M (2003) Lipids on the frontier: a century of cell-membrane bilayers. Nat Rev Mol Cell Biol 4:414–418

Farmer EE, Mueller MJ (2013) ROS-mediated lipid peroxidation and RES-activated signaling. Ann Rev Plant Biol 64:429–450

Garvey CJ, Lenné T, Koster KL, Kent B, Bryant G (2013) Phospholipid membrane protection by sugar molecules during dehydration – insights into molecular mechanisms using scattering techniques. Int J Mol Sci 14:8148–8163

Grille S, Zaslawski A, Thiele S, Plat J, Warnecke D (2010) The functions of steryl glycosides come to those who wait: recent advances in plants, fungi, bacteria and animals. Prog Lipid Res 49:262–288

Hammoudah MM, Nir S, Bentz J, Mayhew E, Stewart TP, HuiSW KRJ (1981) Interactions of La31 with phosphatidylserine vesicles binding, phase transition, leakage, 31P-NMR and fusion. Biochim Biophys Acta 645:102–114

Hansbro PM, Byard SJ, Bushby RJ, Turnbull PJ, Boden N, Saunders MR et al (1992) The conformational behaviour of phosphatidylinositol in model membranes: 2H-NMR studies. Biochim Biophys Acta 1112:187–196

Harwood J (1991) Strategies for coping with low environmental temperatures. Trends Biochem Sci 16:126–127

Heinz E, Roughan PG (1983) Similarities and differences in lipid metabolism of chloroplasts isolated from 18:3 and 16:3 plants. Plant Physiol 72:273–279

Hoekstra FA, Crowe JH, Crowe LM (1992) Germination and ion leakage are linked with phase transitions of membrane lipids during imbibition of Typha latifolia pollen. Physiol Plant 84:29–34

Holzl G, Witt S, Gaude N, Melzer M, Schottler MA, Dormann P (2009) The role of diglycosyl lipids in photosynthesis and membrane lipid homeostasis in Arabidopsis. Plant Physiol 150:1147–1159

Horvath I, Glatz A, Varvasovszki V, Torok Z, Pali T, Balogh G, Kovacs E, Nadasdi L, Benko S, Joo F, Vigh L (1998) Membrane physical state controls the signaling mechanism of the heat shock response in Synechocystis PCC 6803: identification of hsp17 as a 'fluidity gene'. Proc Natl Acad Sci U S A 95:3513–3518

Huang B (2006) Plant-environment interactions, 3rd edn. CRC Press, Boca Raton

Iba K (2002) Acclimative response to temperature stress in higher plants: approaches of gene engineering for temperature tolerance. Ann Rev Plant Biol 53:225–245

Ibrahim AMH, Quick JS (2001) Genetic control of high temperature tolerance in wheat as measured by membrane thermal stability. Crop Sci 41:1405–1407

Jeennor S, Laoteng K, Tanticharoen M, Cheevadhanarak S (2006) Comparative fatty acid profiling of Mucor rouxii under different stress conditions. FEMS Microbiol Lett 259:60–66

Jenkins B,West JA, Koulman A (2015) A review of odd-chain fatty acid metabolism and the role of pentadecanoic acid (C15:0) and heptadecanoic acid (C17:0) in health and disease. Molecules 20:2425–2444

Jouhet J (2013) Importance of the hexagonal lipid phase in biological membrane organization. Front Plant Sci 4:494

Kunst L, Browse J, Somerville C (1988) Altered regulation of lipid biosynthesis in a mutant of Arabidopsis deficient in chloroplast glycerol-3-phosphate acyltransferase activity. Proc Natl Acad Sci U S A 85:4143–4147

Larkindale J, Huang B (2004) Changes of lipid composition and saturation level in leaves and roots for heat-stressed and heat-acclimated creeping bentgrass (Agrostis stolonifera). Environ Exp Bot 51:57–67

Li Q, Zheng Q, Shen W, Cram D, Fowler DB, Wei Y, Zou J (2015) Understanding the biochemical basis of temperature-induced lipid pathway adjustments in plants. Plant Cell 27:86–103

Martineau JR, Specht JE, Williams JH, Sullivan CY (1979) Temperature tolerance in soybeans. I. Evaluation of a technique for assessing cellular membrane thermostability. Crop Sci 19:75–78

Mene-Saffrane L, Dubugnon L, Chetelat A, Stolz S, Gouhier-Darimont C, Farmer EE (2009) Nonenzymatic oxidation of trienoic fatty acids contributes to reactive oxygen species management in Arabidopsis. J Biol Chem 284:1702–1708

Mongrand S, Bessoule J, Cabantous F, Cassagne C (1998) The C16:3/C18:3 fatty acid balance in photosynthetic tissues from 468 plant species. Phytochemistry 49:1049–1064

Murata N, Los DA (1997) Membrane fluidity and temperature perception. Plant Physiol 115:875–879

Muramatsu K, Masumizu T, Maitani Y, Hwang SH, Kohno M, Takayama K, Nagai T (2000) Electron spin resonance studies of dipalmitoylphosphatidylcholine liposomes containing soybean-derived sterylglucoside. Chem Pharm Bull 48:610–613

Narayanan S, Prasad PVV, Fritz AK, Boyle DL, Gill BS (2014) Impact of high night-time and high daytime temperature stress on winter wheat. J Agron Crop Sci 201:206–218

Narayanan S, Tamura PJ, Roth MR, Prasad PVV, Welti R (2016) Wheat leaf lipids during heat stress: I. High day and night temperatures result in major lipid alterations. Plant Cell Environ 39:787–803

Narayanan S, Prasad PVV, Welti R (2018) Alterations in wheat pollen lipidome during high day and night temperature stress. Plant Cell Environ 41:1749–1761

Ohlrogge J, Browse J (1995) Lipid biosynthesis. Plant Cell 7:957–970
Ortiz A, Killian JA, Verkleij AJ, Wilschut J (1999) Membrane fusion and the lamellar-to-inverted-hexagonal phase transition in cardiolipin vesicle systems induced by divalent cations. Biophys J 77:2003–2014
Orvar BL, Sangwan V, Omann F, Dhindsa RS (2000) Early steps in cold sensing by plant cells: the role of actin cytoskeleton and membrane fluidity. Plant J 23:785–794
Prasad PVV, Pisipati SR, Ristic Z, Bukovnik U, Fritz AK (2008) Impact of nighttime temperature on physiology and growth of spring wheat. Crop Sci 48:2372–2380
Quinn PJ (1985) A lipid-phase separation model of low-temperature damage to biological membranes. Cryobiology 22:128–146
Řezanka T, Sigler K (2009) Odd-numbered very-long-chain fatty acids from the microbial, animal and plant kingdoms. Prog Lipid Res 48:206–238
Ristic Z, Bukovnik U, Prasad PVV (2007) Correlation between heat stability of thylakoid membranes and loss of chlorophyll in winter wheat under heat stress. Crop Sci 47:2067–2073
Sairam RK, Deshmukh PS, Shukla DS (1997) Tolerance to drought and temperature stress in relation to increased antioxidant enzyme activity in wheat. J Agron Crop Sci 178:171e177
Seddon JM (1990) Structure of the inverted hexagonal (HII) phase, and non-lamellar phase transitions of lipids. Biochim Biophys Acta 1031:1–69
Shipley GG, Green JP, Nichols BW (1973) The phase behavior of monogalactosyl, digalactosyl, and sulphoquinovosyl diglycerides. Biochim Biophys Acta 311:531–544
Siegel DP, Tenchov BG (2008) Influence of the lamellar phase unbinding energy on the relative stability of lamellar and inverted cubic phases. Biophys J 94:3987–3995
Simon EW (1974) Phospholipids and plant membrane permeability. New Phytol 73:377–420
Simons K, Sampaio JL (2011) Membrane organization and lipid rafts. Cold Spring Harb Perspect Biol 3:a004697
Singer SJ, Nicolson GL (1972) The fluid mosaic model of the structure of cell membranes. Science 175:720–731
Sonnino S, Prinetti A (2013) Membrane domains and the "lipid raft" concept. Curr Med Chem 20:4–21
Sperl W, Murr C, Skladal D, Sass J, Suormala T, Baumgartner R, Wendel U (2000) Odd-numbered long-chain fatty acids in propionic acidaemia. Eur J Pediatr 159:54–58
Suss KH, Yordanov IT (1986) Biosynthetic cause of in vivo acquired thermotolerance of photosynthetic light reactions and metabolic responses of chloroplasts to heat stress. Plant Physiol 81:192–199
Sullivan CY (1972) Mechanisms of heat and drought resistance ingrain sorghum and methods of measurement. In: Raoand NGP, House LR (eds) Sorghum in the seventies. Oxford and IPH Publishing Co, New Delhi
Vikstrom S, Li L, Wieslander A (2000) The nonbilayer/bilayer lipid balance in membranes. Regulatory enzyme in Acholeplasma laidlawii is stimulated by metabolic phosphates, activator phospholipids, and double-stranded DNA. J Biol Chem 275:9296–9302
Voet D, Voet JG, Pratt CW (2008) Fundamentals of biochemistry, 3rd edn. Wiley, New York
Webb MS, Green BR (1991) Biochemical and biophysical properties of thylakoid acyl lipids. Biochim Biophys Acta 1060, 133–158.
Welti R, Rintoul DA, Goodsaid-Zalduondo F, Felder S, Silbert DF (1981) Gel phase phospholipid in the plasma membrane of sterol-depleted mouse LM cells. Analysis by fluorescence polarization and x-ray diffraction. J Biol Chem. 256(14):7528–35
Wendel U (1989) Abnormality of odd-numbered long-chain fatty acids in erythrocyte membrane lipids frompatients with disorders of propionatemetabolism. Pediatr Res 25:147–150
Williams EE (1998) Membrane lipids: what membrane physical properties are conserve during physiochemically-inducedmembrane restructuring? Am Zool 38:280–290
Zheng G, Tian B, Zhang F, Tao F, Li W (2011) Plant adaptation to frequent alterations between high and low temperatures: remodelling of membrane lipids and maintenance of unsaturation levels. Plant Cell and Environment 34:1431–1442

# Current Understanding of Thermotolerance in Wheat

**H. M. Mamrutha, Rinki, Rakesh Kumar, Ankita Pandey, Amandeep Kaur, Gopalareddy K, and Girish Chandra Pandey**

**Abstract** High temperature stress is one of the most common abiotic stresses in many agricultural crops and is highly complex to understand. Though advanced research is going on in understanding heat tolerance, it's very much required to understand the nature of heat tolerance in each crop at field level and how they are coping up with this stress. Hence, the present chapter focuses on understanding the current definition of thermal stress, optimum temperature required in different growth stages, impact of high temperature at different growth stages, phenotyping methods for heat tolerance and the different strategies we need to adapt in mitigating high temperature stress in wheat under field condition.

**Keywords** Wheat · Thermotolerance · Field phenotyping · Terminal heat stress Temperature Induction Response (TIR)

## 1 Introduction

Globally, the astonishing increase in temperature presents an alarming situation to agriculture world. The extreme temperature events are projected to become more frequent, more intense and long lasting than what is being currently observed (Meehl et al. 2007). The Intergovernmental Panel on Climate Change (IPCC) predicts that temperatures in India are likely to rise between 3°C and 4°C by the end of the twenty-first century. In a developing country like India, this situation is more vulnerable in view of the high population depending on agriculture and excessive pressure on natural resources. Cereal productivity is projected to decrease by

---

H. M. Mamrutha (✉) · Rinki · R. Kumar · A. Pandey · A. Kaur · G. K
ICAR-Indian Institute of Wheat and Barley Research, Karnal, Haryana, India
e-mail: mamrutha.M@icar.gov.in

G. C. Pandey
Banasthali Vidyapith, Banasthali, Rajasthan, India

© The Editor(s) (if applicable) and The Author(s), under exclusive license to Springer Nature Switzerland AG 2021

S. H. Wani et al. (eds.), *Physiological, Molecular, and Genetic Perspectives of Wheat Improvement*, https://doi.org/10.1007/978-3-030-59577-7_7

10–40% by 2100, and greater loss is expected in *rabi* due to accelerated thermal stress. By 2080, most cropping areas in the world are likely to be exposed to record average air temperatures (Battisti and Naylor 2009). High average "seasonal" temperatures can increase the risk of drought, limit photosynthesis rates and reduce light interception by accelerating phenological development (Tubiello et al. 2007). Exposure to high temperature, just for few hours, can significantly reduce the production of important food crops (Porter and Semenov 2005). Heat stress damage is particularly severe when high temperatures occur with critical crop development stages, particularly the reproductive period. Because of this, the Fourth Assessment Report of the Intergovernmental Panel on Climate Change (IPCC) has acknowledged heat stress as an important threat to global food supply (IPCC 2007).

Most often thermal stress is defined as the rise in temperature beyond a threshold level for a period of time sufficient to cause irreversible damage to plant growth and development (Hall 2001; Essemine et al. 2010). This rise in temperature is related to both soil and air temperature beyond a critical level. Temperature controls the rate of plant metabolic processes that ultimately influence the production of biomass and grains (Hay and Walker 1989). Greaves 1996 defines high temperature stress as any reduction in growth or induced metabolic, cellular or tissue injury that results in limitations to the genetically determined yield potential, caused as a direct result of exposure to temperatures above or below the thermal thresholds for optimal biochemical and physiological activity or morphological development. In general, a transient elevation in temperature, usually 10–15°C above ambient, is considered heat stress and affects the crop growth, development and particularly the yield. Temperature stress is a multifaceted function of frequency, intensity, duration and rate of increase in temperature. Some researchers reported that night temperature are major limiting factors, while others have believed that the diurnal mean temperature is a better predictor of plant response to high temperature as day and night temperatures do not affect the plant growth and development independently (Sharma et al. 2017). In many cases, there may be adequate soil moisture but the negative water balance created by heat stress lead to withering of plants. Alteration of photorespiratory activities due to diurnal variation in air temperature also impacts normal plant growth and development. The total heat units accumulated during the growth phase of plants influence the physiology, reproduction and maturity of crops. Wheat is presently grown on approx.30 Mha in India, and in this ~3.5 Mha of this area faces heat stress (Joshi et al. 2007). Globally, 36 Mha of the area under wheat (40% of the temperature environment) is subjected to heat stress (terminal heat in irrigated environments) (Sharma et al. 2019).

## 2 Optimum Temperature Required for Different Growth Stages in Spring Wheat

Temperature acts as a critical factor in regulating the developmental and growth stages of wheat ultimately determining the yield and biomass. Wheat is cultivated globally over large geographical regions with a varied temperature range where it can withstand. The lower and higher limits of temperature are −17 ± 1.2°C and 47.5 ± 0.5°C, respectively, at which wheat can survive (Abhinandan et al. 2018). The most optimum temperature for growth and development of the wheat plant is considered to be around 21 to 24°C; the optimal temperature range may vary depending on the prevailing agroclimatic situation at the place where the crop is grown. It requires different temperature at different stages of crop growth and development. Temperature sensitivity varies not only between plant components (Musich et al. 1981) but also changes during the course of development. Thus base and optimum temperature thresholds increase with development (Lumsden 1980; Angus et al. 1981; Slafer and Savin 1991; Slafer and Rawson 1994). Temperature requirement may slightly differ from one variety to another. At the time of germination, the optimum temperature required is 20–25°C. Root growth in upper soil layers is highly variable because of fluctuations in diurnal temperature. The optimal soil temperature for growth of the roots of wheat plants during the vegetative stage is below 20°C (Nielsen and Humphries 1966; MacDowell 1973). Temperatures higher than 35°C have been shown to reduce terminal root growth and accelerate its senescence (Wardlaw and Moncur 1995). Root growth may cease altogether if soil temperatures drop below 2°C (Petr 1991). Cao and Moss 1989 found optimum temp. for leaf emergence ranges from 21.3°C to 24.3°C. Temperatures higher than 25°C have been found to inhibit leaf appearance. For stem growth the optimum temp. requirement is 20–21°C. For culm elongation the optimum temp. requirement is 20°C. Certain winter varieties, which have chilling requirement in order to flower respond positively to cold temperatures. A synthesis of 11 studies revealed that optimum vernalization temperatures lie between 3.8 and 6.0°C. For tillering, the optimum temp. requirement is 6–9°C. For heading and anthesis, optimum temp. requirements are 24°C and 18–24°C (Russell and Wilson 1994), respectively. The temperature sensitivity of the reproductive phase has important implications for grain yield. The number of grains produced is a function of both the number of spikelets and the number of kernels per spikelet. Temperatures above 31°C immediately before anthesis reduce grain yield by inducing pollen sterility, thus reducing grain numbers (Wheeler et al. 1996). For grain filling stage, the optimum temp. requirement is 15–20°C, and for maturity the optimum temp. required is 22–25°C (Table 1).

**Table 1** Optimum temperature requirement in crucial wheat growth stages

| Crop growth stage | Zadoks scale | Optimum temperature range |
|---|---|---|
| Germination to emergence | 0–7 | 20–25°C |
| Crown root initiation/tillering | 21–29 | 6–9°C |
| Heading to anthesis | 51–69 | 18–24°C |
| Grain filling | 71–89 | 20–25°C |

## 3 Effect of High Temperature on Different Growth Stages of Wheat

The reproductive phase is more sensitive to high temperature stress than vegetative stage as it is directly related to grain number and size (Wollenweber et al. 2003). A simulation study predicted the significant loss in wheat yield under tropical and temperate regions with every 2°C rise in temperature (Challinor et al. 2014). However, the effect of temperature is significantly relying on the developmental stage of wheat that is subjected to temperature stress.

### *Germination Stage*

Wheat germination required low temperature and high soil moisture content. The temperature stress during the germination causes delay in germination or no germination of seed because of altered surrounding soil temperature and moisture content. Low germination rate directly reflects the crop density leading to significant yield loss.

### *Vegetative Stage*

The heat stress reduced the vegetative phase (Mishra et al. 2003). Plant height is varietal trait however; it is highly affected by high temperature during plant growth (Begum and Nessa 2014). Wheat flag leaf and yield have positive correlation as shown in other cereal crops such as rice, maize, barley, oat, etc. (Simon 1999; Begum and Nessa 2014). Flag leaf area significantly contributes for supplying photosynthetic product for grain development. A significant decrease in leaf area and leaf dry weight was reported at high temperature (Campbell and Read 1968). At 35°C, the wheat showed 8 to 36% reduction in flag leaf area (Begum and Nessa 2014).

### *Flowering Stage*

Anthesis and panicle emergence are thought to be most prone stages to the temperature stress. However, the priming of wheat plants pre-anthesis results in less severe post-anthesis damage (Wang et al. 2011). The high temperature affects the pollen quality and viability as it perturbs the metabolic regulations for pollen development (Hays et al. 2007). During anthesis and grain filling, with the raise of each degree temperature above optimum causes 6% reduction in wheat production (Asseng et al. 2015; Tewolde et al. 2006).

## *Post-fertilization and Grain Filling Stage*

High temperature shortened the grain filling duration, while grain filling rate is increased. In wheat, a 5°C increase in temperature above 20°C increased the grain filling rate and reduced grain filling duration by 12 days (Yin et al. 2009). If the post-anthesis temperature goes above 30°C, it reduced the grain filling rate (Al-Khatib and Pauben 1984). The yield reduces up to 23% by inducing heat stress of 32°C for 4 days during grain filling period (Stone and Nicolas 1994). However, the yield contributing factors such as spike length, number of spikelet per spike, number of seed per spike and thousand grain weight were found to be significantly reduced even at 28°C. The wheat plant failed to produce seed at 35°C due to sterile florets (Begum and Nessa 2014).

It is reported that each 1°C increase causes reduction in grain-filling duration (GFD) by 2.8 d (Streck 2005), grain number by 4% (Fischer 1985), grain weight by 11% (Mohammadi 2012) and grain yield by 6% (Asseng et al. 2015). The rice (*Oryza sativa* L.) wheat cropping system of the South Asia causes delay in wheat planting. Heat stress resulting from delaying sowing by 1 month can lead to about 20 to 30% loss in grain yield, depending on the climatic conditions (Rane et al. 2007). Delayed sown wheat crop experiences high temperature at the vegetative stage in subtropical countries like India and Pakistan and thus produced few productive tillers (Hossain et al. 2013). Heat stress during the reproductive phase speeds up the development of spikes and can reduce grain number by 63% (Gibson and Paulsen 1999). Under heat stress, the wheat crop completes its life cycle much quicker and can reduce the grain-flling duration by 12 d (Yin et al. 2009). Heat stress also adversely affects biomass (Alam et al. 2014).

## *Grain Number and Weight*

Both grain number and weight are sensitive to elevated temperature. Influence of temperature on each of these components of grain yield depends on the developmental phase at which the elevated temperature occurs. For instance, between spike initiation and anthesis, temperatures above 20°C may substantially reduce grain number per spike. Several events during this phenostage influence grain number including spikelet initiation, floral organ differentiation, male and female sporogenesis, pollination and fertilization. When abiotic stress coincides with meiosis, the first phase of gametogenesis may be further impaired. According to Ferris et al. 1998, warmer maximum temperatures over four consecutive days close to anthesis directly reduces grain number and, as a consequence, grain yield at maturity. Heat stress around floral initiation has severe effects on grain number. For instance, grain number per spike decreased by 4% for every 1°C (from 15–22°C) increase in temperature at 30 days preceding anthesis (Fischer 1985).

### Quality of Wheat Grains at Higher Temperature

The photosynthesis under optimum environmental conditions including temperature assimilates the carbon in the form of energy molecules such as starch, protein, fat, etc. At elevated temperature, the rubisco efficiency reduced and ultimately reduction in the carbon assimilation leads to decrease in the productivity. The elevated temperature at grain filling stage had significant effect on starch and protein content and their composition in the wheat kernels. Starch accounts for up to 70% of wheat grain dry weight. Its accumulation reduces up to 30% during the endosperm development at temperatures between 30°C and 40°C (Stone and Nicolas 1995). B-type starch granules are highly sensitive to high temperature stress as compared to the A-type granules. The B-type granules significantly decreased in response to elevated temperature, whereas A-type granules increased during grain-filling period (Blumenthal et al. 1995). Heat stress increases the grain protein; however, the total grain protein content remained low as heat stress reduces the grain yield (Daniel and Triboi 2000). Hence, the temperature is a critical factor determining the sowing and harvesting time as well as responsible for severe losses in crop yield and quality due to temperature fluctuations during growing season (Yang et al. 2017).

## 4 Adaptive Mechanisms for Heat Stress in Wheat

Wheat have three different mechanisms to adapt for heat stress conditions, namely, heat avoidance, heat tolerance and heat escape mechanisms. These mechanisms are in turn governed by different associated traits (Fig. 1), viz. leaf rolling (e.g. DBW17), waxiness in heat avoidance, production of stress responsive proteins in tolerance (e.g. RAJ3765) and adjusting the phenology of plants once it senses the heat stress to escape (e.g. Halna). Under heat stress condition, early maturation is closely correlated with smaller yield losses, which may be accredited to the engagement of an escape mechanism.

Plants tend to reduce heat-induced damage by leaf rolling, leaf shedding, reducing leaf size, thickening leaves, reducing growth duration, transpirational cooling and other adjustments in morphology and ontogeny (Wahid et al. 2007). Plant responses to heat stress are mediated by an intrinsic capacity to endure basal thermo tolerance and the ability to gain thermo tolerance after acclimation.

Heat tolerance mechanism is commonly known as plant can grow and produce economic yield under heat stress condition. Some main adoptive mechanisms at cellular level includes ion transporters, late embryogenesis abundant proteins, osmoprotectants, ROS defence and many other significant factors involved in cell signaling pathways and transcriptional control (Sairam et al. 2000). Thermotolerance plays a very important role in modification of plant water relations and phytohormones, increase of photosynthetic capacity, pollen tube development and metabolic activities (Almeselmani et al. 2009).

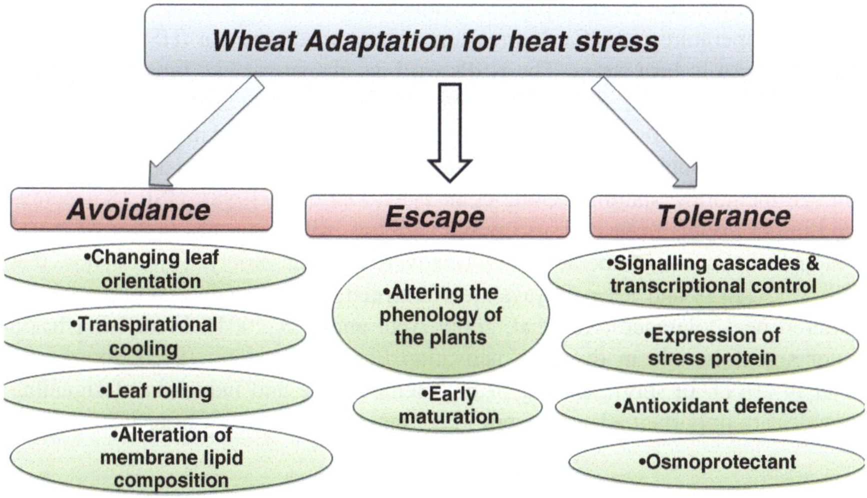


**Fig. 1** Adaptation mechanisms of wheat plants to high temperature

At extreme high temperature stress, plants produce high level of oxidative stress which is inhibited by protective response mechanism. Wheat must be protected from thermo-induced oxidative stress so they can survive under heat tolerance. This tolerance capacity has been linked with the induced antioxidative capacity. ROS such as superoxide radicals, hydroxyl radicals and hydrogen peroxide are formed into the cells in a normal manner, but increase in the production of these compounds can be dangerous to the cells (Srivastava et al. 2012; Esfandiari et al. 2007). Heat stress activates the production and collection of ROS (Sairam et al. 2000; Mittler 2002; Almeselmani et al. 2009). Hence, their detoxification by antioxidant systems is important for protecting plants against heat stress. The antioxidant defence system in wheat involves *enzymatic antioxidants* such as ascorbate peroxidise(APX), dehydroascorbate reductase, glutathione S-transferase, superoxide dismutase(SOD), catalase (CAT), guaiacol peroxidase and glutathione reductase (GR) and *non-enzymatic* antioxidants such as glutathione, ascorbate and tocopherols (Sairam et al. 2000; Mittler 2002). Some more studies shown that defence mechanism like oxidative compounds, which helped to avoid the gathering of ROS, membrane lipid peroxidation and protection of high cell membrane stability, plays a major role. Lipid peroxidation is measured as one of the most destructive processes (Almeselmani et al. 2009). MDA content is considered as the degree of damage at negative environment and is a marker of lipid peroxidation (Mishra et al. 2017). In wheat plant, MDA has been increased threefold under heat stress condition. Cell membrane is the first line of defence mechanism. Cell membrane has many heat-responsive proteins which helps plant to enhance its defence mechanism against heat stress.

Expression of stress proteins is a major adaptation to manage with abiotic stresses like heat stress. Most of the stress proteins are soluble in water (Rodríguez et al. 2005). These stress proteins contribute to stress tolerance apparently via hydration of cellular structures. Normal protein synthesis in wheat is reduced when exposed to high temperature (>35°C). Expression of heat shock protein (HSP) genes is a basic response to heat stress. The HSPs work as chaperone like functions and are concerned with signal transduction in heat stress (Queitsch et al. 2000; Schöffl et al. 1999; Wang et al. 2004; Sharma et al. 2015). The HSPs involved in many physiological phenomena such as photosynthesis, assimilate partitioning, water use efficiency and cell membrane stability (Gong et al. 1997; Dat et al. 1998). HSP 18 accumulated in developing grains in susceptible varieties, whereas it has higher HSP 100 content at increased temperature in a relatively tolerant variety. Studies show that HSP, ABA, ROS and SA pathways are concerned in the growth and protection of acquired heat tolerance (Xu et al. 2006; Apel and Hirt 2004). Methyl-SA has a major signaling role in the gene activation under heat stress (Chakraborty and Pradhan 2011). In short, sensing of high temperature and induction of signaling cascades are important adaptive steps in coping with adversaries of heat stress.

## 5 Phenotyping Methods for Heat Stress

The phenotyping for heat stress is broadly classified into lab-level screening and field-level screening.

### *Lab Screening*

The temperature induction response (TIR) technique is one of the potential screening methods to evaluate genetic variability for intrinsic heat stress tolerance in wheat genotypes. It also explains whether seedling-level tolerance correlates with adult-level tolerance in field. It involves primarily identification of challenging temperature to screen the wheat genotypes. Seedlings are initially exposed to mild temperature followed by a severe challenging temperature. Thus, only the tolerant genotypes will survive during recovery, whereas susceptible ones do not. It has been proved by considering five heat-tolerant (HT) wheat genotypes, namely, Raj3765, WH730, WH147, K7903 (Halna), HD2967, and one heat-sensitive (HS) genotype Raj4014. They were used for identifying the lethal temperature to differentiate between HT and HS wheat genotypes. The wheat seedlings were grown in paper cups in peat for 27 days at 22–24°C with 16/8 h (light/dark) photoperiod, and later they were challenged for increase in temperature for different duration, and finally it has been standardized that a heat treatment of 40°C for 28 h as lethal temperature to differentiate heat-tolerant and heat-sensitive wheat genotypes (Fig. 2, Mamrutha et al. 2015; Rinki et al. 2016). This has been cross-checked with other field trials also, and in all studies seedling tolerance has correlated with adult plant tolerance under field conditions. Hence, this temperature and duration can be used in identifying the true heat-tolerant genotypes in the RILs population developed for heat stress

**Fig. 2** TIR technique standardized in wheat seedlings

at initial seedling stage itself in wheat breeding programme. So that field work load can be reduced.

## *Field Screening*

The phenotyping for heat stress under field condition is routinely done by comparing different traits under timely sowing (mid-November) and late sowing conditions (mid-December). The heat sensitivity index (HSI) is routinely used for identifying heat-tolerant wheat genotypes. The HSI is calculated by the method suggested by Fischer and Maurer 1978 with the following formula, $HSI = (1 - Xh/X)/(1 - Yh/Y)$, where Xh and X are the phenotypic means for each genotype under stress and control conditions, respectively, and Yh and Y are the phenotypic means for all genotypes under stress and control conditions, respectively. The genotypes with HSI score of 1 and below 1 fall under the tolerant category and those with values greater than 1 fall under the susceptible category (ICAR-IIWBR 2019). Multilocation testing in the target hotspot environments is another option towards phenotyping stresses, since they help to obtain response to the stresses under varied natural conditions (Rane et al. 2007).

## *Novel and Precision Field Phenotyping*

Several efforts have been made to decipher traits/genes responsible for imparting high temperature tolerance in wheat. Both controlled and field-based studies have been undertaken in this regard. Lack of sufficient precision in simulating the ambient temperature dynamics and micro-environments prevailing in the field or repeatability of results in the field has been the severe bottlenecks. Hence, at ICAR-IIWBR, Karnal, India, a phenotyping method for screening wheat genotypes under high temperature using state-of-the-art temperature controlled phenotyping facility (TCPF) was developed, which ensures uniform crop stand. This allows screening of several wheat genotypes in a large plot size (simulating the fields) at a desired temperature at any stage of crop growth while allowing plants to grow in the natural environment during rest of the period. To maintain the diurnal cycle during temperature stress treatment, temperature regulation in TCPF is manipulated based on the ambient temperature so that the desired difference between the temperature inside and outside the structure is maintained. A boiler-based heating system is utilized for increasing temperature in which the warm water runs through a network of pipelines hanging from the roof with several inlets and outlets that avoids formation of temperature gradient from one end to another in the structure.

Integrated and automatically governed split air conditioners run through a control panel for cooling purposes. To maintain required humidity levels, a mist system provides fine-water droplets, and the drip system provides irrigation. Once the required temperature for stress treatment is over, the structure gets open and the crop again gets natural environment. There is a clear advantage of this structure in differentiating high temperature response of a large number of genotypes of wheat with greater precision (Sharma et al. 2016, 2019).

In general polyhouses/glasshouses are also used to screen for heat tolerance. Here fixed higher temperature is maintained in the glass house compared to control condition to identify heat tolerant wheat genotype.

## 6 High Temperature Stress Mitigation Strategies Under Field

There are several reports suggesting how we can reduce the effect of high temperature under field conditions. Using sprinkler irrigation to cool down, the canopy in the afternoon whenever the temperature goes beyond 30°C improves productivity. For addressing the early heat stress at tillering stage, need-based light irrigation can be applied. In addition, adopting conservation agriculture practices helps in mitigating the temperature stress by moderating temperature variations, conserving soil moisture and improving soil organic matter status. Early sowing or timely sowing of the crop helps in escaping terminal heat stress and also leads to saving water required for pre-sowing irrigation in wheat crop by utilizing the residual moisture available in soil after harvesting the previous crop (Rane et al. 2007). KCL (0.2%)/ $Cacl_2$ (0.2%) spraying at booting and pre-anthesis stage and whenever the temperature goes beyond 30°C is known to have yield advantage in wheat.

## 7 The Way Forward

If we compare the weather conditions across years in India, there are lot of variations in minimum and maximum temperature across zones. However, still India is harvesting record yield production from past 3 years continuously. Hence, it is still challenging to know how these variations in temperature are contributing for yield. Is there any other factor in combination with temperature is playing role under heat stress needs to be explored. Designated hotspots for screening under high temperature needs to be relooked in view of climate change over years. The advanced technologies like CRISPR/Cas9 also need to be explored to address and improve high temperature stress tolerance in wheat.

## References

Abhinandan K, Skori L, Stanic M, Hickerson NMN, Jamshed M, Samuel MA (2018) Abiotic stress signaling in wheat – an inclusive overview of hormonal interactions during abiotic stress responses in wheat. Front Plant Sci 9:734

Alam MN, Bodruzzaman M, Hossain MM, Sadekuzzaman M (2014) Growth performance of spring wheat under heat stress conditions. Int J Agric Res 4:91–103

Al-Khatib K, Pauben GM (1984) Mode of high temperature injury to wheat during grain development. Physiol Plant 61:363–368

Almeselmani M, Deshmukh PS, Sairam RK (2009) High temperature stress tolerance in wheat genotypes: role of antioxidant defence enzymes. Acta Agronomica Hungarica 57:1–14

Angus JF, MacKenzie DH, Morton R, Schafer CA (1981) Phasic development in field crops. II. Thermal and photoperiodic responses of spring wheat. Field Crops Res 4:269–283

Apel K, Hirt H (2004) Reactive oxygen species: metabolism, oxidative stress and signal transduction. Ann Rev Plant Biol 55:373–399

Asseng S, Ewert F, Martre P, Rötter RP, Lobell DB, Cammarano D (2015) Rising temperatures reduce global wheat production. Nat Clim Chang 5:143–147

Battisti DS, Naylor RL (2009) Historical warnings of future food insecurity with unprecedented seasonal heat. Science 323:240–244

Begum F, Nessa A (2014) Effects of temperature on some physiological traits of wheat. J Bangladesh Acad Sci 38(2):103–110

Blumenthal C, Bekes F, Gras PW, Barlow EW, Wrigley CW (1995) Identification of wheat genotypes tolerant to the effects of heat stress on grain quality. Cereal Chem 72:539–544

Campbell CA, Read DWL (1968) Influence of air temperature, high intensity and soil moisture on the growth, yield and some growth analysis characteristics of wheat grown in the growth chamber. Can J Plant Sci 48:299–311

Chakraborty U, Pradhan D (2011) High temperature-induced oxidative stress in Lens culinaris, role of antioxidants and amelioration of stress by chemical pre-treatments. J Plant Interact 6:43–52

Challinor AJ, Watson J, Lobell DB, Howden SM, Smith DR, Chhetri N (2014) A meta-analysis of crop yield under climate change and adaptation. Nat Clim Chang 4:287–291. https://doi.org/10.1038/nclimate2153

Daniel C, Triboi E (2000) Effects of temperature and nitrogen nutrition on the grain composition of winter wheat: effects on gliadin content and composition. J Cereal Sci 32:45–56

Dat J, Foyer C, Scott I (1998) Change in salicylic acid and antioxidants during induced thermo tolerance in mustard seedlings. Plant Physiol 118:1455–1461

Esfandiari E, Fariborz S, Shekari F, Manouchehr E (2007) The effect of salt stress on antioxidant enzymes' activity and lipid peroxidation on the wheat seedling. Notulae Botanicae Horti Agrobotanici Cluj-Napoca 35

Essemine J, Ammar S, Bouzid S (2010) Impact of heat stress on germination and growth in higher plants: physiological, biochemical and molecular repercussions and mechanisms of defence. J Biol Sci 10(6):565–572

Ferris R, Ellis RH, Wheeler TR, Hadley P (1998) Effect of high temperature stress at anthesis on grain yield and biomass of field-grown crops of wheat. Ann Bot 82:631–639

Fischer RA (1985) Number of kernels in wheat crops and the influence of solar radiation and temperature. J Agric Sci 105:447–461

Fischer RA, Maurer R (1978) Drought resistance in spring wheat cultivars. Grain-yield responses. Aust J Agric Res 29:897–912

Gibson LR, Paulsen GM (1999) Yield components of wheat grown under high temperature stress during reproductive growth. Crop Sci 39

Gong M, Chen SN, Song YQ, Li ZG (1997) Effect of calcium and calmodulin on intrinsic heat tolerance in relation to antioxidant systems in maize seedlings. Aust J Plant Physiol 24:371–379

Greaves JA (1996) Improving sub optimal temperature tolerance in maize – the search for variation. J Exp Bot 47:307–323

Hall AE (2001) Crop responses to environment. CRC Press, Boca Raton

Hay RKM, Walker AJ (1989) An introduction to the physiology of crop yield. Longman Scientific & Technical, New York, p 292

Hays DB, Do JH, Mason RE, Morgan G, Finlayson SA (2007) Heat stress induced ethylene production in developing wheat grains induces kernel abortion and increased maturation in a susceptible cultivar. Plant Sci 172:1113–1123. https://doi.org/10.1016/j.plantsci.2007.03.004

Hossain A, Sarker MAZ, Saifuzzaman M, Teixeira da Silva JA, Lozovskaya MV, Akhter MM (2013) Evaluation of growth, yield, relative performance and heat susceptibility of eight wheat (*Triticum aestivum* L.) genotypes grown under heat stress. Int J Plant Prod 7(3):615–636

ICAR-IIWBR (2019) Progress Report of AICRP on Wheat and Barley 2018-19, Crop Improvement. Ed(s): Gyanendra Singh, Ravish Chatrath, BS Tyagi, SK Singh, Arun Gupta, Satish Kumar, CN Mishra, K Venkatesh, Vikas Gupta, Charan Singh, Gopalareddy K, Hanif Khan, Raj Kumar, AK Sharma, Mamrutha HM, Bhumesh Kumar, Lokendra Kumar, Sindhu Sareen, Ratan Tiwari, Sonia Sheoran, Rinki, Ajay Verma, Suman Lata, Rekha Malik, OP Ahlawat, Rajender Singh, Pradeep Sharma and GP Singh. ICAR-Indian Institute of Wheat and Barley Research, Karnal, Haryana, India. p. 201

IPCC (2007) Climate Change Working Group I: The physical science basis. http://www.ipcc.ch/publications_and_data/ar4/wg1/en/contents.html. Accessed on Feb 2011

Joshi AK, Mishra B, Chatrath R, Ferrara GO, Singh RP (2007) Wheat improvement in India: present status, emerging challenges and future prospects. Euphytica 157:431–446. https://doi.org/10.1007/s10681-007-9385-7

Lumsden ME (1980) The influence of weather on the development of winter wheat. BSc report. University of Bath, UK

MacDowell FDH (1973) Growth kinetics of Marquis wheat. 5. Morphogenic dependence. Can J Bot 51(7):1259–1265

Mamrutha HM, Rinki, Kumar R, Kaur A, Yadav VK (2015) Standardization of temperature induction response technique in wheat. Wheat Barley News Lett 9(1&2):11

Meehl GA, Stocker TF, Collins WD, Friedlingstein, PIERRE, Gaye AT, Gregory JM, Kitoh A, Raper SC (2007) Global climate projections. Climate change 2007: the physical science basis. Contribution of Working Group I to the Fourth Assessment Report of the Intergovernmental Panel on Climate Change 747–845

Mishra V, Mishra RD, Singh M, Verma RS (2003) Dry matter accumulation at pre and post anthesis and yield of wheat (*Triticum aestivum*) as affected by temperature stress and genotypes. Indian J Agron 48(4):27–281

Mishra D, Shekhar S, Agrawal L, Chakraborty S, Chakraborty N (2017) Cultivar-specific high temperature stress responses in bread wheat (Triticum aestivum L.) associated with physico-chemical traits and defense pathways. Food Chem 221:1077–1087

Mittler R (2002) Oxidative stress, antioxidants and stress tolerance. Trends Plant Sci 7(9):405–410

Mohammadi M (2012) Effects of kernel weight and source-limitation on wheat grain yield under heat stress. Afr J Biotech 11:2931–2293

Musich VN, Maistrenko GG, Kolot GA (1981) Dynamics of frost resistance in winter wheat in relation to temperature regime. Nauchno tekhnicheskii Byulleten' Vsesoyuzonogo Selektsionno geneticheskogo Instituta 1:14–19

Nielsen KF, Humphries EC (1966) Effect of temperature on plant growth. Soils Fert 29(1):1–7

Petr J (1991) Weather and yield. Developments in crop science 20: 12–282

Porter J, Semenov M (2005) Crop responses to climatic variation. Philos Trans R Soc B Biol Sci 360:2021–2035

Queitsch C, Hong SW, Vierling E, Lindquist S (2000) Hsp101 plays a crucial role in thermotolerance in Arabidopsis. Plant Cell 12:479–492

Rane J, Pannu RK, Sohu VS, Singh R, Banwari S, Mishra J, Shoran J, Crossa M, Vargas M, Joshi AK (2007) Performance of yield and stability of advanced wheat genotypes under heat stress environments of the Indo-Gangetic plains. Crop Sci 47:1561–1573

Rinki, Mamrutha HM, Kumar R, Tiwari V (2016) Comparison of seedling and adult stage heat stress tolerance in wheat. Wheat Barley Newslett 10(1):9

Rodríguez M, Canales E, Borrás-Hidalgo O (2005) Molecular aspects of abiotic stress in plants. Biotechnol Appl 22:1–10

Russell G, Wilson GW (1994) An agro-pedo-climatological knowledge-base of wheat in Europe. Joint Research Centre European Commission, Luxembourg, p 158

Sairam RK, Srivastava GC, Saxena DC (2000) Increased antioxidant activity under elevated temperatures: a mechanism of heat stress tolerance in wheat genotypes. Biologia Plantarum 43:245–251

Schöffl F, Prand R, Reind A (1999) Molecular responses to heat stress. In: Shinozaki K, Yamaguchi-Shinozaki K (eds) Molecular responses to cold, drought, heat and salt stress in higher plants. RG Landes Co, Austin, pp 81–98

Shrama I, Tyagi BS, Singh G, Venkatesh K, Gupta OP (2015) Enhancing wheat production- a global perspective. Indian J Agril Sci 85:3–13

Sharma D, Singh R, Rane J, Gupta VK, Mamrutha HM, Tiwari R (2016) Mapping quantitative trait loci associated with grain filling duration and grain number under terminal heat stress in bread wheat (*Triticum aestivum L.*). Plant Breed 135:538–545. https://doi.org/10.1111/pbr.12405

Sharma IP, Kumar N, Bhatt S (2017) Heat stress in agriculture – the problem. Indian Farm Digest 50:13–14

Sharma D, Pandey GC, Mamrutha HM, Singh R, Singh NK, Singh GP, Rane J, Tiwari R (2019) Genotype-phenotype relationship for high-temperature tolerance: an integrated method for minimizing phenotyping constraints in wheat. Crop Sci 59

Simon MR (1999) Inheritance of flag leaf angle, flag leaf area and flag angle, leaf area duration in four wheat crosser. Theor Appl Genet 98(2):310–314

Slafer GA, Rawson HM (1994) Sensitivity of wheat phasic development to major environmental factors: a re-examination of some assumptions made by physiologists and modelers. Aust J Plant Physiol 21:393–426

Slafer GA, Savin R (1991) Developmental base temperature in different phenological phases of wheat (*Triticum aestivum*). J Exp Bot 42(241):1077–1082

Srivastava S, Pathak AD, Gupta PS, Shrivastava AK, Srivastava AK (2012) Hydrogen peroxide-scavenging enzymes impart tolerance to high temperature induced oxidative stress in sugarcane. J Environ Biol 33:657–661

Stone PJ, Nicolas ME (1994) Wheat cultivars vary widely in their responses of grain yield and quality to short periods of postanthesis heat stress. Aust J Plant Physiol 21:887–900

Stone PJ, Nicolas ME (1995) A survey of the effects of high temperature during grain filling on yield and quality of 75 wheat cultivars. Aust J Agric Res 46:475–492

Streck NA (2005) Climate change and agroecosystems: the effect of elevated CO2 and temperature on crop growth, development, and yield. Ciênc Rural 35:730–740

Tewolde H, Fernandez CJ, Erickson CA (2006) Wheat cultivars adapted to post-heading high temperature stress. J Agron Crop Sci 192:111–120

Tubiello FN, Soussana JFO, Howden SM (2007) Crop and pasture response to climate change. Proc Natl Acad Sci 104:19686–19690

Wahid AS, Galani S, Ashraf M, Foolad M (2007) Heat tolerance in plants: An overview. Environ and Exp Bot 61: 199–223

Wang W, Vinocur B, Shoseyov O, Altman (2004) A role of plant heat-shock proteins and molecular chaperones in the abiotic stress response. Trends Plant Sci 9:244–252

Wang X, Cai J, Jiang D, Liu F, Dai T, Cao W (2011) Pre-anthesis high-temperature acclimation alleviates damage to the flag leaf caused by postanthesis heat stress in wheat. J Plant Physiol 168:585–593. https://doi.org/10.1016/j.jplph.2010.09.016

Wardlaw IF, Moncur L (1995) The response of wheat to high temperature following anthesis. I. The rate and duration of kernel filling. Aust J Plant Physiol 22:391–397

Wheeler TR, Batts GR, Ellis RH, Hadley P, Morison JI (1996) Growth and yield of winter wheat (*Triticum aestivum*) crops in response to CO2 and temperature. J Agric Sci 127:37–48

Wollenweber B, Porter JR, Schellberg J (2003) Lack of interaction between extreme high temperature events at vegetative and reproductive growth stages in wheat. J Agron Crop Sci 189:142–150

Xu S, Li J, Zhang X, Wei H, Cui L (2006) Effects of heat acclimation pretreatment on changes of membrane lipid peroxidation, antioxidant metabolites, and ultrastructure of chloroplasts in two cool-season turfgrass species under heat stress. Environ Exp Bot 56:274–285

Yang Z, Zhang Z, Zhang T, Fahad S, Cui K, Nie L (2017) The effect of season-long temperature increases on rice cultivars grown in the central and southern regions of China. Front Plant Sci 8:1908. https://doi.org/10.3389/fpls.2017.01908

Yin X, Guo W, Spiertz JH (2009) A quantitative approach to characterize sink-source relationships during grain filling in contrasting wheat genotypes. Field Crop Res 114:119–126

# Advances in Molecular Markers and Their Use in Genetic Improvement of Wheat

**Sachin Kumar, Manoj Kumar, Reyazul Rouf Mir, Rahul Kumar, and Sourabh Kumar**

**Abstract** Molecular markers such as RFLP, RAPD, AFLP, CAPS, DArT, SSR, SNP, etc. have been widely used in wheat genetic studies but have their own limitations. More recent types of molecular makers are the improved versions of some of the already available techniques due to the developments in the areas of next-generation sequencing, high-throughput genotyping, detection procedures and bioinformatics applications. At present, SNP markers have become a good choice due to their abundance in the genome, codominant inheritance, locus specificity, flexibility for high-throughput genotyping/detection formats and being relatively inexpensive. The next-generation sequencing (NGS) technologies have facilitated the discovery of a large number of SNPs for the development of high-resolution genetic maps, QTL/gene discovery and marker-assisted introgression, thereby improving the efficiency in wheat breeding. In this chapter, we attempt to review the recent advancements made in the area of molecular marker technologies such as hybridization-based platforms (fixed array), genotyping-by-sequencing (de novo), KASP genotyping, etc. The molecular markers developed with these advanced technologies in wheat offer easier means to map polymorphic genetic loci at highest density that facilitate enriching chromosomal regions to identify QTLs and candidate genes underlying important traits.

**Keywords** Wheat · Molecular markers · Single-nucleotide polymorphism Array-based markers · Genotyping-by-sequencing · Genetic improvement

---

S. Kumar (✉) · M. Kumar · R. Kumar · S. Kumar
Department of Genetics and Plant Breeding, Chaudhary Charan Singh University, Meerut, UP, India

R. R. Mir
Division of Genetics and Plant Breeding, Sher-e-Kashmir University of Agricultural Sciences and Technology, Sopore, Kashmir, India

© The Editor(s) (if applicable) and The Author(s), under exclusive license to Springer Nature Switzerland AG 2021

S. H. Wani et al. (eds.), *Physiological, Molecular, and Genetic Perspectives of Wheat Improvement*, https://doi.org/10.1007/978-3-030-59577-7_8

## 1 Introduction

Common or bread wheat is an allohexaploid species (*Triticum aestivum*; 2n = 6X = 42; AABBDD genomes) evolved through two separate episodes of polyploidization. In the first episode, two diploid ancestral species, *Triticum urartu* (AA genome) and an unknown *Aegilops* species (BB genome) closely related to *Aegilops speltoides* (SS genome), hybridized to create the allotetraploid wheat (*Triticum turgidum* ssp. *dicoccum*; 2n = 4X = 28; AABB genomes) around 0.5 million years ago. Second episode occurred nearly 10,000 years ago when allotetraploid wheat spontaneously hybridized with an ancestral diploid species, *Ae. tauschii* (DD genome), and therefore genome D was introduced. Hexaploid wheat genome comprises 17 billion nucleotides of DNA packaged into 7 homologous groups. Each homologous group contains one pair of homologous chromosomes from the A, B and D genomes (Sears 1954; Feldman and Levy 2012). In other words, each homologous group of common wheat consists of three closely related chromosomes, one from each of the three (ABD) genomes. For instance, homologous group 3 holds three pairs 3A, 3B and 3D, each derived from common ancestral chromosome. Interestingly, hexaploid wheat behaves like diploids (homologous chromosomes do not pair with each other) during meiosis due to the action of *Ph1* gene located on the long arm of chromosome 5B that ensures meiotic pairing to homologous chromosomes (Riley and Chapman 1958).

Speciation events, domestication and breeding, have created wheat as a productive crop adapted to a variety of climates with increased yield potential. Presently, common wheat alone accounts for some 95% of global wheat production and is most widely cultivated (220 million ha) food crop that feeds >35% of the human population of the world (Shiferaw et al. 2013; IWGSC 2014). Thus, wheat is considered as one of the leading food crops for global food security. Wheat breeding is challenging because it deals with several traits related to abiotic and biotic stresses, grain quality (nutrition), root architecture, etc. The expression of these traits is massively influenced by genetic and environmental factors and their interactions. As a consequence, such traits are known to be quantitative or complex in their phenotypic expression. Achieving genetic improvement of wheat with the traits of quantitative nature remains a challenge for breeders and geneticists. Conventional breeding practices require at least a decade to develop a new variety without any promise that a new breed will be released as a superior variety. Essentially, it is desired to produce new improved varieties as swiftly and economically as possible. Breeders are therefore interested in adopting novel and efficient tools/techniques to achieve such targets.

Since the 1980s, diverse requirements of researchers, continuously emerging technologies, importance of crop species, DNA sequence databases, genomic abundance of polymorphic features, etc. altogether have contributed to the development of new molecular marker systems in plants including wheat. The impact of these developments has already become apparent when looking at the advancements in wheat genomics (Fig. 1). Current advances in DNA-based molecular marker

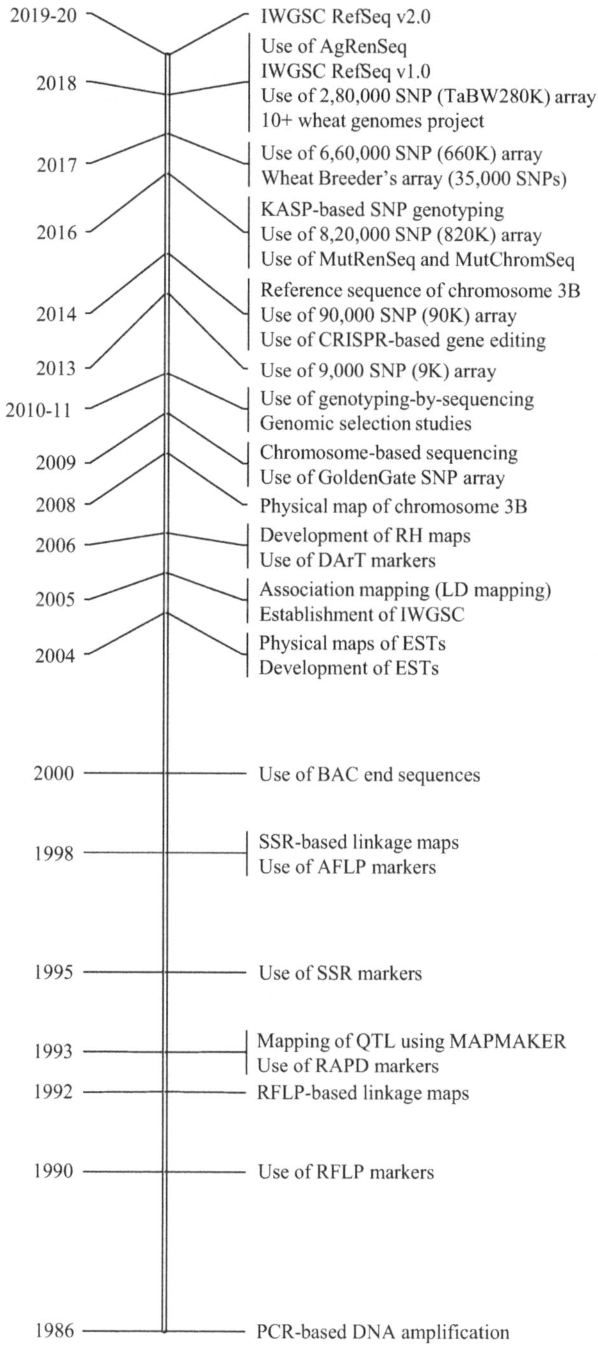


**Fig. 1** Growth of molecular marker systems in wheat. The timeline indicates the important events from the discovery of PCR to reference sequence of wheat genome

technologies, genotyping platforms and reference genome sequence have attracted the attention of practical breeders by providing them an ever-increasing amount of accessible DNA sequence information. The adoption of these advances allows breeders to scale up breeding process and improve precision in selection of plants carrying desirable genes and/or alleles and their favourable combinations. Emerging DNA sequence-based molecular markers can be used to characterize a large number of germplasm for sequence polymorphisms across whole genome in a single run. In this chapter, our main focus is to sum up the recent advancements and progresses in molecular marker technologies and their potential applications (i.e. genomics-assisted breeding) in wheat improvement.

## 2 Molecular Marker Systems in Wheat: An Overview

The three most important and unique features in the genome are about revealing single-nucleotide differences (transitions/transversions), insertions-deletions and variations in the number and size of tandem repeats at a particular locus. Genome-wide distributions of such features provide the key to use them as a molecular tag that can be used to identify an allelic variation of a gene or detect polymorphism in a particular fragment of DNA between two or more individuals (Gupta et al. 1999; Landjeva et al. 2007). Molecular mapping aids in assigning the location of these features/tags/molecular markers on the chromosomes. Simply speaking, DNA-based molecular markers are genetic tools that allow plant breeders and geneticists to identify and tag genomic regions (QTL or gene) for targeted traits within the genome, and their inheritance can be tracked from one generation to the next. Such marker systems have been considered more efficient than conventional methods of plant breeding because these molecular markers have ability to speed up breeding generations in the field and selection efficiency in the laboratory and cut down the cost of labour and phenotyping expenses (Langridge and Chalmers 2004). In order to precisely discriminate alleles of target genes, functional markers (gene-specific) have been developed to accelerate wheat breeding programs (Table 1). The available markers have been proved useful in fingerprinting, trait discovery, genome mapping, genome assembly, comparative mapping, gene cloning, alien gene transfer and marker-assisted selection in wheat breeding (Table 2). In agriculture, genetic improvement of crop plants would not have been possible without the development and use of molecular markers (Fig. 1). In addition, the breakthrough invention of polymerase chain reaction (PCR) technology in 1983 revolutionized the study of DNA profiling till today.

Ideal features of molecular markers and their subsequent applications have been extensively reviewed and discussed elsewhere (Gupta et al. 1999; Prasad et al. 2000; Langridge et al. 2001; Korzun and Ebmeyer 2003; Roder et al. 2004; Varshney et al. 2005; Rustgi et al. 2009; Jiang 2013; Amom and Nongdam 2017; Belete 2018). A genetic marker can be considered perfect if it is of highly polymorphic, codominant inheritance, neutral, uniformly dispersed within genome, reproducible, suitable for

**Table 1** List of available functional molecular markers linked to important genes in wheat

| S. No. | Functional marker | Gene name | Trait | References |
|---|---|---|---|---|
| 1 | *Rht-B1a*, *Rht-B1b*, *Rht-D1a* and *Rht-D1b* | *Rht-B1* and *Rht-D1* | Reduced plant height | Ellis et al. (2002) |
| 2 | *vrn-D1*, *vrn-H1*, *vrn-B3* and *vrn-A1* | *VRN-D1*, *VRN-H1*, *VRN-B3* and *VRN-A1* | Vernalization | Fu et al. (2005) |
| 3 | *PPO18* | *PPO* | Polyphenol oxidase activity | Sun et al. (2005) |
| 4 | SSR (*Pm3a* to *Pmg*) | *Pm3* | Powdery mildew | Tommasini et al. (2006) |
| 5 | *Ppd-D1a*, *Ppd-D1b* | *Ppd-D1* | Photoperiod sensitivity | Beales et al. (2007) |
| 6 | *Ppo-A1a*, *Ppo-A1b* | *Ppo-D1* | Polyphenol oxidase activity | He et al. (2007) |
| 7 | *YP7B-1*, *YP7B-2*, *YP7B-3* and *YP7B-4* | *Psy1* | Yellow pigment content | He et al. (2009) |
| 8 | Five In-Del and one SNP (*cssfr1–cssfr6*) | *Lr34/Yr18/Pm38* | Leaf rust, stripe rust and powdery mildew | Lagudah et al. (2009) |
| 9 | SNP | *Dreb1* | Drought tolerance | Wei et al. (2009) |
| 10 | *gluA3a*, *gluA3b*, *gluA3d*, *gluA3e*, *gluA3f*, *gluA3g* and *gluA3ac* | *Glu-A3* | Gluten content | Wang et al. (2010) |
| 11 | *TaGW2-6A* | *TaGW2-6A* | Grain weight | Su et al. (2011) |
| 12 | *Happa-H* and *Hap-L* | *TaSus2-2B* | Grain weight | Jiang et al. (2011) |
| 13 | *TaZds-D1a* and *TaZds-D1b* | *TaZds-D1* | Yellow pigment content | Zhang et al. (2011) |
| 14 | SNP *LOX16* and *LOX8* | *TaLox-B1* | Lipoxygenase activity | Geng et al. (2012) |
| 15 | *TaZds-A1a* and *TaZds-A1b* | *TaZds-A1* | Yellow pigment content (zeta-carotene) | Dong et al. (2012) |
| 16 | SNP | *TaMYB2* | Dehydration tolerance | Garg et al. (2012) |
| 17 | SNP | *TaAQP* | Drought tolerance | Pandey et al. (2013) |
| 18 | In-Del | *Sr45* | Stripe rust | Periyannan et al. (2014) |
| 19 | *POD-3A1* and *POD-3A2* | *TaPod-A1* | Peroxidase | Wei et al. (2015) |
| 20 | Two SNPs and one In-Del *TaMAMF/TaMAMR* | *TaMOC1-A* | Spikelet number per spike | Zhang et al. (2015) |
| 21 | SNP (*TaGS5-3A-T* and *TaGS5-3A-G*) | *TaGS5-3A* | Kernel size | Ma et al. (2016) |

(continued)

**Table 1** (continued)

| S. No. | Functional marker | Gene name | Trait | References |
|---|---|---|---|---|
| 22 | CAPS-SNP | *TaTGW6-A1* | Thousand grain weight | Hanif et al. (2016) |
| 23 | SNP and SSR *Xbarc62* | *TaELF3-1DL* | Heading | Wang et al. (2016) |
| 24 | *TaPARM1* and *TaPARM2* | *TaPARG* | Plant architecture and yield-related traits | Li et al. (2016) |
| 25 | KASP-SNPs (*S2269949* and *S1077313*) | *CBF-A14* under *Fr-A2* locus | Frost tolerance | Sieber et al. (2016) |
| 26 | *TaTPP6AL1-CAPS-F/R* | *TaTPP-6AL1* | Thousand grain weight | Zhang et al. (2017a, b) |
| 27 | *POD-7D1* and *POD-7D6* | *TaPod-D1* | Peroxidase | Geng et al. (2019) |
| 28 | KASP-SNP | *TaSnRK2.9-5A* | Metabolic regulations signaling (yield related) | Rehman et al. (2019) |
| 29 | *LCY-B1_3765*_SNP | *TaLcy-B1* | Lycopene content | Dong et al. (unpublished) |
| 30 | *PDS-B1*_SNP | *Pds-B1* | Phytoene desaturase | Dong et al. (unpublished) |

*KASP* Kompetitive Allele Specific PCR, *SSR* simple sequence repeat, *SNP* single-nucleotide polymorphism, *In-Del* insertion-deletion, *CAPS* cleaved amplified polymorphic sequences, *F* forward, *R* reverse

a range of applications and user-friendly. However, except in some cases, none of the molecular marker systems would have all the desirable features; depending on the type of study, a marker system can be preferred that would carry the required features. To simplify different marker types, genetic markers can be divided into two broad categories, (a) classical markers and (b) molecular markers. Classical markers comprise (i) morphological, (ii) biochemical or protein/enzyme and (iii) cytological markers. Depending upon features, methods of development, scale of throughput and genotyping/detection procedures, molecular or DNA sequence-based markers include (i) hybridization-based, (ii) PCR-based and (iii) sequencing-based markers. The classification of different genetic markers used in wheat breeding and genomic studies is presented in Fig. 2.

Over the last more than three decades, a continuous progress in the development of DNA-based markers and genotyping methods has provided valuable assistance in the efficient selection of economically important traits (Mir and Varshney 2013). DNA-based markers are considered far better over classical markers because they are abundant, neutral, reliable, convenient to automate and cost-effective. Among hybridization-based markers, restriction fragment length polymorphisms (RFLPs) were the first to be developed and used in wheat for genetic diversity analysis, construction of genetic maps and gene tagging (Chao et al. 1989). Initially, RFLP markers were used for the construction of genetic and physical maps in wheat. Such

**Table 2** List of some cloned genes in wheat. The table incudes linked markers, donor accession, the trait that gene controls and strategy used in cloning

| S. No. | Gene | Linked marker | Donor accession | Trait studied | Cloning strategy | References |
|---|---|---|---|---|---|---|
| 1. | *WX-7A*, *WX-4A* (translocated from 7B), *WX-7D* | Gene-specific primer design from cDNA sequence AB019622, AB019623 and AB019624 | Chinese Spring and nulli-tetrasomic lines (N7A/T7B and N7D/T7B) | Granule bond starch synthase (waxy protein gene) | Map-based cloning (In-Del) | Murai et al. (1999) |
| 2. | *Glu-1* | *Ax1*, *Ax1*, *Bx7*, *Bx17*, *Dx2*, *Dx5*, *By9*, *Dy10* and *Dy12* | GA-250 hexaploid triticale | Dough strength | Map-based cloning | De Bustos et al. (2001) |
| 3. | *pinA* | *Pina-D1a*, *b*, *c*, *d*; *Pinb-D1a*, *b*, *e*, *h*, *k*, *I* and *j* | *Triticum monococcum* and *T. urartu* | Grain hardness | Map-based cloning | Massa et al. (2004) and Guzmán et al. (2012) |
| 4. | *Pinb* | *Pinb-D1* | GaoCheng 8901 | Grain hardness | Map-based cloning | Gautier et al. (1994) and Pan et al. (2004) |
| 5. | *B* | *Xpsr680–7B* and *Xpsr160–7D* | Halberd | Boron toxicity | RFLP marker | Jefferies et al. (2000) |
| 6. | *Cre3* | *Xglk605* and *Xcdo588* | AUS10894 | Cereal cyst nematode resistance | RFLP marker | Ogbonnaya et al. (2001) |
| 7. | *Rlnn1* | *Xpsr121*, *Xpsr680* and *Xcdo347* | Excalibur | Root lesion nematode | RFLP marker | Williams et al. (2002) |
| 8. | *Lr10* | *Xrga1* and *Xrga2* | Hexaploid wheat lineThatcher*Lr10* | Leaf rust | Map-based cloning | Feuillet et al. (2003) |
| 9. | *Pm*, *Pm3b* | *WHS179* RFLP marker | Hexaploid wheat landrace Chul | Powdery mildew | Map-based cloning | Yahiaoui et al. (2004) |
| 10. | *Lr21* | KSUD14 | *Aegilops tauschii* accessions TA1649 and TA1599 | Leaf rust | Map-based cloning | Huang et al. (2003) |
| 11. | *VRN1* | *WG644* | *T. monococcum* ssp. *aegilopoides* accessions G2528 | Wheat vernalization gene | Map-based cloning | Yan et al. (2003) |

(continued)

**Table 2** (continued)

| S. No. | Gene | Linked marker | Donor accession | Trait studied | Cloning strategy | References |
|---|---|---|---|---|---|---|
| 12. | *Nax1* | *Xgwm312* and *Xwmc170* | Durum wheat line149*Tamaroi | Salt tolerance | Synteny-based cloning | Lindsay et al. (2004) |
| 13. | *R genes* | *Tamyb10-A1*, *B1*, *D1* (transcription factors) | AUS1490 | Red grain colour | Expression (transcription factors) | Himi and Noda (2005) |
| 14. | *TaNAM* | *Xuhw106* and *Xucw109* | *T. turgidum* ssp. *durum* cultivar Langdon (LDN) | *Gpc-B1* including Zn and Fe | RNAi expression based | Uauy et al. (2006) and Distelfeld et al. (2007) |
| 15. | *Lr1* | *Xpsr567* | Hexaploid wheat breeding line 87E03-S2B1 | Leaf rust | Map-based cloning | Cloutier et al. (2007) |
| 16. | *Psy1* | *YP7A* | Neixiang188 | Phytoene synthase enzyme (yellow pigment) | Synteny-based cloning | He et al. (2008) |
| 17. | *Ppd-D1 (2D)* | Ppd-D1_F and Ppd-D1_R1/ Ppd-D1_R2 | Ciano 67 2D | Photoperiod insensitive | BAC library | Beales et al. (2007) |
| 18. | *TaVp1* | *TaVp-A1*, *TaVp-B1* and *TaVp-D1* | Minamino | Seed dormancy | Synteny-based cloning | Utsugi et al. (2008) |
| 19. | *TaABI5* | *TaABI5-F/R* and *qTaABFI5-F/R* | SHW-L1 | PHST (ABI signaling) | Map-based cloning | Ohnishi et al. (2008) |
| 20. | *Glu-A1*, *Glu-D1* | *UMN19*, *UMN25* and *UMN26* | Nulli-tetrasomic lines of Chinese Spring | Glutenin content | Map-based cloning | Liu et al. (2008) |
| 21. | *Lr34/Yr18/ Sr57/ Pm38* | *Xgwm1220* and *SWM10* | Hexaploid wheat lines Thatcher *Lr34*, Avocet *Lr34*, Forno and Chinese Spring | Leaf rust, stripe rust, stem rust, powdery mildew | Map-based cloning | Krattinger et al. (2009) |
| 22. | *Yr36* | *Xucw129* and *Xucw148* | Wild emmer wheat accession FA15-3 | Stripe rust | Map-based cloning | Fu et al. (2009) |
| 23. | *Utd1* | *Xgwm234* and *Xgwm443* | D93213 and VIR51658 | Loose smut | Map-based cloning | Randhawa et al. (2009) |

| | | | | | | |
|---|---|---|---|---|---|---|
| 24. | *Tsn1* | *Xfcp623* | Durum wheat cultivar Langdon | *Stagonospora nodorum* blotch, tan spot | Map-based cloning | Faris et al. (2010) |
| 25. | *DOG-1* | *DOG1*-like genes | Norin 61 | Seed dormancy | Synteny-based cloning | Ashikawa et al. (2010) |
| 26. | *TaGW2* | *Xcfd80.2* | Three association panel | Grain weight | Synteny-based cloning | Su et al. (2011) |
| 27. | *TmMla1* | *sbi369* and *sbi314* | *T. monococcum* line DV92 | Powdery mildew | Synteny-based cloning | Jordan et al. (2011) |
| 28. | *Sus2* | *Xgwm122* and *Xgwm328* | 3 diploid and 61 common wheat accession | Sucrose synthase | Map-based cloning | Jiang et al. (2011) |
| 29. | *TaMFT-3A* | *CSZENSSR-F1* and *CSZENSSR-R1* | N61 and Chinese Spring (Zen3A) | Regulation of germination | Map-based cloning | Nakamura et al. (2011) |
| 30. | *TaCwi-A1* | *cwi21* and *cwi22* | Chinese varieties and landraces | Cell wall invertase (CWI) | Synteny-based cloning | Ma et al. (2012) |
| 31. | *Pm8* | *sfr43(Pm8)* | Rye line Petkus, wheat line Kavkaz/4*Federation | Powdery mildew | Synteny-based cloning | Hurni et al. (2013) |
| 32. | *Sr33* | *BE405778* and *BE499711* (EST markers) | *A. tauschii* accession RL5288 | Stem rust | Map-based cloning | Periyannan et al. (2013) |
| 33. | *Sr35* | *AK331487 (0.02 cM)* and *AK332451 (0.98 cM)* | *T. monococcum* line DV92 | Stem rust | Map-based cloning | Saintenac et al. (2013a, b) |
| 34. | *Yr10* | *Xpsp3000* | Hexaploid wheat cultivar Moro | Stripe rust | Map-based cloning | Liu et al. (2014) |
| 35. | *TaSdr-A1*, *TaSdr-B1* and *TaSdr-D1* | *Sdr-2*, *Sdr-3* and *Sdr-4* | Zhongyou 9507, Jing 411, Han 6172 and Heng 7228 | Seed dormancy | Map-based cloning | Zhang et al. (2014) |

(continued)

**Table 2** (continued)

| S. No. | Gene | Linked marker | Donor accession | Trait studied | Cloning strategy | References |
|---|---|---|---|---|---|---|
| 36. | *Sr50* | *Sr50-F1/R1* | Rye cultivar Imperial, wheat introgression line Gabo 1BL.1RS | Stem rust | Map-based cloning | Mago et al. (2015) |
| 37. | *Lr67/Yr46/ Sr55/ Pm46* | *Xgwm165* | Hexaploid wheat line Thatcher+*Lr67* | Leaf rust, stripe rust, stem rust, powdery mildew | Map-based cloning | Moore et al. (2015) |
| 38. | *Snn1* | *Xfcp618* and *Xfcp624* | Chinese Spring-Hope 1B | *S. nodorum* blotch | Map-based cloning | Shi et al. (2016) |
| 39. | *Fhb1* | *STS3B-355* and *STS3B-334* | Hexaploid wheat cultivar Sumai 3 | Fusarium head blight | Map-based cloning | Rawat et al. (2016) |
| 40. | *Phs-A1* | *Xbarc170* and X*wmc420* | Alchemy and Option | PHST | Map-based cloning | Shorinola et al. (2016) |
| 41. | *TaTGW-7A* | *Xbarc174* and *Xbarc222* | Jing 411 | Grain weight | Map-based cloning | Hu et al. (2016) |
| 42. | *Sr13* | *EX24785* | Durum wheat cultivarsLangdon and Kronos | Stem rust | Map-based cloning | Zhang et al. (2017a, b) |
| 43. | *Stb6* | *Xctg8311* and *Xcfn80023* (co-segregated with stb6), *cfn80025* and *cfn80030/ cfn80040* | Hexaploid wheat cultivars Flame, Chinese Spring and Cadenza | *S. tritici* blotch | Map-based cloning | Saintenac et al. (2018) |
| 44. | *Yr15* | *uhw264* and *uhw258* | A set of introgression lines | Yellow rust | Map-based cloning | Klymiuk et al. (2018) |

*In-Del* Insertion-deletion, *RFLP* restriction fragment length polymorphism, *PHST* preharvest sprouting tolerance

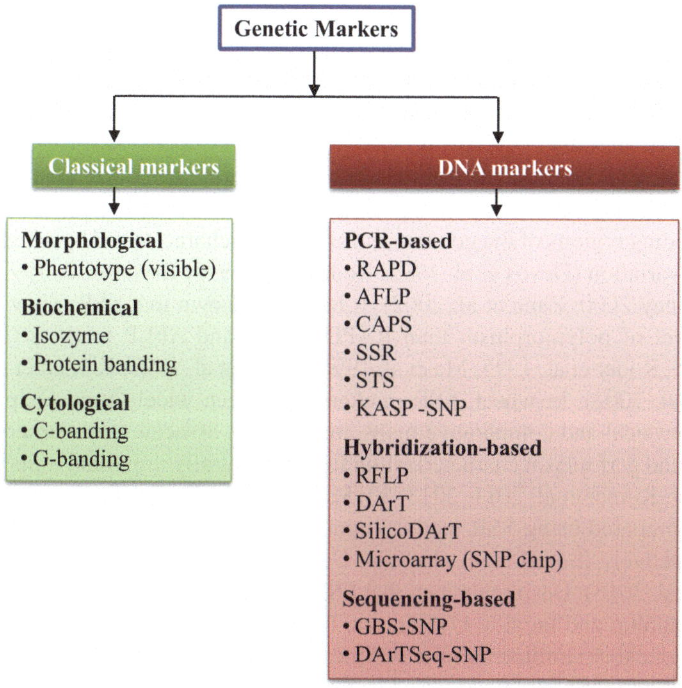


**Fig. 2** Classification of markers frequently used for trait discovery in wheat

efforts in wheat led to the mapping of as many as 2000 RFLP loci in the genetic maps using segregating populations and 1200 RFLP loci in the physical maps using nullisomic-tetrasomic and deletion lines of Chinese Spring (for review, see Gupta et al. 1999, 2008a, b; Hussain and Qamar 2007). RFLP markers have also proved useful in comparative mapping studies because the DNA probes belonging to one species could be readily hybridized to related species (Devos and Gale 1993). Although they have low to medium level of polymorphism, low-throughput nature of detection, high cost of genotyping, etc. discouraged the further use of RFLP markers as a tool in genetic studies of wheat. Subsequent advances in genotyping technology and public genomic database have provided the generation of PCR-based molecular markers such as random amplified polymorphic DNA (RAPD), amplified fragment length polymorphism (AFLP), simple sequence repeat (SSR or microsatellite), diversity arrays technology (DArT) and single-nucleotide polymorphism (SNP). In the beginning, RFLP and RAPD marker systems were not straightforward to use in gene mapping/discovery and marker-based selection. Subsequently, RFLP with the advantage of PCR technique was converted to AFLP marker system (Vos et al. 1995). Therefore, instead of southern hybridization, PCR amplification was performed in AFLP fingerprinting that allowed fractionation of multiple fragments and generated a large number of bands facilitating the detection of polymorphisms. AFLP markers have been used to study genetic diversity, phylogenetic analysis and mapping in wheat (Burkhamer et al. 1998; Parker et al. 1999; Bohn et al. 1999; Gupta et al. 1999). Similarly, the genomic regions amplified by RAPD markers associated with the variation of targeted traits were cloned, sequenced and converted into simple, sturdy and user-friendly PCR-based markers called sequence-characterized amplified regions (SCARs).

In wheat, SSR, DArT and SNP marker systems have been developed in the past and are still in use for an array of applications. The SSRs are located in both coding and noncoding regions of the genome and are usually characterized by a high degree of length variation (Devos et al. 1995; Roder et al. 1995; Bryan et al. 1997; Gupta and Varshney 2000; Zane et al. 2002). It has been shown that SSRs show a much higher level of polymorphism than RAPD, RFLP and AFLP markers (Plaschke et al. 1995; Roder et al. 1995; Ma et al. 1996; Bryan et al. 1997; Korzun et al. 1997; Gupta et al. 2002). In wheat, SSR markers have been widely used for preparing genetic, physical and comparative maps, marker-trait associations, marker-assisted selection and germplasm characterization for agriculturally important traits (Somers et al. 2004; Kumar et al. 2011, 2013, 2015). The available wheat genetic and physical maps prepared using SSR markers contain approximately 3000 and 2000 SSR loci, respectively (Sourdille et al. 2004, Gupta et al. 2008a, b; Goyal et al. 2005; Kumar et al. 2013). Despite the fact that SSRs turned into the most preferred markers for mapping and tagging QTL/genes, there has been a restricted use in wheat genomics due to (i) limited number of SSR motifs in the genome, (ii) uneven distribution, (iii) low-throughput gel-based genotyping and (iv) not being fit to multiplexing. To overcome the above limitations, LGC (https://www.lgcgroup.com/) recently started a new service with state-of-the-art techniques for converting SSR markers into robust, high-throughput and cost-efficient markers (https://www.biosearchtech.com/services/sequencing/microsatellite-ssr-conversion-service#). Different techniques for the development of molecular markers have advantages and disadvantages if compared to each other (Agarwal et al. 2008; Kesawat and Das 2009; Belete 2018), but their reasons for preference vary as per the requirement of users.

At present, SNP markers have rapidly gained the prime position among the available molecular markers not only in wheat but also in other crops due to their abundance in the genomes, flexibility for high-throughput genotyping/detection formats and relatively low-cost. The SNP markers belong to the simplest form of molecular markers that provide a single base pair (nucleotide; DNA building block) variation among alleles of a gene. A single nucleotide (any of A, T, G, C) in gene sequence represents the smallest unit of inheritance; an SNP provides the smallest unit of genetic variation. SNPs encompass variants of four different nucleotides, but as a molecular marker, they are biallelic. According to nucleotide substitution, SNPs can be classified as either transitions (A/G or T/C) or transversions (A/T, A/C, G/T or G/C). For example, an SNP may replace one nucleotide guanine (G) with the nucleotide adenine (A) in a DNA sequence. These SNP variations are present in coding part (exons) as well as noncoding part (introns) of a gene and intergenic regions between genes. SNPs are usually discovered in silico from pre-existing datasets of expressed sequence tags (ESTs) or genome survey sequences (Picoult-Newberg et al. 1999). Lagudah et al. (2009) identified and developed SNP markers in wheat for the disease-resistant genes *Lr34/Yr18/Pm38* that provide resistance to multiple fungal pathogens. Genome-wide distribution of SNP variations aided in development of SNP markers in the close proximity of QTL/genes. Allen et al. (2013) identified 10,251 codominant SNPs from 95,266 putative SNPs following exome sequencing of 8 wheat varieties (Alchemy, Avalon, Cadenza, Hereward, Rialto, Robigus, Savannah and Xi19). These codominant SNP markers and map serve as useful genetic resources for germplasm characterization, QTL studies and marker-assisted selection. Such progresses in wheat genomics and breeding have led to mapping, tagging and cloning of numerous QTL and underlying genes controlling economically important traits (Gupta and Varshney 2000; Lorz and Wenzel 2004; Gale 2005; Landjeva et al. 2007; Mir et al. 2013; Nadeem et al. 2018). The latest information on wheat cultivar development through marker-assisted selection are given in Table 3.

## 3 Current Advances in SNP Marker Development and Genotyping Platform

A prime challenge for wheat geneticists and breeders is to develop robust markers in order to track introgressed segment of QTL/gene in the background of recipient wheat genotype. In this perspective, SNP markers are presently dominating in the genetic analysis due to wide and uniform distribution in the genome, and their discovery relies on the comparison of homologous sequences between genotypes to detect allelic variations at the single-nucleotide level (Ganal et al. 2009; Paux et al. 2011; Wang et al. 2014; Rimbert et al. 2018; Przewieslik-Allen et al. 2019). Next-generation DNA sequencing (NGS) technologies (i.e. Illumina's HiSeq, 454 from Roche Applied Science, SOLiD from Life Technologies, etc.) have significantly accelerated the whole-genome SNPs discovery at ever-reducing cost (Mardis 2008; Berkman et al. 2013; for review see Gupta et al. 2013). These NGS platforms have provided fascinated opportunities for users due to their ability to discover a large number of SNPs from whole genome (Allen et al. 2011; Elshire et al. 2011; Saintenac et al. 2013a, b; Lai et al. 2012; Poland et al. 2012b; Cavanagh et al. 2013; Wang et al. 2014; Winfield et al. 2016; Bajgain et al. 2016). Concurrently with SNP discovery, numerous technologies have been employed for SNP genotyping, from low-throughput to high- and ultra-high-throughput in wheat (Edwards et al. 2009; Cubizolles et al. 2016; Rimbert et al. 2018). In this chapter, two important advanced technologies that are currently being used in wheat for genome-wide SNP discovery and subsequent mapping are discussed in details.

**Table 3** List of improved wheat cultivars or varieties developed through marker-assisted selection (MAS)

| S. No. | Improved cultivar/ variety | Donor line | Trait | Gene or QTL introgressed/ pyramided | Linked marker | References |
|---|---|---|---|---|---|---|
| 1. | Patwin | Madsen (PI 511673) | Resistance to stripe rust, leaf rust, stem rust, strong gluten and good bread-making quality | *Yr17*, *Lr37-Sr38* and *Glu-D1* (5 + 10) | *URIC- LN2* (*ventriup/LN2*) | https://apps.ams.usda.gov/cms/adobeimages/201200476.pdf |
| 2. | Scarlet (WA7994) | Avocet (*Yr15*), Glupro (*Gpc-B1*) | Resistance to stripe rust and high grain protein content | *Yr15* and *Gpc-B1* | *Xwgp34*, *Xgwm934* and *Xgwm33 (Yr15)* | Kidwell et al. (1999) and Carter et al. (2012) |
| 3. | Ben (durum wheat) | 'Sumai 3' and 'Wangshuibai' | Resistance to fusarium head blight | *Qfhs.ndsu-3AS* | *Xgwm533*, *Xgwm493*, *STS3B-66*, *Xbarc133* and *Xbarc180* | Elias et al. (2005) |
| 4. | Lillian | 90B07-AU2B (a breeding line with high GPC) | High grain protein content | *Gpc-B1* | *ND643 (XNor-B2)* | DePauw et al. (2005) |
| 5. | Zentos | AB-lines of Zentos/Syn086 | Grain hardness, test weight, flour yield type 550, sedimentation volume and baking volume | *QGh.Z86-5D.a*, *QHlw.Z86–3B.b*, *QFy550.Z86-5D.a*, *QSed.Z86-1D.a* and *QBvo.Z86-4B.a* | *Xgwm190*, *Xgwm157*, *Xbarc139*, *Xbarc143*, *Xgwm642* and *Xgwm251* | Kunert et al. (2007) |
| 6. | Batis | AB-lines of Batis/Syn022 | Falling number and grain protein content | *QFn.B22–4B.a* and *QGpc.B22-3A.a* | *Xgwm113* and *Xbarc1060* | Kunert et al. (2007) |
| 7. | Stylet | Annuello | Stem, stripe and leaf rusts, semidwarf and glutenin | *Lr34/Yr18*, *Lr46/Yr29*, *Lr24/Sr24*, *Rht-B1*, *Rht-D1*, *Rht8*, *Glu-D1* and *Glu-A3* | *Xgwm295*, *Xgwm140*, *Xwmc44*, *Xgwm3*, *Sr24#50*, *Xcfd36*, *BF-MR1*, *DF-MR2*, *Xgwm261*, *P1 + P2 (Ahmed 2000)* and *Xpsp2999* | Kuchel et al. (2007) |
| 8. | Westmore (durum wheat) | Glupro | Resistant to stripe rust and high grain protein content | *Yr36* and *Gpc-B1* | | Brevis and Dubcovsky (2008) |
| 9. | Haruyokoi | OS21-5 and Leader | Seed dormancy | *Phs-1* | *Xhbg275*, *Xgwm397*, *Xhbe03*, *Xbarc170*, *Xgwm637*, *Xhbe10* and *Xgwm610* | Torada et al. (2008) |

| | | | | | | |
|---|---|---|---|---|---|---|
| 10. | BIOINTA 2004 | *Triticum speltoides* | Resistance to leaf rust | *Lr47* | *Xabc465* (RFLP) | Bainotti et al. (2009) |
| 11. | Goodeve | Clark | Resistance to insect orange blossom wheat midge | *Sm1* | *WM1* | DePauw et al. (2009) |
| 12. | Mace | KS91H184 | Resistance to wheat streak mosaic virus | *Wsm-1* | *STSJ15L* and *STSJ15R* (4DL.4AgS, chromosomal translocation) | Graybosch et al. (2009) |
| 13. | UC1113 (PI638741) | *Thinopyrum elongatum* | Stem rust resistance | *Lr19* and *Sr25* | *Gb* and *STS-Lr19-130* | Yu et al. (2009) |
| 14. | Arina | Tranfer'/6 * Thatcher and Agent/6 * Thatcher | Leaf rust resistance | *Lr9* and *Lr24* | *J13*, *SCS5* and *J09*, *H05* (STS marker) | Moullet et al. (2009) |
| 15. | Nanda2419*Wangshuibai RILs NW66 and N87 (donor) and Mianyang 99-323 | Nanda2419 for *Qfh.nau-2B*, Wangshuibai for *Qfhs.nau-3B*, *Qfhi.nau-4B* and *Qfhi.nau5A* | Scab resistance | *Qfh.nau-2B*, *Qfhs.nau-3B*, *Qfhi.nau-4B* and *Qfhi.nau5A* | *WMC474-WMC499 (Qfh.nau-2B)*, *GWM389-GWM533-BARC147-GWM493 (Qfhs.nau-3B)*, *BARC20-GWM513-GWM192-GWM149-CFD22-WMC349 (Qfhi.nau-4B)* and *BARC180-BARC117-GWM415-GWM304-MAG3794 (Qfhi.nau-5A)* | Xue et al. (2010) |
| 16. | HD2329+*Lr24*+*Lr28* | SPR8198 | PHS tolerance and leaf rust resistace | *QPhs.ccsu-3A.1*, *Lr24* and *Lr28* | *gwm155* and *wmc153* | Kumar et al. (2010) |
| 17. | Sarıçanak-98 | Kyle | γ-Gliadin 45 and LMW-2 glutenin | *γ-gliadin 45* and *LMW-2 glutenin* | *GAG5-6* | Yildirim et al. (2013) |
| 18. | HUW468 | Glu269 | High grain protein content | *Gpc-B1* | *Xucw108* | Vishwakarma et al. (2014) |
| 19. | Kumpa-INIA | CAR3911 | Aluminium tolerance | *TaALMT1* | *ALMT1-4* | Soto-Cerda et al. (2015) |
| 20. | PBW343 | PBW343+*QPhs.ccsu-3A.1* (SPR8198), PBW343+*Lr24/Sr24*+ *QGw.ccsu-1A.3* (Rye Selection 111), PBW343+ *Glu-A1* (*T. dicoccoides* to *T. durum* cv. PBW34), *Lr24/Sr24*+ *Gpc-B1/Yr36* transferred from Yecora Rojo | Leaf rust, stripe rust and yellow rust resistence, preharvest sprouting tolerance and high grain protein content | *QPhs.ccsu-3A.1*, *Lr24/Sr24*, *QGw.ccsu-1A.3*, *Glu-A* and *Gpc-B1/Yr36* | *Xucw108 (GPC)*, *Xgwm155* and *Xwmc153* (PHST), *Xbarc71 (Sr24/Lr24)* and *Xbarc101 (Yr36)* | Tyagi et al. (2014) |

(continued)

**Table 3** (continued)

| S. No. | Improved cultivar/variety | Donor line | Trait | Gene or QTL introgressed/pyramided | Linked marker | References |
|---|---|---|---|---|---|---|
| 21. | HD2932 | HD2687+*Lr19*/*Sr25*, Eagle+*Sr26* and Avocet+*Yr10* | Leaf rust, stripe rust and yellow rust resistance | *Lr19*/*Sr25*, *Sr26* and *Yr10* | *Xwmc221*, *Sr26#43* (presence), *BE518379* (absence) and *Xpsp3000* | Mallick et al. (2015) |
| 22. | DWR162 | PBW343+*Lr24*+*Lr28* | Resistance to leaf rust | *Lr24* and *Lr28* | *SCS1302* and *SCS421* | Yadawad et al. (2015) |
| 23. | HUW234 and HUW468 | PBW343+*Gpc-B1*+Lr24 | High grain protein content | *Gpc-B1* | *Xucw108* | Mishra et al. (2015) |
| 24. | PBW343 Unnat (PBW723) | *Aegilops ventricosa* for *Lr37*/*Yr17* with *Sr38* (2AL) and *Aegilops umbellulata* for *Lr76*/*Yr70* (5DS) | Leaf and yellow rust resistance | *Lr37*/*Yr17* and *Lr76*/*Yr70* | *ventriup*/*LN2* and *CAPS16* | http://dbtindia.gov.in/sites/default/files/uploadfiles/SuccessStories7.pdf |
| 25. | WA8143 | CL0618 Australian Hard Red Wheat | Tolerance to imazamox herbicide | *Als1* and *Als2* | *Als1* and *Als2* (allele-specific primers) | Kumar et al. (2017) |
| 26. | HD2733 | HD2687+*Lr19* and HD2687+*Lr24* | Leaf rust resistance | *Lr19* and *Lr24* | *Xwmc221* and *SCS1302* | Singh et al. (2017) |
| 27. | DBW14 (RAJ3765/PBW343) | Barham, Australian awnless variety | Soft grain | *PinaD1a* | *PinA* and *Pina-N1* | Rai et al. (2019) |
| 28. | Ningchun4, Ningchun47 and Ningchun50 | Spring wheat line CB037 (*D. Villosum* T6AL·6V#2S translocation line containing *Pm21*) | Resistance to powdery mildew | *Pm19* | *6V-4* and *6V-14* | Ye et al. (2019) |
| 29. | McNeal (PI 574642) or Hank (PI 613585) | Pierce | Solid stem (sawfly resistance) | *Qss.msub-3BL* | KASP-SNP marker | Varella et al. (2019) |

### *High-Throughput SNP Genotyping: Array-Based Genotyping*

Development of SNP array requires detection of SNPs from whole-genome and/or transcriptome sequencing using NGS technologies. This collection of DNA/cDNA sequences (NGS reads) serves as an incredible resource for SNP detection. In addition to NGS-based SNP discovery, genomic sequences or EST sequences available in different databases have also been used for SNP identification in the recent past. Clevenger et al. (2015) summarized experimental approaches to SNP calling in polyploid species like wheat. For large-scale SNP genotyping, microarrays that relied on fixed sets of SNP assays have recently been developed by Illumina (Illumina, San Diego, USA) and Affymetrix (Affymetrix Inc., Santa Clara, CA) (for a review see Gupta et al. 2008a, b). For instance, Illumina's BeadArray technology uses Infinium II assay chemistry for genotyping high-density SNPs (Steemers et al. 2006; Steemers and Gunderson 2007). Infinium II assay system, which has designed for one bead type per assay (or SNP), performs whole-genome amplification through single-base extension (SBE) step and discriminates two alleles of a known SNP by incorporating two hapten-labelled dideoxynucleotides (ddNTPs), i.e. dinitrophenol (red fluorescence) for adenosine (A) and thymine (T) and biotin (blue fluorescence) for cytosine (C) and guanine (G). Following SBE, Infinium II assay involves two fluorescence colour assay, and therefore signals contain two intensity values per locus based on allele types (Gunderson 2009). Fluorescence signals of assay matrix are then scanned by the Illumina iScan system for further data visualization in diploid and polyploid versions of GenomeStudio software.

Identification of a large number of SNPs demands their genotyping at high-throughput scale. Illumina presently offers a variety of options for custom genotyping arrays that allow unlimited access of queried SNPs, i.e. Illumina Infinium iSelect HD chip. The high-density Infinium arrays for whole-genome SNP genotyping have been successfully designed and utilized in wheat. For example, the International Wheat SNP Working Group (IWSWG) in collaboration with Illumina developed Infinium 9K and 90K iSelect SNP genotyping arrays (Cavanagh et al. 2013; Wang et al. 2014). Using 9K iSelect SNP array, 7504 SNPs were identified, and a consensus genetic map of wheat was prepared with an average density of 1.9 ± 1.0 SNP/cM (Cavanagh et al. 2013). The 90K iSelect SNP array has been used to map 46,977 SNP markers on wheat chromosomes (Wang et al. 2014; Liu et al. 2016). Afterwards, 90K iSelect SNP array has been used in a range of applications such as phylogenetic analysis (Turuspekov et al. 2015); QTL analysis for preharvest sprouting tolerance (Cabral et al. 2014), loose smut resistance (Kumar et al. 2018), leaf rust resistance (Gao et al. 2016a, b; Kumar et al. 2019), physiological traits (Gao et al. 2016a, b) and agronomic traits (Zou et al. 2016); and genome-wide association analysis (Liu et al. 2017, 2018; Li et al. 2019; Garcia et al. 2019; Alomari et al. 2019). Recently, Gao et al. (2017) identified 7989 iSelect SNP loci involved in the domestication and improvement and constructed first-generation map of selection loci for evolutionary studies and breeding in wheat. A year after, Rimbert et al. (2018) identified 3.3 million SNPs in the genic, repetitive and non-repetitive intergenic fractions of 8 wheat lines. They developed TaBW280K high-throughput SNP genotyping array. The TaBW280K SNP array has been used to genotype a biparental population derived from a cross between Chinese Spring and Renan and generated an ultra-high-density genetic map comprising 83,721 SNP markers (Rimbert et al. 2018).

In addition, a large selection of high-density wheat genotyping SNP arrays has also been developed at the Affymetrix Axiom platform. Earlier, 1.57 million SNPs were identified by Jordan et al. (2015) targeting 107 Mb sequences from nonredundant low-copy genic regions in 62 wheat genotypes. Following exome sequencing, Winfield et al. (2016) captured ~57 Mb of coding sequences in 43 hexaploid wheat accessions and identified 921,705 (921K) putative SNPs. Of which, 820K high-quality SNPs were included in an array and used for the genotyping of 475 accessions of wheat and relatives. Subsequently, a set of 35,143 highly polymorphic and evenly distributed SNP markers from 820K SNP array were picked, and a 35K SNP genotyping array (also known as Wheat Breeder's Array) was developed at the Affymetrix GeneTitan platform (Allen et al. 2017). This Wheat Breeder's Array contains informative SNP markers that were used to characterize 2713 wheat genotypes, including landraces, elite lines and five mapping populations (Allen et al. 2017). In addition, another important SNP array named Wheat660K SNP array, designed by Chinese Academy of Agricultural Sciences (https://wheat.pw.usda.gov/ggpages/topics/Wheat660_SNP_array_developed_by_CAAS.pdf) and synthesized by Affymetrix Axiom, became available for a wide range of potential applications in wheat. Cui et al. (2017) prepared an ultra-high-density genetic map consisting 119,566 SNP loci by genotyping Affymetrix Wheat660K SNP array on 188 recombinant inbred lines (RILs) derived from a cross between KN2904 and J411. A major stable QTL (*qKnps-4A*) for kernel number per spike was identified by using high-density SNP map along with phenotypic data (Cui et al. 2017). Using mapped SNP flanking sequences and corresponding contig sequences of wheat, comparative genomic analysis has also been carried out with the genomic sequences of rice (*Oryza sativa*), thale cress (*Brachypodium distachyon*), sorghum (*Sorghum bicolor*) and maize (*Zea mays*). Furthermore, as many as 53,063 SNP sequence tags were carefully selected from the Wheat660K SNP array, and a new Affymetrix Wheat55K SNP array was developed. The SNP tags included in the Wheat55K array were uniformly distributed in all the 21 wheat chromosomes (~2600 SNPs per chromosome) with an average distance of 0.1 cM and corresponding average physical distance of approximately 300 kb (Ren et al. 2018). In order to verify the SNPs in Wheat55K array, a high-density SNP map was developed using 371 RILs derived from a cross between Chuan-Nong18 and T1208. Using phenotyping data from multiple environments, seven stable QTLs for tiller number around different growth stages were identified (Ren et al. 2018).

The two ultra-high-density genetic maps of SNP markers developed and communicated by Cui et al. (2017) and Rimbert et al. (2018) serve as useful genomic resources for mapping and dissecting complex traits in hexaploid and tetraploid wheat. To date, none of the publicly available genetic maps of wheat have the same SNP density as presented by Cui et al. (2017) and Rimbert et al. (2018). It is also worth noting that the International Wheat Genome Sequencing Consortium

(IWGSC) has considered these two SNP maps as the reference genetic maps for anchoring and ordering the wheat genome reference sequence.

The studies exemplified above demonstrate in wheat the value and power of array-based SNP genotyping. The array-based SNP genotyping technologies have achieved popularity among users due to a number of advantages like variation detection at the nucleotide level, flexibility, speed, cost-effectiveness, etc. (Thomson 2014). By assessing the current SNP genotyping platforms, researchers are able to perform genetic and physical mapping, marker-trait associations and investigations into evolutionary relationships. However, the genotyping data using the assays may have an ascertainment bias due to use of a limited set of wheat germplasm for developing SNP arrays (Wang et al. 2014). Such limitations can be overcome by employing genotyping-by-sequencing (GBS), a more advanced alternative to genotyping technology. The identification of a large collection of SNPs and development of Infinium and Axiom arrays currently present the wheat community with valuable resources and tools that could transform wheat breeding in a more specific way.

## *High-Throughput SNP Genotyping: Genotyping-by-Sequencing*

The increasing adaptability and affordability of next-generation DNA sequencing (NGS) for genetic analysis has crossed the borders from a small set of loci to hundreds of thousands of SNPs. The reduced-representation sequencing (RRS) approach holds secret of reducing genome complexity by capturing only specific DNA regions flanked by restriction enzymes prior to sequencing. The family of RRS approach comprises at least 13 different methods (Scheben et al. 2017); one of them is genotyping-by-sequencing (GBS). The GBS, introduced first time in maize by Elshire et al. (2011) and later in barley and wheat (Poland et al. 2012a), is becoming an attractive method of genotyping due to its simple, rapid and robust nature. In GBS, entire DNA is sampled for sequencing with some average sequence depth (depending on the species being used) of genome as an alternative to genotyping system where a specific polymorphism is targeted. GBS mainly relies on the nucleotide sequencing of complexity-reduced fraction of the genome, which employs one or more restriction endonucleases to capture only the portion of the genome flanked by restriction sites (Elshire et al. 2011; He et al. 2014). A comparison between the features of GBS and array-based genotyping technologies are presented in Table 4. Genotyping with GBS method requires good-quality genomic DNA at appropriate concentration for library preparation (Davey et al. 2011). The usefulness of GBS has been demonstrated to predict breeding values through genomic selection in wheat (Poland et al. 2012a) and genetic analyses in other crop plants (Bhatia et al. 2013). Besides the detection of SNPs, GBS also allows detection of polymorphisms due to presence/absence variations (PAVs) (Deschamps et al. 2012). The GBS for high-throughput genotyping has been widely used in wheat for preparation of high-density genetic maps, marker-trait association and genomic selection for days to heading, thousand grain weight and yield (Poland and Rife 2012; Poland et al.

**Table 4** Comparison of high-throughput SNP genotyping technologies: array-based genotyping and genotyping-by-sequencing

| Features | Array-based genotyping | | Genotyping-by-sequencing (GBS) |
|---|---|---|---|
| | Illumina Infinium iSelect HD | Affimatrix Axiom | |
| Genotyping method | Hybridization (fixed array) | Hybridization (fixed array) | Restriction enzyme-based |
| Number of SNPs (range) | 3K–700K | 50K–650K | Variable (range 1K–100K) |
| Number of samples | 24 | 96, 384 | 48, 96, 384 |
| Compensation | Highly multiplexed | Highly multiplexed | Massive amount of sequence data relative to cost |
| Discovery of new SNP variants | Not possible | Not possible | Always possible |
| Proportion of missing data | Low | Low | High |
| Data imputation | May or may not be required | May or may not be required | Required |
| Accuracy in prediction of missing data | Low | Low | High |
| Ascertainment bias | Yes | Yes | No |
| Reduction in genome complexity | No | No | Yes |
| Cost per sample | Moderately high | Moderately high | Low to moderate |
| Requirement of reference genome | Not required | Not required | May or may not be required |
| Bar coding | Required to tag SNP probe | Required to tag SNP probe | Required to tag sequence variant |

2012a, b; He et al. 2014; Gao et al. 2017; Bhatta et al. 2018a; Jamil et al. 2019). GBS has also been used in wheat for mapping genes/QTLs for preharvest sprouting tolerance and disease and insect resistance (Forrest et al. 2014, Gao et al. 2015, Li et al. 2015a, b; Lin et al. 2015; Bhatta et al. 2018b; Zhao et al. 2019). Despite the fact that GBS has potential to identify several million SNPs, higher amount of missing data (incomplete SNP data) largely due to insufficient sequencing coverage often set limits to the number of usable SNPs for downstream analysis (Elshire et al. 2011). The missing data in big datasets like GBS come under a situation when some of the experimental lines are missing a genotype value at a particular locus but it is correctly detected and called in the remaining lines. High-quality genomic DNA, optimized sequence depth, efficient GBS library preparation and accuracy in sequencing can minimize the amount of missing data. The improved version of the GBS protocol in wheat and other cereals has also been developed and used in order to increase informative SNPs at affordable cost (Poland et al. 2012a, b; Huang et al. 2014). There has been an interest to deal with missing data using imputation methods. Genotype imputation is a process of predicting missing genotypes with the help of some computational algorithms like IMPUTE and fastPHASE; thereby any value for missing data can be estimated with logical values according to the available reference genome sequence (Torkamaneh and Belzile 2015); however accuracy in predicted missing data may rely on the completeness of reference genome. Recently, Alipour et al. (2019) demonstrated the use of reference genome to impute missing genotype data generated by GBS in wheat and barley. Authors showed that, among the four reference genomes (wheat reference genomes of CSSS, W7984 and IWGSC RefSeq v1.0 and barley reference genome), IWGSC RefSeq v1.0 imputed the maximum number of missing SNP data points with adequate imputation accuracy. Combined with data imputation, GBS provides a simple, fast and effective technology of choice for simultaneous detection and genotype SNPs for genomic-assisted breeding in wheat improvement.

## 4 Utility and Achievement of High-Throughput Genotyping Approaches in Wheat

It has been critical for identification of genome-wide SNPs using NGS technologies in a polyploid species like common wheat, which is known for large and complex genome. With the help of modern fast sequencing technologies (NGS) and suitable computer software, it is now possible to scan the whole genome for SNP discovery and variation. Accessibility of high-quality reference genome of Chinese Spring wheat has further accelerated the re-sequencing of germplasm accessions and population lines to accurately detect SNP variations even in highly similar breeding lines. At present, SNP markers hold promise in wheat breeding and genomic research and are contributing towards the analysis of complex traits in all modern breeding programs. For instance, SNP markers have provided greater insight into genetically complex trait such as drought and heat tolerance. Being a complex trait, drought tolerance in wheat is governed by numerous QTLs (or polygenes) with small effect. The drought-responsive traits include water-use efficiency (WUE), root system architecture (RSA), coleoptile length, stomatal conductance, canopy temperature (CT), carbon isotope discrimination (CID), plant phenology, grain yield and related traits (Ahmad et al. 2017; Gupta et al. 2017). Although a number of QTL identified for the above traits have been mapped, these QTLs are most often placed in large intervals between the flanking markers due to low-density genetic maps. The large distance between the QTLs and flanking markers has discouraged the deployment of the QTLs in breeding through MAS in wheat. This has led to use high-throughput SNP genotyping methods (array-based and GBS) for generating a large number of useful SNP markers that are closely associated with the QTL/genes of targeted traits. Recently, Infinium 90K SNP genotyping assay and a panel of 123 wheat cultivars from Pakistan (released from 1947 to 2015) were used to conduct genome-wide association study (GWAS) for yield and related traits under rain-fed conditions (Ain et al. 2015). This allowed identification of 14,960 polymorphic

SNPs permitting identification of 44 marker-trait associations (MTAs) for 9 yield-related traits. Of which nine multi-trait MTAs were mapped on seven different wheat chromosomes. Gene annotation of the 44 MTAs and their syntenic relationship to the genes in rice, brachypodium and sorghum, allowed detection of genes underlying 14 MTAs, which encode proteins that are expressed in response to stress environments (Ain et al. 2015).

Genotyping-by-sequencing has been used in the study of 1423 spring wheat accessions for various important traits, including drought and heat tolerance at CIMMYT under Seeds of Discovery (SeeD) program (Sehgal et al. 2015). They identified 1273 GBS-SNPs in the landraces adapted to drought and 4473 SNPs in the landraces adapted to heat stress environments. In order to utilize the marker information, >200 landraces and synthetic wheat were selected to exploit their potential use in pre-breeding and for the allele mining of possible candidate genes for drought and heat stress tolerance. The mean diversity index indicated that accessions representing synthetic wheat were relatively more diverse than landraces and elite cultivars. Study infers that the characterization of unexploited genetic variation in landraces and synthetic hexaploid wheat accessions can be mobilized into well-adapted popular cultivars (Sehgal et al. 2015).

While a number of studies have been conducted, only a few reports have focused on the comparison of genotypic datasets from array- and GBS-based methods (Torkamaneh and Belzile 2015; Elbasyoni et al. 2018). The SNPs obtained from array-based genotyping are of high quality, but per sample cost they are considerably higher. Conversely, SNP data obtained from GBS platform are larger in amount but contain a high proportion of missing calls. Array-based genotyping does not allow the detection of new SNPs, which is not the case in GBS. Nevertheless, based on the available SNP genotyping data, both the genotyping technologies are complementary for detecting and mapping of important QTL/genes (Negro et al. 2019). Recently, Elbasyoni et al. (2018) compared the SNP genotyping data scored from 90K SNP array and from GBS in winter wheat for estimating population structure and genomic kinship. The authors highlighted that GBS-scored SNPs are comparable to or better than 90K SNP array-scored SNPs for genomic prediction application. The options of genotyping technologies should be considered carefully in keeping with desired purposes and objectives.

## 5 Conversion of Trait-Linked SNPs to User-Friendly Markers

As we have already discussed, array-based genotyping and GBS are the most preferred technologies for multiplexing and high-throughput SNP analysis in trait discovery. They do not provide flexibility and become expensive if a small number of selected SNPs need to be genotyped on a large collection of germplasm and breeding lines. It is important to look at a suitable SNP assay that can be flexible, cost-effective, user-friendly and time-saving and generate good-quality data. The LGC Genomics (http://www.lgcgroup.com/) provided the solution of such scientific problem and introduced uniplex SNP genotyping platform such as KASP (KBiosciences Competitive Allele-Specific PCR [also named as Kompetitive Allele Specific PCR]) (Neelam et al. 2013; Mir et al. 2013; Semagn et al. 2014). The KASP genotyping system, developed earlier by KBiosciences and later acquired by the LGC Genomics in 2011, is a homogeneous fluorescent, endpoint genotyping technology. Among the available uniplex systems (Semagn et al. 2014), KASP offers easier, cheapest and flexible way to determine both SNP and insertion-deletion genotypes. Trait-associated SNP flanking sequences (50 bp upstream and 50 bp downstream around the SNP variant position), derived from GBS or array-based systems, can be used to design KASP assays using SNiPlay3 (Dereeper et al. 2015). A comprehensive procedure/protocol of KASP genotyping chemistry, requirement of equipment, software and reagents, designing KASP primers, data output and data scoring can be found here (He et al. 2014; Smith and Maughan 2015; https://www.cerealsdb.uk.net/cerealgenomics/CerealsDB/PDFs/KASP_SNP_Genotyping_Manual.pdf). Allen et al. (2011) demonstrated first time the application of KASP SNP genotyping in hexaploid wheat for constructing a genetic linkage map of 548 loci using an Avalon/Cadenza doubled haploid mapping population. LGC Genomics directly as well as through the Generation Challenge Program (GCP) and Integrated Breeding Platform (IBP) offers SNP genotyping services for several crops including wheat. Information on KASP assays and their mapping to wheat chromosomes can be found at following websites/databases:

(i) LGC Genomics wheat panel (http://www.lgcgroup.com/wheat/#.VfMk3q10y70)

(ii) CerealsDB KASP SNPs database (http://www.cerealsdb.uk.net/cerealgenomics/CerealsDB/indexNEW.php; https://www.cerealsdb.uk.net/cerealgenomics/CerealsDB/kasp_mapped_snps.php)

(iii) Integrated Breeding Platform (https://www.integratedbreeding.net/482/communities/genomics-crop-info/crop-information/gcp-kasparsnp-marker)

(iv) LGC's online wheat genotyping (https://biosearch-cdn.azureedge.net/assetsv6/Wheat-poster-Key-trait-screening.pdf; https://www.researchgate.net/institution/LGC_Biosearch_Technologies2/post/58458fbfdc332d599f0c2991_KASPR_Genotyping_Markers_for_Key_Wheat_Traits)

Recently, Rasheed et al. (2016) utilized 70 KASP-based assays of functional markers for agronomic, disease resistance, drought tolerance, preharvest sprouting tolerance and end-use quality traits in wheat and validate them using a panel of 300 diverse cultivars and 4 recombinant inbred line (RIL) populations. The validated KASP assays related to (i) agronomic traits including *Ppd-B1*, *Ppd-D1*, *VRN-A1*, *VRN-B1*, *VRN-D1*, *Rht-B1*, *Rht-D1*, *TaCwi-5D*, *TaGS-D1*, *TaTGW6-3A*, *TaGASR-A1*, *TaSus2-2B*, *TaCKX-D1* and *TaMoc1-7A*; (ii) disease resistance including Lr34TCCIND and Lr34jagger for *Lr34*; (iii) drought tolerance including *TaDreb-B1*, *1-feh w3* and *TaCwi-4A*; (iv) preharvest sprouting tolerance including *TaPHS1*, *TaSdr-B1*, *TaVp-1B* and *TaMFT-A1*; and (v) end-use quality comprising *Glu-A1*, *Glu-B1*, *Glu-D1*, *Pina-D1*, *Pinb-D1*, *Pinb-B2*, *Ppo-A1*, *Ppo-D1*, *Psy-A1*, *Psy-B1*,

*Psy-D1* and *Zds-A1* were used for function polymorphism (Rasheed et al. 2016). After validation, KASP-based SNP markers can be used to pyramid favourable genes/alleles following marker-assisted selection in wheat genetic improvement. Furthermore, the KASP marker system has also been applied in other crop plants, including pigeon pea (Saxena et al. 2012), chickpea (Hiremath et al. 2012), Indica rice (Pariasca-Tanaka et al. 2015; Steele et al. 2018) and Japonica rice (Cheon et al. 2018), for genetic analysis. The KASP platform provides an opportunity to customize a set of trait-linked SNPs for genotyping on a panel of wheat germplasm and further validation.

In addition to KASP genotyping system, TaqMan assay (Woodward 2014), semi-thermal asymmetric reverse PCR (STARP) (Long et al. 2016), Amplifluor SNP genotyping system (Jatayev et al. 2017) and RNase H2 enzyme-based amplification (rhAmp) (https://eu.idtdna.com/pages/products/qpcr-and-pcr/genotyping/rhamp-snp-genotyping) are also available in the market that have emerged as promising techniques of SNP genotyping. All the five techniques are offering allele-specific uniplex genotyping platforms with exceptional chemistry and scalable flexibility without compromising cost and data throughput (Rasheed et al. 2017; Broccanello et al. 2018; Ayalew et al. 2019).

# 6 Conclusions and Future Directions

It is apparent that DNA sequencing and genotyping technologies have evolved rapidly and become one of the most promising breeding tools to discover useful alleles contributing to trait variation. With the continuous support of modern technologies, wheat genome sequence data are being produced at a faster and cheaper rate. Identification of additional numbers of genome-wide SNPs is likely to have the greatest impact for revealing hidden variations particularly near centromeric regions of the chromosomes. Current challenges are likely to move from the wheat genome analysis to the association of sequence variation with economically important traits. The development of ultra-high-density SNP map (over 100K markers) will speed up high-resolution mapping and cloning of major QTLs. Moreover, co-localization of similar QTLs from different studies and projecting a meta-QTL can also refine QTL position and corresponding genes. Hence, candidate gene-based user-friendly functional assays can be used to target alleles in wheat marker-assisted breeding projects. Future research is likely to appear with continued advancements in molecular marker technology to make them more useful and effective breeding tool.

**Acknowledgement** The work on this chapter was supported by the SERB (Science and Engineering Research Board), New Delhi, India, through ECR/2016/001774.

**Author Contributions** SK designed the outline and wrote the first draft of this manuscript. MK, RRM, RK and SoK contributed to the writing of the manuscript. SK, MK, RRM, RK and SoK made critical revisions and approved final version. All authors read and approved the final manuscript.

## References

Agarwal M, Shrivastava N, Padh H (2008) Advances in molecular marker techniques and their applications in plant sciences. Plant Cell Rep 27:617–631

Ahmad I, Ali N, Ahmad H (2017) Association mapping of root traits for drought tolerance in bread wheat. In: Wanyera R, Owuoche J (eds) Wheat improvement, management and utilization. InTech, London, pp 39–57

Ain QU, Rasheed A, Anwar A, Mahmood T, Imtiaz M, Mahmood T, Xia X et al (2015) Genome-wide association for grain yield under rainfed conditions in historical wheat cultivars from Pakistan. Front Plant Sci 6:743

Alipour H, Bai G, Zhang G, Bihamta MR, Mohammadi V, Peyghambari SA (2019) Imputation accuracy of wheat genotyping-by-sequencing (GBS) data using barley and wheat genome references. PLoS One 14(1):e0208614

Allen AM, Barker GLA, Berry ST, Coghill JA, Gwilliam R, Kirby S, Robinson P, Brenchley RC, D'Amore R, McKenzie N (2011) Transcript-specific, single nucleotide polymorphism discovery and linkage analysis in hexaploid bread wheat (*Triticum aestivum* L.). Plant Biotechnol J 9:1086–1099

Allen AM, Barker GL, Wilkinson P, Burridge A, Winfield M, Coghill J, Uauy C et al (2013) Discovery and development of exome-based, codominant single nucleotide polymorphism markers in hexaploid wheat (*Triticum aestivum* L.). Plant Biotechnol J 11:279–295

Allen AM, Winfield MO, Burridge AJ, Downie RC, Benbow HR, Barker GL et al (2017) Characterization of a Wheat Breeders' Array suitable for high-throughput SNP genotyping of global accessions of hexaploid bread wheat (*Triticum aestivum* L.). Plant Biotechnol J 15:390–401

Alomari DZ, Eggert K, von Wirén N, Polley A, Plieske J, Ganal MW, Liu F, Pillen K, Röder MS (2019) Whole-genome association mapping and genomic prediction for iron concentration in wheat grains. Int J Mol Sci 20(1):76

Amom T, Nongdam P (2017) The use of molecular marker methods in plants: a review. Int J Cur Res Rev 9(17):1–7

Ashikawa I, Abe F, Nakamura S (2010) Ectopic expression of wheat and barley *DOG1*-like genes promotes seed dormancy in Arabidopsis. Plant Sci 179(5):536–542

Ayalew H, Tsang PW, Chu C, Wang J, Liu S, Chen C et al (2019) Comparison of TaqMan, KASP and rhAmp SNP genotyping platforms in hexaploid wheat. PLoS One 14(5):e0217222

Bainotti CT, Fraschina J, Salines JH, Nisi JE, Dubcovsky J, Lewis SM, Bullrich L, Vanzetti L, Cuniberti M, Campos P, Formica MB (2009) Registration of 'BIOINTA 2004' wheat. J Plant Regist 3(2):165–169

Bajgain P, Rouse MN, Tsilo TJ, Macharia GK, Bhavani S et al (2016) Nested association mapping of stem rust resistance in wheat using genotyping-by-sequencing. PLoS One 11:e0155760

Beales J, Turner A, Griffiths S, Snape JW, Laurie DA (2007) A pseudo-response regulator is misexpressed in the photoperiod insensitive Ppd-D1a mutant of wheat (*Triticum aestivum* L.). Theor Appl Genet 115(5):721–733

Belete T (2018) Improvement of crop plants by molecular markers. JOJ Hortic Arboric 1(5):555572

Berkman PJ, Visendi P, Lee HC, Stiller J, Manoli S, Lorenc MT et al (2013) Dispersion and domestication shaped the genome of bread wheat. Plant Biotechnol J 11:564–571

Bhatia D, Wing RA, Singh K (2013) Genotyping-by-sequencing, its implications and benefits. Crop Improv 40(2):101–111

Bhatta M, Morgounov A, Belamkar V, Baenziger PS (2018a) Genome-wide association study reveals novel genomic regions for grain yield and yield-related traits in drought-stressed synthetic hexaploid wheat. Int J Mol Sci 19:3011

Bhatta M, Morgounov A, Belamkar V, Yorgancılar A, Baenziger PS (2018b) Genome-wide association study reveals favorable alleles associated with common bunt resistance in synthetic hexaploid wheat. Euphytica 214:200

Bohn M, Utz HF, Melchinger AE (1999) Genetic similarities among winter wheat cultivars determined on the basis of RFLPs, AFLPs, and SSRs and their use for predicting progeny variance. Crop Sci 39:228–237

Brevis JC, Dubcovsky J (2008) Effect of the *Gpc-B1* region from *Triticum turgidum* ssp. *dicoccoides* on grain yield, thousand-grain weight and protein yield. In: Appels R, Eastwood R, Lagudah E, Langridge P, Mackay M, McIntyre L, Sharp P (eds) Proceedings of 11th international wheat genetic symposium, Brisbane, Australia, August 24–29, 2008, Sydney University Press, pp. 1–3

Broccanello C, Chiodi C, Funk A, McGrath JM, Panella L, Stevanato P (2018) Comparison of three PCR-based assays for SNP genotyping in plants. Plant Methods 14(1):28

Bryan GJ, Collins AJ, Stephenson P, Orry A, Smith JB, Gale MD (1997) Isolation and characterization of microsatellites from hexaploid bread wheat. Theor Appl Genet 94:557–563

Burkhamer RL, Lanning SP, Martens RJ, Martin JM, Talbert LE (1998) Predicting progeny variance from parental divergence in hard red spring wheat. Crop Sci 38(1):243–248

Cabral AL, Jordan MC, McCartney CA, You FM, Humphreys DG, MacLachlan R, Pozniak CJ (2014) Identification of candidate genes, regions and markers for pre-harvest sprouting resistance in wheat (*Triticum aestivum* L.). BMC Plant Biol 14(1):340

Carter AH, Santra DK, Kidwell KK (2012) Assessment of the effects of the *Gpc-B1* allele on senescence rate, grain protein concentration and mineral content in hard red spring wheat (*Triticum aestivum* L.) from the Pacific Northwest Region of the USA. Plant Breed 131(1):62–68

Cavanagh CR, Chao S, Wang S, Huang BE, Stephen S, Kiani S, Forrest K, Saintenac C, Brown-Guedira GL, Akhunova A, See D, Bai G, Pumphrey M, Tomar L, Wong D, Kong S, Reynolds M, da Silva ML, Bockelman H, Talbert L, Anderson JA, Dreisigacker S, Baenziger S, Carter A, Korzun V, Morrell PL, Dubcovsky J, Morell MK, Sorrells ME, Hayden MJ, Akhunov E (2013) Genome-wide comparative diversity uncovers multiple targets of selection for improvement in hexaploid wheat landraces and cultivars. Proc Natl Acad Sci U S A 110(20):8057–8062

Chao S, Sharp PJ, Worland AJ, Warham EJ, Koebner RMD, Gale MD (1989) RFLP-based genetic maps of wheat homoeologous group 7 chromosomes. Theor Appl Genet 78:495–504

Cheon KS, Baek J, Cho Y, Jeong Y-M, Lee Y-Y et al (2018) Single nucleotide polymorphism (SNP) discovery and kompetitive allele-specific PCR (KASP) marker development with Korean japonica rice varieties. Plant Breed Biotech 6(4):391–403

Clevenger J, Chavarro C, Pearl SA, Ozias-Akins P, Jackson SA (2015) Single nucleotide polymorphism identification in polyploids: a review, example, and recommendations. Mol Plant 8:831–846

Cloutier S, McCallum BD, Loutre C, Banks TW, Wicker T, Feuillet C, Keller B, Jordan MC (2007) Leaf rust resistance gene *Lr1,* isolated from bread wheat (*Triticum aestivum* L.) is a member of the large *psr567* gene family. Plant Mol Biol 65:93–106

Cubizolles N, Rey E, Choulet F, Rimbert H, Laugier C, Balfourier F, Bordes J, Poncet C, Jack P, James C, Gielen J, Argillier O, Jaubertie J-P, Auzanneau J, Rohde A, Ouwerkerk PBF, Korzun V, Kollers S, Guerreiro L, Hourcade D, Robert O, Devaux P, Mastrangelo A-M, Feuillet C, Sourdille P, Paux E (2016) Exploiting the repetitive fraction of the wheat genome for high-throughput single nucleotide polymorphism discovery and genotyping. Plant Genome 9:1–11

Cui F, Zhang N, Fan XL, Zhang W, Zhao CH, Yang LJ, Pan RQ, Chen M, Han J, Zhao XQ, Ji J (2017) Utilization of a Wheat660K SNP array-derived high-density genetic map for high-resolution mapping of a major QTL for kernel number. Sci Rep 7(1):3788

Davey JW, Blaxter ML (2011) RADSeq: next-generation population genetics. Brief Funct Genomics 9:416–423

De Bustos A, Rubio P, Jouve N (2001) Characterisation of two gene subunits on the 1R chromosome of rye as orthologs of each of the *Glu-1* genes of hexaploid wheat. Theor Appl Genet 103(5):733–742

DePauw RM, Townley-Smith TF, Humphreys G, Knox RE, Clarke FR, Clarke JM (2005) Lillian hard red spring wheat. Can J Plant Sci 85(2):397–401

DePauw RM, Knox RE, Thomas JB, Smith M, Clarke JM, Clarke FR, McCaig TN, Fernandez MR (2009) Goodeve hard red spring wheat. Can J Plant Sci 89(5):937–944

Dereeper A, Homa F, Andres G et al (2015) SNiPlay3: a web-based application for exploration and large scale analyses of genomic variations. Nucleic Acids Res 43:295–300

Deschamps S, Llaca V, May GD (2012) Genotyping-by-sequencing in plants. Biology 1(3):460–483

Devos KM, Gale MD (1993) Extended genetic maps of the homoeologous group 3 chromosomes of wheat, rye and barley. Theor Appl Genet 85:649–652

Devos KM, Moore G, Gale MD (1995) Conservation of marker synteny during evolution. Euphytica 85:367–372

Distelfeld A, Cakmak I, Peleg Z, Ozturk L, Yazici AM, Budak H, Saranga Y, Fahima T (2007) Multiple QTL-effects of wheat *Gpc-B1* locus on grain protein and micronutrient concentrations. Physiol Plant 129(3):635–643

Dong CH, Xia XC, Zhang LP, He ZH (2012) Allelic variation at the *TaZds-A1* locus on wheat chromosome 2A and development of a functional marker in common wheat. J Integr Agric 11(7):1067–1074

Edwards KJ, Reid AL, Coghill JA, Berry ST, Barker GL (2009) Multiplex single nucleotide polymorphism (SNP)-based genotyping in allohexaploid wheat using padlock probes. Plant Biotechnol J 7:375–390

Elbasyoni IS, Lorenz AJ, Guttieri M, Frels K, Baenziger PS, Poland J, Akhunov E (2018) A comparison between genotyping-by-sequencing and array-based scoring of SNPs for genomic prediction accuracy in winter wheat. Plant Sci 270:123–130

Elias EM, Manthey FA, Stack RW, Kianian SF (2005) Breeding efforts to develop Fusarium head blight resistant durum wheat in North Dakota. Paper presented at the proceedings 2005 National Fusarium Head Blight Forum, Milwaukee, WI

Ellis M, Spielmeyer W, Gale K, Rebetzke G, Richards R (2002) Perfect markers for the *Rht-B1b* and *Rht-D1b* dwarfing genes in wheat. Theor Appl Genet 105:1038–1042

Elshire RJ, Glaubitz JC, Sun Q et al (2011) A robust, simple genotyping-by-sequencing (GBS) approach for high diversity species. PLoS One 6:e19379

Faris JD, Zhang Z, Lu H, Lu S, Reddy L, Cloutier S, Fellers JP, Meinhardt SW, Rasmussen JB, Xu SS, Oliver RP (2010) A unique wheat disease resistance-like gene governs effector-triggered susceptibility to necrotrophic pathogens. Proc Natl Acad Sci U S A 107(30):13544–13549

Feldman M, Levy AA (2012) Genome evolution due to allopolyploidization in wheat. Genetics 192(3):763–774

Feuillet C, Travella S, Stein N, Albar L, Nublat A, Keller B (2003) Map-based isolation of the leaf rust disease resistance gene *Lr10* from the hexaploid wheat (*Triticum aestivum* L.) genome. Proc Natl Acad Sci U S A 100(25):15253–15258

Forrest K, Pujol V, Bulli P, Pumphery M, Wellings C, Herrera-Foessel S, Huerta-Espino J, Singh R, Lagudah E, Hayden M, Spielmeyer W (2014) Development of a SNP marker assay for the *Lr67* gene of wheat using a genotyping-by-sequencing approach. Mol Breeding 34:2109–2118

Fu D, Szucs P, Yan L, Helguera M, Skinner JS, Von Zitzewitz J, Hayes PM, Dubcovsky J (2005) Large deletions within the first intron in *VRN-1* are associated with spring growth habit in barley and wheat. Mol Genet Genomics 273(1):54–65

Fu D, Uauy C, Distelfeld A, Blechl A, Epstein L, Chen X, Sela H, Fahima T, Dubcovsky J (2009) A kinase-START gene confers temperature-dependent resistance to wheat stripe rust. Science 323:1357–1360

Gale KR (2005) Diagnostic DNA markers for quality traits in wheat. J Cereal Sci 41:181–192

Ganal MW, Altmann T, Roder MS (2009) SNP identification in crop plants. Curr Opin Plant Biol 12:211–217

Gao L, Kielsmeier-Cook J, Bajgain P, Zhang X, Chao S, Rouse MN, Anderson JA (2015) Development of genotyping-by-sequencing (GBS)- and array-derived SNP markers for stem rust resistance gene *Sr42*. Mol Breeding 35:207

Gao F, Liu J, Yang L, Wu X, Xiao Y, Xia X et al (2016a) Genome-wide linkage mapping of QTL for physiological traits in a Chinese wheat population using the 90K SNP array. Euphytica 209:789–804

Gao L, Turner MK, Chao S, Kolmer J, Anderson JA (2016b) Genome-wide association study of seedling and adult plant leaf rust resistance in elite spring wheat breeding lines. PLoS One 11:e0148671

Gao L, Zhao G, Huang D, Jia J (2017) Candidate loci involved in domestication and improvement detected by a published 90K wheat SNP array. Sci Rep 7:44530

Garcia M, Eckermann P, Haefele S, Satija S, Sznajder B, Timmins A et al (2019) Genome-wide association mapping of grain yield in a diverse collection of spring wheat (*Triticum aestivum* L.) evaluated in southern Australia. PLoS One 14(2):e0211730

Garg B, Lata C, Prasad M (2012) A study of the role of gene *TaMYB2* and an associated SNP in dehydration tolerance in common wheat. Mol Biol Rep 39(12):10865–10871

Gautier MF, Aleman ME, Guirao A, Marion D, Joudrier P (1994) *Triticum aestivum* puroindolines, two basic cystine-rich seed proteins: cDNA sequence analysis and developmental gene expression. Plant Mol Biol 25(1):43–57

Geng H, Xia X, Zhang L, Qu Y, He Z (2012) Development of functional markers for a lipoxygenase gene *TaLox-B1* on chromosome 4BS in common wheat. Crop Sci 52(2):568–576

Geng H, Shi J, Fuerst EP, Wei J, Morris CF (2019) Physical mapping of peroxidase genes and development of functional markers for *TaPod-D1* on bread wheat chromosome 7D. Front Plant Sci 10:523

Goyal A, Bandopadhyay R, Sourdille P, Endo TR, Balyan HS, Gupta PK (2005) Physical molecular maps of wheat chromosomes. Funct Integr Genomics 5:260–263

Graybosch RA, Peterson CJ, Baenziger PS, Baltensperger DD, Nelson LA, Jin Y, Kolmer J, Seabourn B, French R, Hein G, Martin TJ (2009) Registration of 'Mace' hard red winter wheat. J Plant Regist 3(1):51–56

Gunderson KL (2009) Whole-genome genotyping on bead arrays. Methods Mol Biol 529:197–213

Gupta PK, Varshney RK (2000) The development and use of microsatellite markers for genetic analysis and plant breeding with emphasis on bread wheat. Euphytica 113:163–185

Gupta PK, Varshney RK, Sharma PC, Ramesh B (1999) Molecular markers and their applications in wheat breeding. Plant Breed 118(5):369–390

Gupta PK, Balyan HS, Edwards KJ, Isaac P, Korzun V, Roder MS, Gautier MF, Joudrier P, Schlatter AR, Dubcovsky J, De La Pena RC, Khairallah M, Penner G, Hayden MJ, Sharp P, Keller B, Wang RCC, Hardouin JP, Jack P, Leroy P (2002) Genetic mapping of 66 new microsatellite (SSR) loci in bread wheat. Theor Appl Genet 105:413–422

Gupta PK, Mir RR, Mohan A, Kumar J (2008a) Wheat genomics: present status and future prospects. Int J Plant Genomics 2008:896451

Gupta PK, Rustgi S, Mir RR (2008b) Array-based high-throughput DNA markers for crop improvement. Heredity 101:5–18

Gupta PK, Rustgi S, Mir RR (2013) Array-based high-throughput DNA markers and genotyping platforms for cereal genetics and genomics. In: Gupta PK, Varshney RK (eds) Cereal Genomic II. Springer, Dordrecht, pp 11–55

Gupta PK, Balyan HS, Gahlaut V (2017) QTL analysis for drought tolerance in wheat: present status and future possibilities. Agronomy 7:5

Guzmán C, Caballero L, Martín MA, Alvarez JB (2012) Molecular characterization and diversity of the *Pina* and *Pinb* genes in cultivated and wild diploid wheat. Mol Breeding 30(1):69–78

Hanif M, Gao F, Liu J, Wen W, Zhang Y, Rasheed A, Xia X, He Z, Cao S (2016) *TaTGW6-A1*, an ortholog of rice *TGW6*, is associated with grain weight and yield in bread wheat. Mol Breeding 36(1):1

He XY, He ZH, Zhang LP, Sun DJ, Morris CF, Fuerst EP, Xia XC (2007) Allelic variation of *polyphenol oxidase* (PPO) genes located on chromosomes 2A and 2D and development of functional markers for the *PPO* genes in common wheat. Theor Appl Genet 115(1):47–58

He XY, Zhang YL, He ZH, Wu YP, Xiao YG, Ma CX, Xia XC (2008) Characterization of *phytoene synthase 1 gene* (*Psy1*) located on common wheat chromosome 7A and development of a functional marker. Theor Appl Genet 116(2):213–221

He XY, He ZH, Ma W, Appels R, Xia XC (2009) Allelic variants of *phytoene synthase 1* (*Psy1*) genes in Chinese and CIMMYT wheat cultivars and development of functional markers for flour colour. Mol Breeding 23(4):553–563

He C, Holme J, Anthony J (2014) SNP genotyping: the KASP assay. In: Fleury D, Whitford R (eds) Crop breeding. Methods in molecular biology (methods and protocols), vol 1145. Humana Press, New York, pp 75–86

Himi E, Noda K (2005) Red grain colour gene (R) of wheat is a Myb-type transcription factor. Euphytica 143(3):239–242

Hiremath PJ, Kumar A, Penmetsa RV, Farmer A, Schlueter JA, Chamarthi SK et al (2012) Large-scale development of cost-effective SNP marker assays for diversity assessment and genetic mapping in chickpea and comparative mapping in legumes. Plant Biotechnol J 10:716–732

Hu MJ, Zhang HP, Liu K, Cao JJ, Wang SX, Jiang H, Wu ZY, Lu J, Zhu XF, Xia XC, Sun GL (2016) Cloning and characterization of *TaTGW-7A* gene associated with grain weight in wheat via SLAF-seq-BSA. Front Plant Sci 7:1902

Huang L, Brooks SA, Li W, Fellers JP, Trick HN, Gill BS (2003) Map-based cloning of leaf rust resistance gene *Lr21* from the large and polyploid genome of bread wheat. Genetics 164(2):655–664

Huang YF, Poland JA, Wight CP, Jackson EW, Tinker NA (2014) Using genotyping-by-sequencing (GBS) for genomic discovery in cultivated oat. PLoS One 9:e102448

Hurni S, Brunner S, Buchmann G, Herren G, Jordan T, Krukowski P, Wicker T, Yahiaoui N, Mago R, Keller B (2013) Rye *Pm8* and wheat *Pm3* are orthologous genes and show evolutionary conservation of resistance function against powdery mildew. Plant J 76(6):957–969

Hussain SS, Qamar R (2007) Wheat genomics: challenges and alternative strategies. Proc Pakistan Acad Sci 44:305–327

International Wheat Genome Sequencing Consortium (2014) A chromosome-based draft sequence of the hexaploid bread wheat (*Triticum aestivum*) genome. Science 345:1251788

Jamil M, Ali A, Gul A, Ghafoor A, Napar AA, Ibrahim AM, Naveed NH, Yasin NA, Mujeeb-Kazi A (2019) Genome-wide association studies of seven agronomic traits under two sowing conditions in bread wheat. BMC Plant Biol 19(1):149

Jatayev S, Kurishbayev A, Zotova L, Khasanova G, Serikbay D, Zhubatkanov A, Botayeva M, Zhumalin A, Turbekova A, Soole K, Langridge P (2017) Advantages of Amplifluor-like SNP markers over KASP in plant genotyping. BMC Plant Biol 17(2):254

Jefferies SP, Pallotta MA, Paull JG, Karakousis A, Kretschmer JM, Manning S, Islam AKMR, Langridge P, Chalmers KJ (2000) Mapping and validation of chromosome regions conferring boron toxicity tolerance in wheat (*Triticum aestivum*). Theor Appl Genet 101:767–777

Jiang GL (2013) Molecular markers and marker-assisted breeding in plants, plant breeding from laboratories to fields. In: Andersen SB (ed) Plant breeding from laboratories to fields. InTech. Publishers, Rijeka, p 52583

Jiang Q, Hou J, Hao C, Wang L, Ge H, Dong Y, Zhang X (2011) The wheat (*Triticum aestivum*) *sucrose synthase 2* gene (*TaSus2*) active in endosperm development is associated with yield traits. Funct Integr Genomics 11(1):49–61

Jordan T, Seeholzer S, Schwizer S, Toller A, Somssich IE, Keller B (2011) The wheat *Mla* homologue *TmMla1* exhibits an evolutionarily conserved function against powdery mildew in both wheat and barley. Plant J 65(4):610–621

Jordan K, Wang S, Lun Y, Gardiner L-J, MacLachlan R, Hucl P et al (2015) A haplotype map of allohexaploid wheat reveals distinct patterns of selection on homoeologous genomes. Genome Biol 16:48

Kesawat MS, Das BK (2009) Molecular markers: it's application in crop improvement. J Crop Sci Biotech 12(4):168–178

Kidwell KK, Shelton GS, Morris CF, Line RF, Miller BC, Davis MA, Konzak CF (1999) Registration of 'Scarlet' wheat. Crop Sci 39(4):1255

Klymiuk V, Yaniv E, Huang L, Raats D, Fatiukha A, Chen S, Feng L, Frenkel Z, Krugman T, Lidzbarsky G, Chang W (2018) Cloning of the wheat *Yr15* resistance gene sheds light on the plant tandem kinase-pseudokinase family. Nat Commun 9(1):3735

Korzun V, Ebmeyer E (2003) Molecular markers and their applications in wheat breeding. In: Pogna NE, Romano M, Pogna EA, Galterio G (eds) Proceedings 10th international wheat genetics symposium, Istituto Sperimentrale per la Cerealicoltura, Rome, Italy, Vol 1, pp. 140–143

Korzun V, Roder MS, Worland AJ, Borner A (1997) Application of microsatellite markers to distinguish inter-varietal chromosome substitution lines of wheat (*Triticum aestivum* L.). Euphytica 95:149–155

Krattinger SG, Lagudah ES, Spielmeyer W, Singh RP, Huerta-Espino J, McFadden H, Bossolini E, Selter LL, Keller B (2009) A putative ABC transporter confers durable resistance to multiple fungal pathogens in wheat. Science 323:1360–1363

Kuchel H, Fox R, Reinheimer J, Mosionek L, Willey N, Bariana H, Jefferies S (2007) The successful application of a marker-assisted wheat breeding strategy. Mol Breeding 20(4):295–308

Kumar J, Mir RR, Kumar N, Kumar A, Mohan A, Prabhu KV, Balyan HS, Gupta PK (2010) Marker-assisted selection for pre-harvest sprouting tolerance and leaf rust resistance in bread wheat. Plant Breed 129(6):617–621

Kumar J, Jaiswal V, Kumar A, Kumar N, Mir RR, Kumar S et al (2011) Introgression of a major gene for high grain protein content in some Indian bread wheat cultivars. Field Crop Res 123:226–233

Kumar S, Goyal A, Mohan A, Balyan HS, Gupta PK (2013) An integrated physical map of simple sequence repeats in bread wheat. Aust J Crop Sci 7(4):460–468

Kumar S, Knox R, Clarke F, Pozniak C, Depauw R, Cuthbert R et al (2015) Maximizing the identification of QTL for pre-harvest sprouting resistance using seed dormancy measures in a white-grained hexaploid wheat population. Euphytica 205:287–309

Kumar N, Randhawa HS, Higginbotham RW, Chen X, Murray TD, Gill KS (2017) Targeted and efficient transfer of value-added genes into a wheat variety. Mol Breeding 37(5):68

Kumar S, Knox RE, Singh AK, DePauw RM, Campbell HL, Isidro-Sanchez J, Clarke FR, Pozniak CJ, N'Daye A, Meyer B, Sharpe A, Ruan Y, Cuthbert RD, Somers D, Fedak G (2018) High-density genetic mapping of a major QTL for resistance to multiple races of loose smut in a tetraploid wheat cross. PLoS One 13(2):e0192261

Kumar S, Phogat BS, Vikas VK, Sharma AK, Saharan MS, Singh AK et al (2019) Mining of Indian wheat germplasm collection for adult plant resistance to leaf rust. PLoS One 14(3):e0213468

Kunert A, Naz AA, Dedeck O, Pillen K, Léon J (2007) AB-QTL analysis in winter wheat: I. Synthetic hexaploid wheat (*T. turgidum* ssp. *dicoccoides* × *T. tauschii*) as a source of favourable alleles for milling and baking quality traits. Theor Appl Genet 115(5):683–695

Lagudah ES, Krattinger SG, Herrera-Foessel S et al (2009) Gene-specific markers for the wheat gene *Lr34/Yr18/Pm38*, which confers resistance to multiple fungal pathogens. Theor Appl Genet 119(5):889–898

Lai K, Duran C, Berkman PJ, Lorenc MT, Stiller J, Manoli S, Hayden MJ, Forrest KL, Fleury D, Baumann U, Zander M, Mason AS, Batley J, Edwards D (2012) Single nucleotide polymorphism discovery from wheat next-generation sequence data. Plant Biotechnol J 10:743–749

Landjeva S, Korzun V, Borner A (2007) Molecular markers: actual and potential contributions to wheat genome characterization and breeding. Euphytica 156:271–296

Langridge P, Chalmers K (2004) Biotechnology in agriculture and forestry. In: Lorz H, Wenzel G (eds) Molecular marker systems, vol 55. Springer-Verlag, Berlin Heidelberg

Langridge P, Lagudah ES, Holton TA, Appels A, Sharp P, Chalmers KJ (2001) Trends in genetic and genome analyses in wheat: a review. Aust J Agr Res 52:1043–1077

Li H, Vikram P, Singh RP, Kilian A, Carling J, Song J, Burgueno-Ferreira JA, Bhavani S, Huerta-Espino J, Payne T, Sehgal D, Wenzl P, Singh S (2015a) A high density GBS map of bread wheat and its application for dissecting complex disease resistance traits. BMC Genomics 16(1):216

Li G, Wang Y, Chen MS, Edae E, Poland J, Akhunov E, Chao S, Bai G, Carver BF, Yan L (2015b) Precisely mapping a major gene conferring resistance to hessian fly in bread wheat using genotyping-by-sequencing. BMC Genomics 16(1):108

Li B, Li Q, Mao X, Li A, Wang J, Chang X, Hao C, Zhang X, Jing R (2016) Two novel *AP2/EREBP* transcription factor genes *TaPARG* have pleiotropic functions on plant architecture and yield-related traits in common wheat. Front Plant Sci 7:1191

Li F, Wen W, Liu J et al (2019) Genetic architecture of grain yield in bread wheat based on genome-wide association studies. BMC Plant Biol 19:168

Lin M, Cai S, Wang S, Liu S, Zhang G, Bai G (2015) Genotyping-by-sequencing (GBS) identified SNP tightly linked to QTL for pre-harvest sprouting resistance. Theor Appl Genet 128(7):1385–1395

Lindsay MP, Lagudah ES, Hare RA, Munns R (2004) A locus for sodium exclusion (*Nax1*), a trait for salt tolerance, mapped in durum wheat. Funct Plant Biol 31(11):1105–1114

Liu S, Chao S, Anderson JA (2008) New DNA markers for high molecular weight glutenin subunits in wheat. Theor Appl Genet 118(1):177

Liu W, Frick M, Huel R, Nykiforuk CL, Wang X, Gaudet DA, Eudes F, Conner RL, Kuzyk A, Chen Q, Kang Z (2014) The stripe rust resistance gene *Yr10* encodes an evolutionary-conserved and unique CC–NBS–LRR sequence in wheat. Mol Plant 7(12):1740–1755

Liu S, Assanga SO, Dhakal S, Gu X, Tan CT, Yang Y et al (2016) Validation of chromosomal locations of 90K array single nucleotide polymorphisms in US wheat. Crop Sci 56(1):364–373

Liu J, He Z, Rasheed A et al (2017) Genome-wide association mapping of black point reaction in common wheat (*Triticum aestivum* L.). BMC Plant Biol 17:220

Liu J, Xu Z, Fan X, Zhou Q, Cao J, Wang F, Ji G, Yang L, Feng B, Wang T (2018) A genome-wide association study of wheat spike related traits in China. Front Plant Sci 9:1584

Long YM, Chao WS, Ma GJ, Xu SS, Qi LL (2016) An innovative SNP genotyping method adapting to multiple platforms and throughputs. Theor Appl Genet 130:597–607

Lorz H, Wenzel G (2004) Molecular marker systems in plant breeding and crop improvement. Springer Verlag, Germany, p 476

Ma ZQ, Roder M, Sorrells ME (1996) Frequencies and sequence characteristics of di-, tri- and tetra- nucleotide microsatellites in wheat. Genome 39:123–130

Ma D, Yan J, He Z, Wu L, Xia X (2012) Characterization of a cell wall invertase gene *TaCwi-A1* on common wheat chromosome 2A and development of functional markers. Mol Breeding 29(1):43–52

Ma L, Li T, Hao C, Wang Y, Chen X, Zhang X (2016) *Ta GS 5-3A*, a grain size gene selected during wheat improvement for larger kernel and yield. Plant Biotechnol J 14(5):1269–1280

Mago R, Zhang P, Vautrin S, Simkova H, Bansal U, Luo MC, Rouse M, Karaoglu H, Periyannan S, Kolmer J, Jin Y (2015) The wheat *Sr50* gene reveals rich diversity at a cereal disease resistance locus. Nat Plants 1(12):15186

Mallick N, Sharma JB, Tomar RS, Sivasamy M, Prabhu KV (2015) Marker-assisted backcross breeding to combine multiple rust resistance in wheat. Plant Breed 134(2):172–177

Mardis ER (2008) Next-generation DNA sequencing methods. Annu Rev Genomics Hum Genet 9:387–402

Massa AN, Morris CF, Gill BS (2004) Sequence diversity of puroindoline-a, puroindoline-b, and the grain softness protein genes in *Aegilops tauschii* Coss. Crop Sci 44(5):1808–1816

Mir RR, Varshney RK (2013) Future prospects of molecular markers in plants. In: Henry RJ (ed) Molecular markers in plants. Blackwell Publishing Ltd, Oxford, pp 169–190

Mir RR, Hiremath PJ, Riera-Lizarazu O, Varshney RK (2013) Evolving molecular marker technologies in plants: from RFLPs to GBS. In: Lubberstedt T, Varshney RK (eds) Diagnostics in plant breeding. Springer, New York, pp 229–247

Mishra VK, Gupta PK, Arun B, Chand R, Vasistha NK, Vishwakarma MK, Yadav PS, Joshi AK (2015) Introgression of a gene for high grain protein content (*Gpc-B1*) into two leading cultivars of wheat in eastern Gangetic Plains of India through marker assisted backcross breeding. J Plant Breed Crop Sci 7(8):292–300

Moore JW, Herrera-Foessel S, Lan C, Schnippenkoetter W, Ayliffe M, Huerta-Espino J, Lillemo M, Viccars L, Milne R, Periyannan S, Kong X (2015) A recently evolved hexose transporter variant confers resistance to multiple pathogens in wheat. Nat Genet 47(12):1494

Moullet O, Fossati D, Mascher F, Guadagnolo R, Schori A (2009) Use of marker-assisted selection (MAS) for pyramiding two leaf rust resistance genes,(*Lr9* and *Lr24*) in wheat. In Intl Conf "Conventional and Molecular Breeding of Field and Vegetable Crops", November 24–27, 2008, Novi Sad, Serbia, pp. 24–27

Murai J, Taira T, Ohta D (1999) Isolation and characterization of the three waxy genes encoding the granule-bound starch synthase in hexaploid wheat. Gene 234(1):71–79

Nadeem MA, Nawaz MA, Shahid MQ, Dogan Y, Comertpay G, Yildiz M, Hatipoglu R, Ahmad F, Alsaleh A, Labhane N, Ozkan H, Chung G, Baloch FS (2018) DNA molecular markers in plant breeding: current status and recent advancements in genomic selection and genome editing. Biotechnol Biotechnol Equip 32(2):261–285

Nakamura S, Abe F, Kawahigashi H, Nakazono K, Tagiri A, Matsumoto T, Utsugi S, Ogawa T, Handa H, Ishida H, Mori M (2011) A wheat homolog of MOTHER OF FT AND TFL1 acts in the regulation of germination. Plant Cell 23(9):3215–3229

Neelam K, Brown-Guedira G, Huang L (2013) Development and validation of a breeder-friendly KASPar marker for wheat leaf rust resistance locus *Lr21*. Mol Breeding 31(1):233–237

Negro SS, Millet EJ, Madur D et al (2019) Genotyping-by-sequencing and SNP-arrays are complementary for detecting quantitative trait loci by tagging different haplotypes in association studies. BMC Plant Biol 19:318

Ogbonnaya FC, Subrahmanyam NC, Moullet O, De Majnik J, Eagles HA, Brown JS, Eastwood RF, Kollmorgen J, Appels R, Lagudah ES (2001) Diagnostic DNA markers for cereal cyst nematode resistance in bread wheat. Aust J Agr Res 52(12):1367–1374

Ohnishi N, Himi E, Yamasaki Y, Noda K (2008) Differential expression of three ABA-insensitive five binding protein (AFP)-like genes in wheat. Genes Genet Syst 83(2):167–177

Pan Z, Song W, Meng F, Xu L, Liu B, Zhu J (2004) Characterization of genes encoding wheat grain hardness from Chinese cultivar GaoCheng 8901. Cereal Chem 81(2):287–289

Pandey B, Sharma P, Pandey DM, Sharma I, Chatrath R (2013) Identification of new aquaporin genes and single nucleotide polymorphism in bread wheat. Evol Bioinformatics Online 9:437–452

Pariasca-Tanaka J, Lorieux M, He C, McCouch S, Thomson MJ, Wissuwa M (2015) Development of a SNP genotyping panel for detecting polymorphisms in *Oryza glaberrima*/ *O. sativa* interspecific crosses. Euphytica 201:67–78

Parker GD, Chalmers KJ, Rathjen AJ, Langridge P (1999) Mapping loci associated with milling yield in wheat (*Triticum aestivum* L). Mol Breeding 5:561–568

Paux E, Sourdille P, Mackay I, Feuillet C (2011) Sequence-based marker development in wheat: advances and applications to breeding. Biotechnol Adv 30:1071–1088

Periyannan S, Moore J, Ayliffe M, Bansal U, Wang X, Huang L, Deal K, Luo M, Kong X, Bariana H, Mago R (2013) The gene *Sr33,* an ortholog of barley *Mla* genes, encodes resistance to wheat stem rust race Ug99. Science 341:786–788

Periyannan S, Bansal U, Bariana H, Deal K, Luo MC, Dvorak J, Lagudah E (2014) Identification of a robust molecular marker for the detection of the stem rust resistance gene *Sr45* in common wheat. Theor Appl Genet 127(4):947–955

Picoult-Newberg L, Ideker TE, Pohl MG, Taylor SL, Donaldson MA, Nickerson DA, Boyce-Jacino M (1999) Mining SNPs from EST databases. Genome Res 9(167):174

Plaschke J, Ganal M, Roder MS (1995) Detection of genetic diversity in closely related bread wheat using microsatellite markers. Theor Appl Genet 91:1001–1007

Poland JA, Rife TW (2012) Genotyping-by-sequencing for plant breeding and genetics. Plant Genome 5:92–102

Poland JA, Brown PJ, Sorrells ME, Jannink J-L (2012a) Development of high-density genetic maps for barley and wheat using a novel two-enzyme genotyping-by-sequencing approach. PLoS One 7:e32253

Poland JA, Endelman J, Dawson J, Rutkoski J, Wu S, Manes Y et al (2012b) Genomic selection in wheat breeding using genotyping-by-sequencing. Plant Genome 5:103–113

Prasad M, Varshney RK, Roy JK, Balyan HS, Gupta PK (2000) The use of microsatellites for detecting DNA polymorphism, genotype identification and genetic diversity in wheat. Theor Appl Genet 100:584–592

Przewieslik-Allen AM, Burridge AJ, Wilkinson PA, Winfield MO, Shaw DS, McAusland L, King J, King IP, Edwards KJ, Barker GLA (2019) Developing a high-throughput SNP-based marker system to facilitate the introgression of traits from *Aegilops* species into bread wheat (*Triticum aestivum*). Front Plant Sci 9:1993

Rai A, Mahendru-Singh A, Raghunandan K, Kumar TPJ, Sharma P, Ahlawat AK, Singh SK, Ganjewala D, Shukla RB, Sivasamy M (2019) Marker-assisted transfer of *PinaD1a* gene to develop soft grain wheat cultivars. 3 Biotech 9(5):183

Randhawa HS, Popovic Z, Menzies J, Knox R, Fox S (2009) Genetics and identification of molecular markers linked to resistance to loose smut (*Ustilago tritici*) race T33 in durum wheat. Euphytica 169(2):151–157

Rasheed A, Wen W, Gao F, Zhai S, Jin H, Liu J, He Z (2016) Development and validation of KASP assays for genes underpinning key economic traits in bread wheat. Theor Appl Genet 129(10):1843–1860

Rasheed A, Hao Y, Xia X, Khan A, Xu Y, Varshney RK, He Z (2017) Crop breeding chips and genotyping platforms: progress, challenges, and perspectives. Mol Plant 10:1047–1064

Rawat N, Pumphrey MO, Liu S, Zhang X, Tiwari VK, Ando K, Trick HN, Bockus WW, Akhunov E, Anderson JA, Gill BS (2016) Wheat *Fhb1* encodes a chimeric lectin with agglutinin domains and a pore-forming toxin-like domain conferring resistance to Fusarium head blight. Nat Genet 48(12):1576

Rehman SU, Wang J, Chang X, Zhang X, Mao X, Jing R (2019) A wheat protein kinase gene TaSnRK2.9-5A associated with yield contributing traits. Theor Appl Genet 132(4):907–919

Ren T, Hu Y, Tang Y, Li C, Yan B, Ren Z, Tan F, Tang Z, Fu S, Li Z (2018) Utilization of a Wheat55K SNP array for mapping of major QTL for temporal expression of the tiller number. Front Plant Sci 9:333

Riley R, Chapman V (1958) Genetic control of the cytologically diploid behavior of hexaploid wheat. Nature 182:713–715

Rimbert H, Darrier B, Navarro J, Kitt J, Choulet F, Leveugle M et al (2018) High throughput SNP discovery and genotyping in hexaploid wheat. PLoS One 13(1):e0186329

Roder MS, Plaschke J, Konig SU, Borner A, Sorrels ME (1995) Abundance, variability and chromosomal location of microsatellite in wheat. Mol Gen Genet 246:327–333

Roder MS, Huang X-Q, Ganal MW (2004) Wheat microsatellites: potential and implications. In: Lörz H, Wenzel G (eds) Molecular marker systems in plant breeding and crop improvement. Biotechnology in agriculture and forestry, vol 55. Springer, Berlin Heidelberg, pp 255–266

Rustgi S, Bandopadhyay R, Balyan HS, Gupta PK (2009) EST-SNPs in bread wheat: discovery, validation, genotyping and haplotype structure. Czech J Genet Breed 45:106–116

Saintenac C, Jiang D, Wang S, Akhunov E (2013a) Sequence-based mapping of the polyploid wheat genome. G3 (Bethesda) 3:1105–1114

Saintenac C, Zhang W, Salcedo A, Rouse MN, Trick HN, Akhunov E, Dubcovsky J (2013b) Identification of wheat gene *Sr35* that confers resistance to Ug99 stem rust race group. Science 341:783–786

Saintenac C, Lee WS, Cambon F, Rudd JJ, King RC, Marande W, Powers SJ, Bergès H, Phillips AL, Uauy C, Hammond-Kosack KE (2018) Wheat receptor-kinase-like protein Stb6 controls gene-for-gene resistance to fungal pathogen Zymoseptoria tritici. Nat Genet 50(3):368

Saxena RK, Penmetsa RV, Upadhyaya HD, Kumar A, Carrasquilla-Garcia N, Schlueter JA et al (2012) Large-scale development of cost-effective single-nucleotide polymorphism marker assays for genetic mapping in pigeonpea and comparative mapping in legumes. DNA Res 19:449–461

Scheben A, Batley J, Edwards D (2017) Genotyping-by-sequencing approaches to characterize crop genomes: choosing the right tool for the right application. Plant Biotechnol 15:149–161

Sears ER (1954) The aneuploids of common wheat. Missouri Agric Exp Stn Res Bull 572:1–5

Sehgal D, Vikram P, Sansaloni CP, Ortiz C, Pierre CS, Payne T et al (2015) Exploring and mobilizing the gene bank biodiversity for wheat improvement. PLoS One 10(7):e0132112

Semagn K, Babu R, Hearne S et al (2014) Single nucleotide polymorphism genotyping using Kompetitive Allele Specific PCR (KASP): overview of the technology and its application in crop improvement. Mol Breeding 33:1–14

Shi G, Zhang Z, Friesen TL, Raats D, Fahima T, Brueggeman RS, Lu S, Trick HN, Liu Z, Chao W, Frenkel Z (2016) The hijacking of a receptor kinase–driven pathway by a wheat fungal pathogen leads to disease. Sci Adv 2(10):e1600822

Shiferaw B, Smale M, Braun HJ, Duveiller E, Reynolds M, Muricho G (2013) Crops that feed the world: past successes and future challenges to the role played by wheat in global food security. Food Sec 5:291–317

Shorinola O, Bird N, Simmonds J, Berry S, Henriksson T, Jack P, Werner P, Gerjets T, Scholefield D, Balcárková B, Valárik M (2016) The wheat *Phs-A1* pre-harvest sprouting resistance locus delays the rate of seed dormancy loss and maps 0.3 cM distal to the *PM19* genes in UK germplasm. J Exp Bot 67(14):4169–4178

Sieber AN, Longin CFH, Leiser WL, Würschum T (2016) Copy number variation of *CBF-A14* at the *Fr-A2* locus determines frost tolerance in winter durum wheat. Theor Appl Genet 129(6):1087–1097

Singh M, Mallick N, Chand S, Kumari P, Sharma JB, Sivasamy M, Jayaprakash P, Prabhu KV, Jha SK (2017) Marker-assisted pyramiding of Thinopyrum-derived leaf rust resistance genes *Lr19* and *Lr24* in bread wheat variety HD2733. J Genet 96(6):951–957

Smith SM, Maughan PJ (2015) SNP genotyping using KASPar assays. In: Batley J (ed) Plant genotyping. Methods in molecular biology (methods and protocols), vol 1245. Humana Press, New York

Somers DJ, Isaac P, Edwards K (2004) A high-density microsatellite consensus map for bread wheat (*Triticum aestivum* L.). Theor Appl Genet 109:1105–1114

Soto-Cerda BJ, Inostroza-Blancheteau C, Mathias M, Penaloza E, Zuniga J, Munoz G, Rengel Z, Salvo-Garrido H (2015) Marker-assisted breeding for *TaALMT1*, a major gene conferring aluminium tolerance to wheat. Biol Plantarum 59(1):83–91

Sourdille P, Singh S, Cadalen T, Gina L, Brown-Guedira GG, Qi L, Gill BS, Dufour P, Murigneux A, Bernard M (2004) Microsatellite-based deletion bin system for the establishment of genetic-physical map relationships in wheat (*Triticum aestivum* L). Funct Integr Genomics 4:12–25

Steele KA, Quinton-Tulloch MJ, Amgai RB, Dhakal R, Khatiwada SP, Vyas D et al (2018) Accelerating public sector rice breeding with high-density KASP markers derived from whole genome sequencing of Indica rice. Mol Breed 38:38

Steemers FJ, Gunderson KL (2007) Whole genome genotyping technologies on the BeadArray platform. Biotechnol J 2:41–49

Steemers F, Chang W, Lee G, Shen R, Barker DL, Gunderson KL (2006) Whole genome genotyping (WGG) using single base extension (SBE). Nat Methods 3(1):31–33

Su Z, Hao C, Wang L, Dong Y, Zhang X (2011) Identification and development of a functional marker of *TaGW2* associated with grain weight in bread wheat (*Triticum aestivum* L.). Theor Appl Genet 122(1):211–223

Sun DJ, He ZH, Xia XC, Zhang LP, Morris CF, Appels R, Ma WJ, Wang H (2005) A novel STS marker for polyphenol oxidase activity in bread wheat. Mol Breeding 16(3):209–218

Thomson MJ (2014) High-throughput SNP genotyping to accelerate crop improvement. Plant Breed Biotechnol 2:195–212

Tommasini L, Yahiaoui N, Srichumpa P, Keller B (2006) Development of functional markers specific for seven *Pm3* resistance alleles and their validation in the bread wheat gene pool. Theor Appl Genet 114(1):165–175

Torada A, Koike M, Ikeguchi S, Tsutsui I (2008) Mapping of a major locus controlling seed dormancy using backcrossed progenies in wheat (*Triticum aestivum* L.). Genome 51(6):426–432

Torkamaneh D, Belzile F (2015) Scanning and filling: ultra-dense SNP genotyping combining genotyping-by-sequencing, SNP array and whole genome resequencing data. PLoS One 10(7):e0131533

Turuspekov Y, Plieske J, Ganal M, Akhunov E, Abugalieva S (2015) Phylogenetic analysis of wheat cultivars in Kazakhstan based on the wheat 90 K single nucleotide polymorphism array. Plant Genet Resources 15:29–35

Tyagi S, Mir RR, Kaur H, Chhuneja P, Ramesh B, Balyan HS, Gupta PK (2014) Marker-assisted pyramiding of eight QTLs/genes for seven different traits in common wheat (*Triticum aestivum* L.). Mol Breeding 34(1):167–175

Uauy C, Distelfeld A, Fahima T, Blechl A, Dubcovsky J (2006) A NAC gene regulating senescence improves grain protein, zinc, and iron content in wheat. Science 314:1298–1301

Utsugi S, Nakamura S, Noda K, Maekawa M (2008) Structural and functional properties of Viviparous1 genes in dormant wheat. Genes Genet Syst 83(2):153–166

Varella AC, Weaver DK, Blake NK, Hofland ML, Heo HY, Cook JP, Lamb PF, Jordan KW, Akhunov E, Chao S, Talbert LE (2019) Analysis of recombinant inbred line populations derived from wheat landraces to identify new genes for wheat stem sawfly resistance. Theor Appl Genet 132:2195–2207

Varshney R, Graner A, Sorrells ME (2005) Genic microsatellite markers in plants: features and applications. Trends Biotechnol 23:48–55

Vishwakarma MK, Mishra VK, Gupta PK, Yadav PS, Kumar H, Joshi AK (2014) Introgression of the high grain protein gene *Gpc-B1* in an elite wheat variety of Indo-Gangetic Plains through marker assisted backcross breeding. Curr Plant Biol 1:60–67

Vos P, Hogers R, Bleeker M, Reijans M, van de Lee T, Hornes M, Frijters A, Pot J, Peleman J, Kuiper M, Zabeau M (1995) AFLP: a new technique for DNA fingerprinting. Nucleic Acids Res 23:4407–4414

Wang L, Li G, Peña RJ, Xia X, He Z (2010) Development of STS markers and establishment of multiplex PCR for *Glu-A3* alleles in common wheat (*Triticum aestivum* L.). J Cereal Sci 51(3):305–312

Wang S, Wong D, Forrest K, Allen A et al (2014) Characterization of polyploid wheat genomic diversity using a high-density 90000 single nucleotide polymorphism array. Plant Biotechnol J 12:787–796

Wang J, Wen W, Hanif M, Xia X, Wang H, Liu S, Liu J, Yang L, Cao S, He Z (2016) *TaELF3-1DL*, a homolog of *ELF3*, is associated with heading date in bread wheat. Mol Breeding 36(12):161

Wei B, Jing R, Wang C, Chen J, Mao X, Chang X, Jia J (2009) *Dreb1* genes in wheat (*Triticum aestivum* L.): development of functional markers and gene mapping based on SNPs. Mol Breeding 23(1):13–22

Wei J, Geng H, Zhang Y, Liu J, Wen W, Zhang Y, Xia X, Chen X, He Z (2015) Mapping quantitative trait loci for peroxidase activity and developing gene-specific markers for *TaPod-A1* on wheat chromosome 3AL. Theor Appl Genet 128(10):2067–2076

Williams K, Taylor S, Bogacki P, Pallotta M, Bariana H, Wallwork H (2002) Mapping of the root lesion nematode (*Pratylenchus neglectus*) resistance gene *Rlnn1* in wheat. Theor Appl Genet 104(5):874–879

Winfield MO, Allen AM, Burridge AJ, Barker GL, Benbow HR, Wilkinson PA et al (2016) High-density SNP genotyping array for hexaploid wheat and its secondary and tertiary gene pool. Plant Biotechnol J 14:1195–1206

Woodward J (2014) Bi-allelic SNP genotyping using the TaqMan® assay. Methods Mol Biol 1145:67–74

Xue S, Li G, Jia H, Lin F, Cao Y, Xu F, Tang M, Wang Y, Wu X, Zhang Z, Zhang L (2010) Marker-assisted development and evaluation of near-isogenic lines for scab resistance QTLs of wheat. Mol Breeding 25(3):397–405

Yadawad A, Hanchinal RR, Nadaf HL, Desai SA, Biradar S, Rudranaik V (2015) Improvement of leaf rust resistance in bread wheat variety DWR162 (*Triticum aestivum* L.) through marker assisted selection. Indian J Genet 75(2):256–259

Yahiaoui N, Srichumpa P, Dudler R, Keller B (2004) Genome analysis at different ploidy levels allows cloning of the powdery mildew resistance gene *Pm3b* from hexaploid wheat. Plant J 37(4):528–538

Yan L, Loukoianov A, Tranquilli G, Helguera M, Fahima T, Dubcovsky J (2003) Positional cloning of the wheat vernalization gene *VRN1*. Proc Natl Acad Sci U S A 100(10):6263–6268

Ye X, Zhang S, Li S, Wang J, Chen H, Wang K, Lin Z, Wei Y, Du L, Yan Y (2019) Improvement of three commercial spring wheat varieties for powdery mildew resistance by marker-assisted selection. Crop Prot 125:104889

Yildirim A, Sonmezoglu OA, Sayaslan A, Koyuncu M, Guleç T, Kandemir N (2013) Marker-assisted breeding of a durum wheat cultivar for γ-gliadin and LMW-glutenin proteins affecting pasta quality. Turk J Agric For 37(5):527–533

Yu L-X, Abate Z, Anderson JA, Bansal UK, Bariana HS, Bhavani S, Dubcovsky J, Lagudah ES, Liu S, Sambasivam PK, Singh RP, Sorrells ME (2009) Developing and optimizing markers for stem rust resistance in wheat. The Borlaug Global Rust Initiative 2009 Technical Workshop, March 17–20, 2009, Ciudad Obregon, Mexico

Zane L, Bargelloni L, Patarnello T (2002) Strategies for microsatellite isolation: a review. Mol Ecol 11:1–16

Zhang C, Dong C, He X, Zhang L, Xia X, He Z (2011) Allelic variants at the *TaZds-D1* locus on wheat chromosome 2DL and their association with yellow pigment content. Crop Sci 51(4):1580–1590

Zhang Y, Miao X, Xia X, He Z (2014) Cloning of seed dormancy genes (*TaSdr*) associated with tolerance to pre-harvest sprouting in common wheat and development of a functional marker. Theor Appl Genet 127(4):855–866

Zhang B, Liu X, Xu W, Chang J, Li A, Mao X, Zhang X, Jing R (2015) Novel function of a putative *MOC1* ortholog associated with spikelet number per spike in common wheat. Sci Rep 5:12211

Zhang W, Chen S, Abate Z, Nirmala J, Rouse MN, Dubcovsky J (2017a) Identification and characterization of *Sr13*, a tetraploid wheat gene that confers resistance to the Ug99 stem rust race group. Proc Natl Acad Sci USA 114(45):9483–9492

Zhang P, He Z, Tian X, Gao F, Xu D, Liu J, Wen W, Fu L, Li G, Sui X, Xia X (2017b) Cloning of *TaTPP-6AL1* associated with grain weight in bread wheat and development of functional marker. Mol Breeding 37(6):78

Zhao J, Abdelsalam NR, LKhalaf L, Chuang W-P, Zhao L, Smith MC, Carver B, Bai G (2019) Development of single nucleotide polymorphism markers for the wheat curl mite resistance gene *Cmc4*. Crop Sci 59:1567–1575

Zou J, Semagn K, Iqbal M, N'diaye A, Chen H, Asif M et al (2016) Mapping QTLs controlling agronomic traits in the 'Attila' × 'CDC go' spring wheat population under organic management using 90K SNP array. Crop Sci 57:365–377

# Genomic Selection for Wheat Improvement

**Neeraj Kumar, Maneet Rana, Brijesh Kumar, Subhash Chand, Aalok Shiv, Shabir H. Wani, and Satish Kumar**

**Abstract** Recent advances in plant breeding and agronomic practices have contributed significantly to the annual genetic gain in crop productivity to the tune of 0.8–1.2%. However, the present rate of gain is insufficient to meet out the fast-growing food demand of the expected global population of 2050. Till 1980s genetic enhancement of crop plants was primarily based on conventional plant breeding approaches. Although conventional breeding is continued to be breeder's choice, faster genetic gain is hampered particularly for complex traits. Increasing the rate of genetic gain through modern breeding technologies is essential for food and nutritional security. Genomic selection (GS) is one such proven technology in animal breeding and recently incorporated in plant breeding programmes, especially large-scale private sector. GS is a promising approach for the rapid selection of superior genotypes and accelerating the breeding cycle. A comprehensive review of the existing GS literature in crop plants may provide insights for integrating GS in crop breeding programmes. Incorporation and effective use of GS in breeding programme depend upon several factors such as breeding method, genetic architecture and heritability number of targeted traits, statistical models, availability of genotyping and phenotyping facilities and the budget of breeding program. In this chapter, we discuss GS in wheat while highlighting various studies carried for improvement of grain yield, biotic and abiotic stresses, disease resistance and grain quality parameters. Also discussed are the challenges and key considerations to be followed for successful implementation of GS in varietal development programmes. Most of the GS studies are used to predict the additive genetic value and lag behind for non-additive and Genotype X Environment Interaction (GEI). Multi-trait and multi-environment modelling is essential for improving the prediction accuracy for

N. Kumar · M. Rana · B. Kumar · S. Chand
ICAR-Indian Grassland and Fodder Research Institute, Jhansi, India

A. Shiv
ICAR-Indian Institute of Sugarcane Research, Lucknow, India

S. H. Wani
Mountain Research Centre for Field Crops, Sher-e-Kashmir University of Agricultural Sciences and Technology of Kashmir, Anantnag, Jammu and Kashmir, India

S. Kumar (✉)
ICAR-Indian Institute of Wheat & Barley Research, Karnal, India

© The Editor(s) (if applicable) and The Author(s), under exclusive license to Springer Nature Switzerland AG 2021

S. H. Wani et al. (eds.), *Physiological, Molecular, and Genetic Perspectives of Wheat Improvement*, https://doi.org/10.1007/978-3-030-59577-7_9

environment-sensitive traits. Another potential of GS is mining of genes in gene bank accessions to access unexplored diversity into breeding programmes. '

**Keywords** Genomic selection · Wheat · Next-generation phenotyping

## 1 Introduction

Wheat (*Triticum aestivum* L.) considered as one of the most important cereal crop and staple food for half of population worldwide (CRP 2018). Its annual production is about 722.4 million metric tons with gross cultivated area of 220 million hectares globally (FAO 2018). Genetic gain in wheat is restricted by its low annual growth of 0.9% (Ray et al. 2013) which can be attributed to stagnating yields (Ray et al. 2012), impact of diseases (Singh et al. 2016) and drought and heat stresses (Zampieri et al. 2017). Traditional wheat breeding and yield improvement efforts are inadequate to cope the 2% annual increment rate in global population and feed estimated ten billion population by 2050 (Hickey et al. 2019). Traditional breeding methodologies basically rely on evaluating phenotypic merit along with pedigree information (Rasmusson and Phillips 1997) prompting the lower accuracy and efficiency for trait selections that are modulated by prevailing environmental conditions (Heffner et al. 2009) and hindering precision in selection. To overcome these challenges and to sustain production, modification and upgradation of conventional breeding techniques are prerequisite for meeting out the production to feed the increasing population.

Hence, to hasten rate of genetic gain for higher yield and stress resilience, incorporation of genomic tools can help to achieve precise and accurate selection which thus facilitates in saving time, resources and labour in wheat breeding programmes. To surmount the shortcomings of conventional wheat breeding, new breeding approaches with combination of phenotyping and genotyping approaches have led to accelerated genetic gains. Numerous marker-assisted breeding strategies like MAS (marker-assisted selection), MABB (marker-assisted backcross breeding) and MARS (marker-assisted recurrent selection) can assist the selection of favourable alleles for desired traits in early generations (Howes and Woods 1998; Bonnett et al. 2005). However, the requirements of marker identification and overestimation of marker effects with small phenotypic variance explained are the significant limitations of marker-assisted breeding programme (Hayes and Goddard 2001; Meuwissen et al. 2001).

However, the availability of wheat genome reference sequence, RefSeq v.1.0 (IWGSC 2018), and advances in high-throughput genotyping platform has potential to hasten the process of marker identification and prediction of accurate marker effect on phenotype which are highly useful for mapping trait, gene discovery and advance molecular breeding process. Identification of genome-wide distributed single-nucleotide polymorphisms (SNPs) markers can provide new possibilities and opportunities for genetic improvement of bread wheat in addition to enhancing the rate of productivity gains (Juliana et al. 2019a). With accurate genotypic and phenotypic information, genomic selection (GS) can facilitate the rapid selection and identification of desired genotypes by utilizing genome-wide distributed markers to estimate the effects of all loci and predict the genomic estimated breeding values (GEBV) in order to achieve more reliable selection. Linear models like G-BLUP and machine learning algorithms are used in understanding the complex patterns of data to make correct decisions. These prediction models can be effectively utilized in exploiting positive G × E interactions. Modelling multi-trait and multi-environment is prerequisite for improving the prediction accuracy and performance of newly developed lines. The main advantages of GS over phenotype-based selection breeding are significant as it can facilitate accuracy in selection, breeding time and phenotyping costs in developing a variety (Fig. 1), especially for complex traits with low heritability (Heffner et al. 2009; Crossa et al. 2017).

GS schemes are being implemented to attain genetic gains of economically important and low heritable traits which are otherwise very difficult to improve genetically by using conventional breeding principles. Incorporation and effective use of GS in breeding programme depend upon several factors such as breeding method, genetic architecture and heritability number of targeted traits, statistical models, availability of genotyping and phenotyping facilities and the budget of breeding programme (Heffner et al. 2009). Effective GS strategy utilizes an

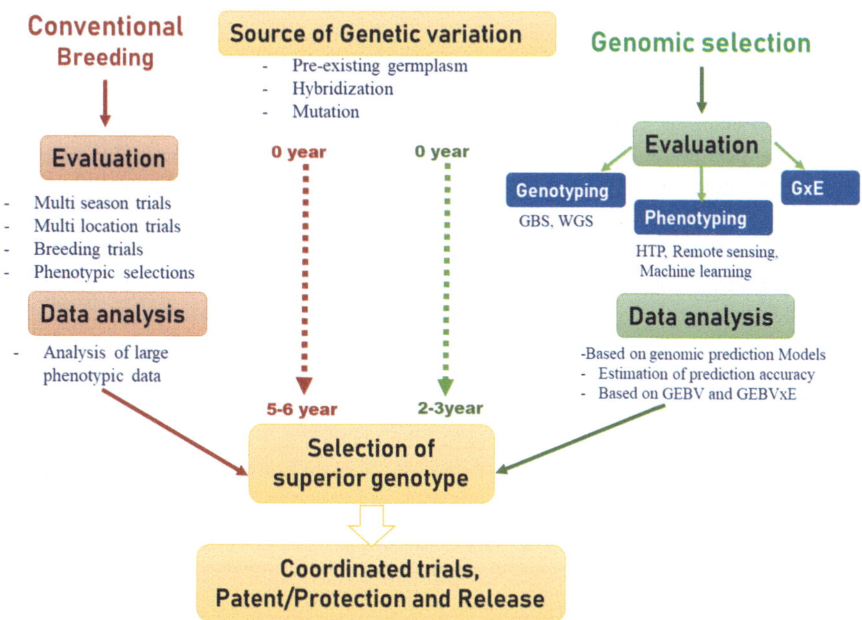


**Fig. 1** Schematic representation of conventional breeding and genomic selection approaches in wheat

extensively genotyped and phenotyped population called as training population, which is used to optimize statistical prediction model, with the help of which breeding values of un-phenotyped population called as breeding population are calculated called as genomic estimated breeding value (GEBV) purely on the basis of genotyping data, which results in cutting down the breeding cycle and eliminating unnecessary multi-location and multi-environmental phenotyping trials. Thus GS breeding scheme has edge over traditional and marker-assisted breeding methods to increase genetic gains rapidly for complex traits. The practical application and implementation of genomic selection schemes are gaining momentum in wheat breeding programmes, with accurate assessments of the genomic predictabilities and lowering cost of genotyping (Juliana et al. 2019a; Charmet et al. 2020). The capacity of GS to deliver more genetic gain for complex end-use traits in wheat breeding will help to transform its production and breeding methodologies in coming years. In this article, we highlight the genomic selection advancement for important traits like grain yield, disease resistance, grain quality traits and drought/heat stress tolerance for wheat in the post-genomics era and finally discussed the future prospects of this emerging technology for the improvement of genetic gains in wheat.

## 2 Advances in Phenomics Platform for Efficient Genomic Selection

Next-generation phenotyping (NGP) is gaining momentum nowadays with aiming for higher genetic gains in plant breeding programmes by employing automated, precise and accurate monitoring and evaluation of plant traits in large-scale plant breeding experiments (Araus et al. 2018; Rutkoski et al. 2016). These new phenotyping tools generate huge data points on plant physiological and morphological attributes with high temporal and spatial resolution. With advances in high-throughput genotyping platform, genomic selection (GS) has become popular breeding method (Desta and Ortiz 2014) which has its core in genotype to phenotype prediction models on the basis of genome-wide markers (Martin et al. 2013). These models are successful only if extensive genomic information is correlated with accurate phenotypic information (van Eeuwijk et al. 2019), which can be effective if large-scale and accurate phenotyping strategies are adopted in plant breeding. Thus, high-throughput phenotyping (HTP) data is also one of the essential requirements for the success of any genomic selection-assisted breeding programme (Desta and Ortiz 2014).

Further, in contrast to conventional phenotyping with naked eye and plant breeders' perspective, NGP includes automation, data analytics and computational capacity to improve selection efficiency and overcome individual breeder's biasedness including its effective incorporation in genomic selection (Cooper et al. 2014). NGP includes phenotyping capabilities in both control environments (Furbank and Tester

2011) and field environments (Araus and Cairns 2014). In modern and viable plant breeding programmes, phenotyping at field environment is critical and referred as high-throughput field phenotyping (HTFP). Advances in HTFP have enabled monitoring crop traits remotely in non-destructive manner for large breeding populations (White et al. 2012; Araus and Cairns 2014). Its wide scale applicability has been demonstrated for various complex traits like leaf area, biomass, canopy architecture, leaf senescence and grain yield (Jimenez-Berni et al. 2018) including ground cover display for single leaf features (Kipp et al. 2014) and aerial-based thermal imaging for crop transpiration (Deery et al. 2016; Rutkoski et al. 2016). Various phenomics platforms are available for high-throughput phenotyping of desirable traits of interest in wheat, and the same has been extensively summarized in Table 1.

HTFP is supported by various latest tools and gadgets whose application depends on image acquisition, resolution, costs and time. Although satellite imaging provides multispectral spatial resolution but to support breeder's decisions higher-resolution images are required in case of smaller plots in plant breeding experiments (Tattaris et al. 2016), thus satellite imaging can only be useful for evaluating large sized yield trial plots at later stages of breeding experiments. While for early generation breeding experiments and small plot evaluation, unmanned aerial vehicles (UAVs) and proximal phenotyping platforms were proved to be the best alternatives. UAVs are being extensively used in plant breeding trials for various traits like height in sugarcane (De Souza et al. 2017), normalized difference vegetation index in sunflower (Vega et al. 2015), ground cover and plant height in sorghum (Watanabe et al. 2017) lodging and grain yield in wheat (Lelong et al. 2008; Singh et al. 2019; Hu et al. 2020). These studies suggest immense potential of UAVs in HTFP in plant breeding experiments, but its high cost and operation hurdles are restricting its efficient applications. In contrast to this, proximal phenotyping includes ground-based vehicles and sensors for phenotyping the large-scale breeding trials irrespective of cost and operational hurdles. Its application has been demonstrated as mobile vehicle platforms for estimating early vigour, leaf area index, plant height, maturity and biomass. These ground vehicles can maneuverer manually or motorized and aided with high-throughput sensors including IR and multispectral sensors, camera along with navigation satellite system (Crain et al. 2016). Sensing cart has been successfully used in breeding trials of wheat, barley and cotton with aided sensors like ultrasonic transducer, infrared thermometer, GPS receiver, multispectral reflectance sensor, RGB cameras, data logger and weather station for estimating the correlation among grain yield with canopy temperature and plant health (White and Conley 2013; Thompson et al. 2018). Advances in sensing carts have led to development of PhenoMobile vehicles like PhenoTrac 4 and GPhenoVision which have been demonstrated in wheat, barley and cotton breeding trails for evaluating yield, canopy architecture and plant growth and developments (Jiang et al. 2018; Barmeier and Schmidhalter 2016). This allows greater convenience with advance hydraulic field drive system with high-clearance vehicle.

These HTFP tools can serve as beneficial phenotyping platform for evaluation of multiple traits with reduced errors as compared to conventional phenotyping which differs from breeders to breeder's perspective. However, these motorized

**Table 1** High-throughput phenotyping platforms utilized for improving genetic gain in wheat

| S. no. | Platform | Trait | Condition | Accuracy | Reference |
|---|---|---|---|---|---|
| 1. | Vehicle-based multispectral active sensor | Early plant vigour index | Field | $r^2 = 0.98$ | Kipp et al. (2014) |
| 2. | Thermal and hyperspectral camera mounted to manned aircraft | Canopy temperature and NDVI | Field | $r^2 = 0.78$ | Rutkoski et al. (2016) |
| 3. | Structure from motion photogrammetry using UAV with a mounted RGB camera | Plant height and growth rate | Field | $r^2 = \geq 0.92$ | Holman et al. (2016) |
| 4. | Airborne thermography with manned helicopter | Canopy temperature | Field | – | Deery et al. (2016) |
| 5. | Multi-spectral satellite sensors | Diseases and insects | Field | $r^2 = 0.74$ | Yuan et al. (2017) |
| 6. | Unmanned aerial systems (IRIS+ and eBee Ag) | Large-scale phenotyping | Field | $r^2 = 0.64$–0.76 | Haghighattalab et al. (2016) |
| 7. | Automatic disease diagnosis system | Disease | Field | 95.12–97.95 | Lu et al. (2017) |
| 8. | Near-infrared or nuclear magnetic resonance predictions | Grain end-use quality traits | Lab | $r^2 = 0.69$ | Hayes et al. (2017) |
| 9. | PhenoMobile Lite containing light detection and ranging (LiDAR) | Plant height, ground cover and aboveground biomass | Field | $r^2 = 0.92$–0.99 | Jimenez-Berni et al. (2018) |
| 10. | Hyperspectral camera using airplane | Grain yield | Field | $r^2 = 0.42$–0.58 | Krause et al. (2018) |
| 11. | Unmanned aerial or drone-based system | Lodging | Field | $r^2 = 0.19$–0.55 | Singh et al. (2019) |
| 12. | ArduCrop wireless IR thermometer and airborne thermography using manned helicopter | Canopy temperature | Field | $r^2 = 0.96$–0.98 | Deery et al. (2019) |
| 13. | Mobile field vehicle | Flowering time | Field | 99.7–100 | Wang et al. (2019) |
| 14. | Multi-spectral UAV | NDVI for grain yield | Field | $r^2 = 0.38$–0.90 | Hassan et al. (2019) |
| 15. | Unmanned aerial vehicle | Grain yield | Field | 0.65–0.96 | Hu et al. (2020) |

phenotyping platforms are costly and require technical knowledge to operate and calibrate for different breeding experiments. To optimize the cost of phenotyping technologies, Reynolds et al. (2019) describes various scenarios with which cost can be significantly reduced to normal. Moreover, with these HTP platforms, enormous amount of phenotypic data is generated, which is tedious to handle and manage; hence emphasis should also be there on its easy interpretation using automatic analysis pipeline which not only provides learning data interface but also helps in visualization of plant growth trends (Araus and Cairns 2014; Araus et al. 2018; Lee et al. 2018), and the data should meet the FAIR criteria (findable, accessible, interoperable and reusable) so that it can be accessed by anyone across the globe (Watt et al. 2020). Further, in order to improve the prediction accuracy and effectiveness of the phenotyping technologies, envirotype parameters along with different modelling strategies should also be included in the analysis (van Eeuwijk et al. 2019).

Keeping in view the population explosion, climate change scenario and limiting factors like water and land, there is an urgent requirement of combinatorial technologies like high-throughput phenotyping along with high-throughput genotyping for improvement of genetic gain of crop plants so that yields must be increased with the faster rates. To achieve this high-end goal, several advancements have been made in this high-throughput phenotyping technologies like *phenomic selection* (Rincent et al. 2018) whereby near-infrared spectroscopy was incorporated into the system for indirectly capturing the endophenotypic variants; *functional phenomics* (York 2019) whereby significant knowledge gaps that are still existing between plant phenotype and its function can be filled; and ultimately *PANOMICS platform* (Weckwerth et al. 2020), whereby various datasets which arise from various omics technologies can be integrated with phenomics approaches. Ultimately, these phenotyping platforms provide a new window to plant breeders to have a look into the plant and most importantly without any invasive methods being employed (Watt et al. 2020).

## 3 Genomic Selection for Grain Yield Improvement

Wheat breeding programme emphasizes strongly towards the grain yield improvement. Reports of stagnation or decline in yield are nowadays a major concern as more than 37% of areas under wheat cultivation have been affected by this (Ray et al. 2012). Genetic gain for yield is directly correlated with other component traits like spike length, thousand kernel weight, kernel length, kernel width and fertile spikelet, which ultimately affect grain yield. Various studies have been undertaken to identify QTLs related to yield and its contributing traits in different mapping populations (Deng et al. 2011; Heidari et al. 2011; Liu et al. 2014). The grain yield is complex quantitative trait having low heritability and greatly affected by environmental factors. So, employing GS breeding strategy has potential to lower the investment in phenotyping and minimizing replication and data collection costs in wheat breeding programmes (Crossa et al. 2014; Juliana et al. 2018). Studies of genomic predictabilities of key end-use quality and yield traits along with genomic selection were performed to identify significant marker-trait associations. A reference map of wheat genotype-phenotype was built to explore dynamics of allele frequency over time from 44,624 lines generating about 7.6 million data points, which can serve a valuable resource to the wheat breeders to aid in GS to improve productivity and stress resilience (Juliana et al. 2019b). Further, BWGS (breed wheat genomic selection) R library has been created to facilitate easy computation of GEBV for genomic selection which is a simplified powerful tool for wheat breeding programmes (Charmet et al. 2020). These tools and resources are encouraging their effective utilization in robust GS breeding schemes in wheat breeding programmes.

## *Successful Application of GS in Grain Yield Improvement*

Genomic selection is getting popularity among wheat breeders, and there are many successful examples employing GS breeding strategies in wheat improvement (Table 2). Studies have done to evaluate the gains for yield from over 2200 winter wheat breeding lines. GEBV indicated the presence of alleles responsible for enhancing yield and most significant SNPs from genome-wide association resulting high observed estimated breeding values for yield (Lozada et al. 2019). Similar study was conducted for improving yield under CIMMYT's semiarid wheat breeding programme which employed genotyping by sequencing approach to identify 41,371 SNPs in 254 breeding line sets. GEBV prediction accuracies were ranging from 0.28 to 0.45 for grain yield making GBS an excellent marker identification platform for GS breeding (Poland et al. 2012). GS technique has also been

**Table 2** Genomic selection studies in wheat for improving grain yield

| S. no | Population size | Genotyping | Statistical model | Trait | Accuracy | Reference |
|---|---|---|---|---|---|---|
| 1. | 599 lines | 1279 DArTs | RR-BLUP, RKHS, Bayes-LASSO | Grain yield (GY) | 0.48–0.61 | Crossa et al. (2010) |
| 2. | 94 lines | 234 DArTs | Bayes-LASSO, RKHS | GY | 0.43–0.79 | Crossa et al. (2011) |
| 3. | 374 lines | 1158 DArTs | RR BLUP Bayes-A, B, C | GY (0.87) | 0.48 | Heffner et al. (2011a) |
| | | | | Flour yield (0.21) | 0.76 | |
| 4. | 599 lines | 1279 DArTs | Bayes-LASSO | GY | 0.5–0.6 | Burgueno et al. (2012) |
| | | | | Flowering time (0.84) | Av. 0.43 | |
| 5. | 306 lines | 1717 DArTs | RR-BLUP, Bayes-A, B, LASSO, RKHS, RBFNN, BRNN | Date to heading (0.92) | 0.5–0.6 | Perez-Rodriguez et al. (2012) |
| | | | | GY (0.67) | 0.6–0.7 | |
| 6. | 254 lines | 1726 DArTs | GBLUP | GY (0.62) | 0.2–0.4 | Poland et al. (2012) |
| | | 34749 SNPs | | 1000-kernel weight (0.95) ~0.3 | | |
| 7. | 90 hybrids from 35 elite parental lines | 1201 SNPs | RR-BLUP, Bayes-A, B, C | GY (0.56) | 0.3–0.6 | Zhao et al. (2013) |

implemented in simultaneous breeding for protein content along with grain yield which is otherwise very difficult in traditional breeding due to negative correlation among these traits. Employing GS strategy overcomes the negative trade-off between these two traits leading to significant genetic gain and selection response for grain yield with high protein, i.e. total seed nitrogen content, which suggests its feasibility to develop varieties having multiple superior traits (Michel et al. 2019).

# 4 Genomic Selection for Improving Disease Resistance

Based on inheritance pattern, disease resistance in plants is broadly classified into two groups, qualitative and quantitative (Vanderplank 2012). Genetically qualitative resistance is controlled by a single resistance gene with large effects and follows the gene-for-gene mechanism of resistance against a particular race of a known pathogen species (race specificity), while quantitative resistance is governed by large number of genes with small effects and does not involve race specificity. Qualitative resistance is often overcome by pathogen through rapid evolution of new race virulent over the deployed resistance gene, while quantitative resistance provides durable resistance since the overcoming of multiple modes of resistance is very difficult for the pathogen until super-race is evolved (Parlevliet 2002).

## *Conventional Breeding for Disease Resistance*

For developing any successful improved variety, the breeding programmes must have the efficient strategy to combine disease resistance with high yield and better agronomic performance without altering the end-use qualities (Chakradhar et al. 2017). The decision on breeding strategy dealing with disease resistance largely depends on the disease reaction whether qualitative or quantitative. The most commonly used methods for selection for disease resistance have been reviewed by Poland and Rutkoski (2016). Breeding programmes for qualitative resistance often start with the screening of large number of lines in initial cycles of selection where susceptible plants are discarded through creating epiphytotic conditions. The resistant lines or plants are then selected and further advanced in breeding programmes. Backcrossing is another breeding strategy, which involves the crossing of resistant parent to an elite but susceptible parent followed by continuous backcrossing of progenies with susceptible parent until desired level of susceptible parent genome is recovered. In qualitative resistance breeding, homozygous resistant lines can be obtained in relatively less number of breeding cycles, and no further selection of resistance allele is required (Frisch and Melchinger 2005). Breeding strategy for quantitative resistance involves selection for multiple highly heritable traits including disease resistance (Henderson 1963). However, the genetic gain per cycle for each trait is usually less compared to selection of one trait only. Breeding for quantitative resistance thus necessitates multiple selection cycles to improve other important traits like yield along with disease resistance (Poland and Rutkoski 2016).

## *Molecular Breeding in Disease Resistance*

The advances in molecular breeding have enabled identification of molecular markers linked with resistance loci which can be employed in any breeding programme for direct selection of disease resistance (Moose and Mumm 2008). The resistance genes/QTLs for many diseases such as stem rust (Haile et al. 2012; Saintenac et al. 2013; Saccomanno et al. 2018), leaf rust (Krattinger et al. 2009; Kthiri et al. 2019; Zhang et al. 2019; Gebrewahid et al. 2020), stripe rust (Yuan et al. 2018; Zhang et al. 2019; Gebrewahid et al. 2020), powdery mildew (Chantret et al. 2001; Asad et al. 2014; Yu et al. 2018), spot blotch (Singh et al. 2018) and common bunt (Bokore et al. 2019) have been mapped in wheat. However, in applied plant breeding, the execution of marker-assisted selection (MAS) for disease resistance has not been successful on large scale in wheat due to many constraints such as availability of very few linked markers, insufficient linkage with markers, prevalence of QTL background effect and high economic investment (Miedaner and Korzun 2012). Besides, the less durability of monogenic resistance and small effect of resistance QTLs further limit the practical application of MAS (Poland and Rutkoski 2016).

In the scope of genomics-assisted breeding, MAS is efficient for monogenic resistance. However, race-specific resistance shows differential reaction response to varying environments leading to greater genotype × environment (G × E) interaction (Poland and Rutkoski 2016). By comparison, race non-specific resistance displays much less G × E as it does not involve pathogen-host genes interaction. Thus, breeding for quantitative resistance is one of the approaches to achieve yield stability by minimizing G × E, particularly in epidemics prone areas (Jannink et al. 2010). In addition, application of quantitative genetic and genomic prediction models in breeding programmes also minimize G × E given that there is some genetic correlation between environments (Kelly et al. 2007; Burgueno et al. 2012). Therefore, as the breeding for resistance shifts from qualitative to quantitative durable resistance, the molecular breeding approach needs to shift from MAS to genomic selection (Bekele et al. 2019).

## *Genomic Selection in Disease Resistance*

GS is a powerful tool for designing novel breeding programmes for disease resistance using existing molecular marker sets and to identify new markers set for predicting breeding value of selection candidates (Bekele et al. 2019). GS uses whole genome markers rather than the subset of markers in case of MAS for prediction of models; therefore over MAS, GS have greater power to (i) capture small effect resistance QTLs (Meuwissen et al. 2001), (ii) to account of whole available additive genetic variance (Jannink et al. 2010) and (iii) to evaluate large number of breeding lines for disease resistance (Poland and Rutkoski 2016). Considering the above points, GS thus provides greater opportunity to achieve maximum genetic gain for disease resistance by reducing the breeding cycles and increasing the selection accuracy and selection intensity.

## Genomic Selection Models for Disease Resistance

The statistical models applied to estimate breeding values in GS for disease resistance have been described in detail by Poland and Rutkoski (2016). Genomic best linear unbiased prediction (G-BLUP) and ridge-regression BLUP (RR-BLUP) are the two most commonly used models for purely quantitative traits. When there is a large effect QTL, the selected number of markers can be considered as fixed effects in G-BLUP or RR-BLUP models to increase the prediction accuracy, and this has been shown for resistance to stem rust resistance in wheat (Rutkoski et al. 2014). For the traits that fall between quantitative and qualitative resistance, Bayesian models such as BayesA, BayesB (Meuwissen et al. 2001), BayesCπ (Habier et al. 2011), BayesR (Daetwyler et al. 2014) and Bayesian LASSo (Park and Casella, 2008) are the most suitable models. Some authors have reported equal performance of RR-BLUP and Bayesian LASSO in their studies on Fusarium head blight resistance in wheat (Lorenz et al. 2012; Rutkoski et al. 2012; Rutkoski et al. 2014) and BayesCπ (Lorenz et al. 2012; Mirdita et al. 2015; Rutkoski et al. 2014; Sallam et al. 2015). However, Ornella et al. (2012) reported slightly higher performance of Bayesian LASSO than RR-BLUP for wheat rust resistance. In contrast Arruda et al. (2015) observed higher prediction abilities of Bayesian LASSO for FHB resistance compared to RR-BLUP. Two other models, Reproducing Kernel Hilbert Space (RKHS) (Gianola and Van Kaam 2008) and Random Forest (RF) (Breiman 2001) can capture both additive and non-additive effects and are useful for predicting total genetic value. Some studies of GS for FHB resistance in wheat have reported that RKHS and RF models performed better than linear models (Rutkoski et al. 2012; Mirdita et al. 2015).

## Success of GS in Wheat Disease Resistance Breeding

In the last decade, numerous studies have been conducted on GS for disease resistance in wheat (Table 3). These studies have proven the utility of the current GS models and GS approaches for predicting the breeding values of disease resistance in wheat, particularly quantitative resistance

**Table 3** Compiled studies demonstrating the utility of genomic selection for improving disease resistance in wheat

| Disease type | Models used | Validation type | Prediction accuracy | Reference |
|---|---|---|---|---|
| Stem rust | LGM, RR-BLUP, BL, SVR | Cross-validation within full-sib family | 0.39–0.85 | Ornella et al. (2012) |
| | | Cross-validation across family | 0.14–0.67 | |
| Yellow rust | LGM, RR-BLUP, BL, SVR | Cross-validation within full-sib family | 0.14–0.63 | |
| | | Cross-validation across family | 0.14–0.63 | |
| All the three rusts | G-BLUP, BayesR | Fivefold cross-validation | 0.27–0.44 | Daetwyler et al. (2014) |
| Stem rust | MLR, G-BLUP, BL, BayesCπ | k-fold cross-validation | 0.56–0.62 | Rutkoski et al. (2014) |
| Stem rust | G-BLUP, BayesR | Forward validation | 0.20–0.40 | Rutkoski et al. (2015) |
| | | Cross-validation | 0.55 | |
| Stripe rust | RR-BLUP | Cross-validation | 0.45–0.65 | Muleta et al. (2017) |
| All the three rusts | G-BLUP, LS, three RKHS models | Tenfold cross-validation | 0.12–0.78 | Juliana et al. (2017a) |
| FHB | RR-BLUP, BL, RKHS, RF, MLR | Fivefold cross-validation | 0.59–0.64 | Rutkoski et al. (2012) |
| | | Single cross-validation across years | 0.19–0.41 | |
| FHB | RKHS, G-BLUP, RR-BLUP, BayesCπ | k-fold cross-validation | 0.46–0.64 | Mirdita et al. (2015) |
| *Septoria tritici* blotch | RKHS, G-BLUP, RR-BLUP, BayesCπ | k-fold cross-validation | 0.36–0.59 | |
| FHB | RR-BLUP, BL | k-fold cross-validation | 0.40–0.64 | Arruda et al. (2015) |
| FHB | GS model (RR-BLUP | Fourfold cross-validation | 0.4–0.9 | Arruda et al. (2016) |
| | MAS models | Fourfold cross-validation | <0.3 | |
| FHB | GS models (RR-BLUP, RKHS, BayesCπ) | Cross-validation | 0.72–0.74 | Jiang et al. (2017) |
| | | Independent validation | 0.64–0.69 | |
| | MAS models | Cross-validation | 0.01–0.62 | |
| | | Independent validation | −0.01–0.46 | |

(continued)

**Table 3** (continued)

| Disease type | Models used | Validation type | Prediction accuracy | Reference |
|---|---|---|---|---|
| FHB | RR-BLUP | Forward validation (bi-parental families) | 0.72 | Herter et al. (2019) |
| *Septoria tritici* blotch | RR-BLUP | Cross-validation (bi-parental families) | 0.15 | |
| *Septoria tritici* blotch | LS, G-BLUP, Bayesian models, RKHS | Tenfold cross-validation | 0.19–0.57 | Juliana et al. (2017b) |
| *Stagonospora nodorum* blotch | LS, G-BLUP, Bayesian models, RKHS | Tenfold cross-validation | 0.43–0.60 | |
| Tan spot | LS, G-BLUP, Bayesian models, RKHS | Tenfold cross-validation | 0.28–0.77 | |
| Powdery mildew | RR-BLUP | Cross-validation | 0.36–0.57 | Sarinelli et al. (2019) |

*FHB* Fusarium head blight, *LGM* linear genetic model, *RR-BLUP* ridge-regression genomic best linear unbiased prediction, *BL* Bayesian Lasso, *SVR* support vector regression, *G-BLUP* genomic best linear unbiased prediction, *BayesR* Bayesian regression, *MLR* multiple linear regression, *LS* least-squares, *RKHS* Reproducing Kernel Hilbert Space

## Genomic Selection for Wheat Rusts

Wheat rusts are the most studied crop diseases for demonstrating the utility of GS in disease resistance breeding. Three common forms of rusts which occur in wheat are stem rust (*Puccinia graminis*), yellow or stripe rust (*Puccinia striiformis*), brown or leaf rust (*Puccinia triticina*), stripe/yellow rust (*Puccinia striiformis*) and leaf/brown rust (*Puccinia triticina*). Resistance response to rust pathogens can be of qualitative or quantitative. Qualitative resistance expressed at seedling stage, while quantitative resistance usually detected at adult plant stage, therefore also known as adult plant resistance (APR) (Poland and Rutkoski 2016). Concentrated efforts towards breeding for major resistance genes resulted in development of highly resistant wheat varieties, but they are short lived as rapid evolving rust pathogens often overcome major genes. Therefore, breeding for minor genes (quantitative) resistance much needed for developing wheat cultivars with durable resistance. Moreover, being race non-specific, quantitative resistance sometimes provides protection against more than one rust pathogen species (William et al. 2003; Krattinger et al. 2009; Herrera-Foessel et al. 2011).

The first study on GS for rusts in wheat was conducted by Ornella et al. (2012). They reported moderate to high prediction accuracies (range: 0.3–0.8) over within and between bi-parental validation populations. They also found that linear models performed better than non-linear models, and Bayesian LASSO was slightly superior to RR-BLUP. Daetwyler et al. (2014) performed GS for all the three rusts using 206 diverse landrace collections of wheat and confirmed the feasibility of GS models in wheat. They also reported that inclusion of gene-based diagnostic markers in GS models improves the prediction accuracies. Rutkoski et al. (2014) incorporated *sr2*-linked markers and seedling phenotype score as fixed effects in classical GS models for improving their predictability for APR in wheat. They found that GS models containing *sr2*-linked markers along with whole genome markers (GBS) performed better than classical models which contain only GBS markers. However, incorporation of seedling phenotype score as fixed effects did not improved prediction accuracies of classical models. This study also reported higher prediction accuracies for G-BLUP compared to MLR and Bayesian models. Rutkoski et al. (2015) compared selection efficacy of one cycle phenotypic selection to two cycles of GS and observed equal realized gain per unit time from both GS and PS for stem rust resistance in spring wheat. They also reported that GS more rapidly reduced genetic variance and increased inbreeding compared to PS. Muleta et al. (2017) applied GS for stripe rust APR in a diverse set of accessions of spring wheat. They demonstrated that prediction accuracies can be increased by (i) increasing the training population size, (ii) selecting genetically related populations for validation and (iii) increasing marker density. They also found that the use of a subset of markers instead of whole genome markers can efficiently predict the genetic gain. In a similar way, Juliana et al. (2017a) predicted selection accuracies of different GS models for all the three rusts using few markers as fixed effects. They observed highest prediction accuracy for RKHS-MP model and lowest prediction accuracy for least-squares (LS) model.

### Genomic Selection for Fusarium Head Blight

FHB caused by a fungal pathogen (*Fusarium graminearum*) is a serious disease of wheat throughout the world, particularly in humid regions. It infects the spikes leading to shriveled and discoloured grain which result in significant yield loss and reduction of grain quality (Dexter et al. 1996). The FHB infection deposits a mycotoxin, deoxynivalenol (DON) in the grain that has poor correlation with visible disease symptoms, and creates difficulties in phenotyping. The FHB resistance is governed by a single large effect QTL with additive genetic variation (Jin et al. 2013). Considering the phenotyping evaluation and quantitative nature of inheritance, GS could accelerate FHB resistance breeding.

Rutkoski et al. (2012) applied GS for first time in wheat for predicting selection accuracies of different linear and non-linear models for FHB resistance traits including DON. They observed that GS models, especially RF and RHKS, outperformed MLR model, and for DON, marker plus trait RF model showed high prediction accuracy than all other models. This study also compared QTL-linked markers with genome-wide markers and observed higher accuracy for QTL targeted markers for DON, whereas for other resistance traits QTL-linked markers performed equal to genome-wide markers. Mirdita et al. (2015) incorporated main and epistatic effects in different GS models and compared their predictability for FHB resistance in wheat. They observed high prediction ability for epistatic models amounting to 0.6

with RKHS. Arruda et al. (2015) trained GS models for FHB in diverse set of breeding lines and observed moderate to high prediction accuracies. In contrast to Rutkoski et al. (2012), Arruda et al. (2015) found high prediction accuracy for RR-BLUP with genome-wide markers compared to QTL targeted markers. Arruda et al. (2016) compared GS models with MAS models for FHB in soft red winter wheat and found higher accuracies for GS models compared to MAS models. They also observed higher prediction accuracies for GS models containing QTL markers as fixed effects compared to classical GS models. Jiang et al. (2017) predicted the accuracies of independent and cross-validated GS and MAS models for FHB in wheat. They reported that the cross-validated MAS models overestimated the prediction accuracies compared to independently validated models, whereas prediction accuracies of cross-validated GS models found similar to independently validated models. Herter et al. (2019) trained GS for FHB in winter wheat lines derived from 14 bi-parental families. They compared the prediction accuracies of genomic selected populations with random sampled populations and observed higher prediction accuracies for genomic selected populations compared to the randomly chosen populations.

### Genomic Selection for Other Diseases

Mirdita et al. (2015) applied the same GS approach in predicting selection accuracies for *Septoria tritici* blotch (STB) as for FHB in wheat. They observed that GS models covering epistatic effects outperformed GS models covering main effects. Juliana et al. (2017b) applied LS, G-BLUP, Bayesian and RKHS models in GS for STB, *Stagonospora nodorum* blotch (SNB) and tan spot (TS) in wheat. They observed moderate to high prediction accuracies across the traits and models. RKHS models gave the high prediction accuracies, while LS model gave lowest accuracy. They also compared GBS and diversity array technology (DArT) sequencing approaches and found slightly higher accuracies for GBS than DArT sequencing. Herter et al. (2019) followed the same GS approach for STB as for FHB in winter wheat and reported very little advantage (2.14 %) of GS for improving STB resistance. Sarinelli et al. (2019) investigated the effects of selection methods and size of training population on predictability of powdery mildew resistance in wheat. The results found increase in prediction ability as population size increases, and higher prediction ability was observed for prediction error variance methods than for random and clustering training population selection methods. The authors also reported improved prediction ability of GS models when gene linked markers were used as fixed effects.

One of the major challenges for utility of GS in wheat is high heritability of resistance for many diseases which makes the GS hard to overcome phenotypic selection in term of both breeding cycles required and per cycle genetic gain achieved (Rutkoski et al. 2015; Arruda et al. 2015). Therefore, to beat phenotypic selection, GS for quantitative disease resistance must be focused on increasing selection intensity along with decreasing breeding cycles. For disease resistances with low heritability, markers plus phenotype GS can considerably improve selection accuracy (Heffner et al. 2010). Another challenge of GS is low direct cost-to-benefit ratio of GBS or markers compared to phenotypic selection. Furthermore, GS for quarantined disease resistances is also challenging because phenotyping is possible only in disease occurring countries or states or in expensive biocontainment facilities. The advances in genomics-assisted breeding opens new horizon for developing durable resistance through combined selection for major resistance (R) genes and quantitative resistance. The GS scheme adopted for combined selection of R-genes and quantitative resistance has been extensively summarized by Poland and Rutkoski 2016.

## 5 Genomic Selection for Wheat Quality Improvement

Quality trait improvement is considered as second priority in breeding programmes than yield per se because most of the market does not value their nutritional aspects and does not fetch good selling price. Wheat grain yield with its grain nutritional quality decides the economic value in the man-made markets. Qualitative and quantitative traits determine wheat quality such as grain protein content and grain, governed by quantitative and qualitative traits, respectively, and determine market acceptance and wheat milling quality (Battenfield et al. 2016). Starch-protein concentration is higher in hard wheat than soft wheat. These starch-protein attachments break in milling process by additional use of energy and produce high amount of mutilated starch granules, and leavened product tends to produce high baking property because of more water absorption. Soft wheat is most preferably used as pastries, cakes and cookies because of production of less damaged starch. Soft wheat tends to have low protein, while hard wheat tends to have high protein. There are some additional milling quality traits in wheat such as kernel colour and weight, flour colour and yield, flour protein, protein quality and protein quality that determines preference to plant breeders as well as consumers. The primary wheat storage proteins, glutenins and gliadins, provide unique viscoelastic property to wheat flour during baking (Battenfield et al. 2016). The wheat dough elasticity, tolerance, excellent mixing time and strength are determined by mixing wheat flour with water, and these dough traits are conferred by the concentration of protein components such as gliadin and glutenin. Different chemical and analytic approaches such as alveograph, mixographs and farinographs are being used to estimate various dough properties, but these are not cost-effective and need high flour quantity and more time consuming. It may not be feasible to test baking quality component traits, i.e. loaf volume and texture in early generation breeding trials due to unavailability and or high cost of inputs.

## *Milling and Flour Quality*

GS-based prediction models are in trend for quantitative as well as quantitative attributes for their more prediction accuracy and cost-effectiveness. Heffner et al. (2011a, b) studied two wheat quality attributes, viz. milling as well as flour quality to know the effect of prediction models for accurate selection and also compared GS models with direct phenotypic selection. They concluded that performance of phenotypic prediction accuracy was higher for GS models for quality attributes in wheat. Bayes-Cπ and RR-BLUP models of GS were studied and simultaneously compared and revealed that the performance of Bayes-Cπ model was high in elite × elite mapping population, while performance of RR-BLUP model was high in QTL bi-parental mapping population (Heffner et al. 2011a, b).

Prediction accuracies were higher from multifamily panel predictions rather than bi-parental panel predictions for wheat quality attributes (Heffner et al. 2011a, b). But this is contradictory to general assumption that bi-parental mapping population contains maximum prediction accuracy (Lehermeier et al. 2014). Only 96 individuals (Heffner et al. 2011a) in each family and about 450 SSR markers were used in bi-parental training population while 288 individuals (Heffner et al. 2011b) and about 1158 SSR genetic markers were applied during study of multifamily mapping population. Hoffstetter et al. (2016) studied flour softness and flour yield of soft winter wheat in advanced $F_4$-derived breeding population, and prediction accuracy was improved by subsetting the marker sets and training population. They found that for wheat quality traits, genotypes associated with high genotype-by-environment interdependence can be removed without changing the prediction accuracy inside the specific environment but use of subsetting markers, i.e. markers closely associated with trait of interest boosted prediction accuracy compare to complete marker set for flour softness and flour yield. This approach not only saves time but also cost-effective.

Battenfield et al. (2016), used multiple GS-based models while analysing the breeding population panel datasets in CIMMYT multiyear spring wheat, and all the data were used with great precision, processing and phenotypic quality for prediction analysis in subsequent year. As years progress from 2011 to 2015, forward prediction accuracies increase due to addition of more years into the model. However, cross-validation with above years reveals outperformance of all forward prediction models in studied quality traits. It was outperformed due to not including genotype-by-phenotype interactions and informative details from their relatives over the years in forward prediction. The genetic gain from GS increased more than 100% for wheat grain quality traits such as grain protein, flour yield and protein and other dough-related attributed (Battenfield et al. 2016). They used GS-based models as an accompaniment rather than substitute of direct phenotypic selection at CIMMYT in their breeding programme and revealed genetic gain through selection might be higher by considering that more than 10,000 individuals can be genotyped at the same cost as phenotyping of 1000 individuals. The use of near-infrared (NIR) and/or nuclear magnetic resonance (NMR) high-throughput phenotyping (HTP) of large training population into multi-traits GS models increases the prediction accuracy. For example, Hayes et al. (2017), phenotyped 44 quality traits, by using combination of NIR and NMR HTP, in training set for more than 2 years, and validation test was conducted more than 3 years by using training subset and Australian National Variety Trails (NVTs) system. Prediction accuracy was increased for grain, milling and baking attributes, but dough rheology traits could not be improved as such. Prediction accuracies for selection were higher, reliable and effective at early generations for quality attributes. These accuracies help in selection of elite suitable breeding material from large available elite line for genetic gain in wheat quality traits improvement.

## *Pre-harvest Sprouting*

It happens when physiologically matured crop is bared to excess moisture, prior to harvest, in field for substantially long time and germination process of seeds starts and forms coleoptiles and rootlets on the spike itself. There are few ways and means to measure pre-harvest sprouting (PHS) damages like germination test of soft grains and mist test of wheat spike at individual level (direct method), while falling dough properties, an indirect method for measurement of α-amylase enzymatic activity. Heffner et al. (2011a, b), firstly, studied the impact of GS for PHS tolerance in wheat. In Fist study, they used population size of 96 segregating individuals, derived from single bi-parental mating system, for PHS tolerance/resistance to train prediction model. Interestingly, the accuracy of phenotypic selection was significantly superior to GS models. GS-based models, viz. RR-BLUP, etc., overshadow the impact of marker-assisted selection (MAS) – *mostly used for genomic selection of different plant attributes*, and its performance was higher over MAS. The superiority in accuracy of PS over both GS models for PHS was due to high heritability, small size of training population, polygenic nature of PHS, low-density marker and low genotype-by-phenotype interaction. In the second study, they used training population, to predict forward prediction accuracy by using different GS models along with models related to association, from multifamily breeding lines in soft winter wheat. They concluded that GS model-based prediction accuracy for PHS was higher for association models. However, it can be concluded from both the studies that GS-based models reduce breeding cycle time and cost for quality traits including PHS and improved quality traits than phenotypic selection. Training population, composed of 1118 breeding lines of hard white and red winter wheat, was used by Moore et al. (2017), and concluded that hard red wheat lines are comparatively more tolerant than white wheat lines. They used SNP genotypic markers, derived from GBS, for study the level of resistance for PHS attribute by using germination test. Surprisingly, when grain colour was used as fixed effect, they could not get improved prediction accuracy. Although prediction accuracy was improved by using five significant markers from GWAS, these findings were previously supported by Arruda et al. (2016).

## GS for Nutritional Quality Traits

Globally, burgeoning population mostly depends on cereal grain crops, viz. rice and wheat, as a primary source of calorie, to mitigate their food and nutritional demand. Micronutrient deficiency leads to thriving of malnutrition, particularly in developing world along with scatter form in developed nations; thus development of biofortified varieties, by using amalgamation of different breeding approaches with advance tool such as GS, can alleviate the nutrient deficiency and malnutrition, particularly in women and children. Globally, iron (Fe) and zinc (Zn), essential micronutrients, deficiency jointly affects nearly two billion individuals (HarvestPlus 2017). Wheat grain for micronutrients components is genetically polygenic and quantitative in nature and makes it suitable for GS (Velu et al. 2014). Wheat grain Zn and Fe concentration was studied by Velu et al. (2016) by using CIMMYT association panel of HarvestPlus as training population for various prediction models. Association panel is made by landraces and progenies derived synthetically. The prediction accuracies were low to medium for these traits across multi-locations for 2 years. Prediction accuracy was observed by using G-BLUP models, by adding of both G × E interaction and pedigree × environment kernel, with both genetic and pedigree matrix relationship. It was evident that inclusion of G × E interactions from multi-environments across the years in GS-based models increases (Lopez-Cruz et al. 2015) the prediction efficiency than single environment. Although, the wheat grain Zn and Fe concentration (Ortiz-Monasterio et al. 2007) highly influenced by Fe and Zn status of native soil and, major factor for high genotype-by-environmental interactions. Two similar sets, composed by Afghan wheat landraces, were sown for grain phosphorus, potassium, manganese, iron, magnesium and zinc in Afghanistan and Japan (Manickavelu et al. 2017). They analysed each nutritional trait by GWAS and found only single significant association for Zn. Mostly, wheat grain nutritional traits showed quantitative inheritance. Prediction accuracies were medium to high for every nutritional grain attribute in both the countries when they used multiple genomic prediction models. Prediction accuracy was higher for macronutrients (P, K and Mg) than micronutrients (Zn, Fe and Mn). As compare to Afghanistan, prediction accuracy was higher in Japan due to more genotype-by-environment interactions and soil edaphic conditions in Afghanistan. There is need of the hour to do research specially on models based on G × E interaction to enhance prediction accuracies for accurate selection in breeding program.

## GS for Grain Protein Content

Wheat grain yield – a complex and quantitative attribute – and its protein content, a qualitative attribute, are negatively associated with each other and are potent for improving grain quality in wheat breeding program. Rapp et al., 2018, evaluated two panels consisting 189 and 159 lines of durum wheat and were assessed across the Europe and genotyped by GBS approach. Forward prediction accuracy was relatively low for grain protein content as compared to grain yield in durum wheat panels when they used RR-BLUP-based GS model in Southern and Central Europe. The same result of prediction accuracies was found by He et al. (2016), for grain yield and by Würschum et al. (2016), for protein content in bread wheat. However, there was decline in prediction accuracy when the genotypic value of one panel was used to predict the accuracy of other panel. These findings are in line with Crossa et al. (2014) where they found high variability in germplasm leads to high prediction accuracy when used as training and prediction set because they show large genomic coverage. Zhao et al. (2015) also concluded that the close genetic relationship of genotypes in prediction and training set affects prediction accuracy. Zhao et al. (2012), also used handful amount of genotypes over the breeding cycles to stabilize breeding prediction abilities. Thus, it is interesting to know how genetic closeness between prediction and training sets improve the prediction accuracies in future.

## *GS for Semolina Quality*

Durum wheat grain quality traits such as semolina quality and baking quality, etc. are very crucial and potent in breeding programmes for its commercialization and high market value. Conventional phenotypic selection hinders the effective selection efficiency of breeding programme by promoting the undesirable entries from one phase to another phase. Each trial needs huge natural and mane-made resources for testing of grain quality traits and grain yield per se. Genome-wide association mapping was used by Fiedler et al. (2017) while studying 5 important durum wheat quality characters in 1184 breeding lines at North Dakota State University. RR-BLUP method was used for estimation of genomic prediction precision for quality traits. The sedimentation volume and semolina colour confer highest prediction accuracy, while semolina protein content confers lowest prediction accuracy. Genomic selection improves the selection efficiency by increasing selection intensity and shortens the total time period of breeding cycle and thus enhances the genetic gain per unit cost, time and space. Pre-selection based on predicted breeding values allows eliminating poor lines into the high cost multi-locational and multi-year yield trials. They used full sib lines in testing and training population. It showed adjacent relatedness between these populations during cross-validation, while those testing and training mapping population derived from new breeding cycle showed distant relationship (Crossa et al. 2014). Close relatedness among testing and training population tends to increase prediction accuracies. Forward prediction accuracy was analysed by using testing and training population and evaluated by using different GS-based models such as RR-BLUP and Bayesian Lasso (BL), etc. Durum wheat semolina and grain quality attributes showed forward prediction accuracy of 0.27–0.66. Forward prediction accuracy was high (0.44) for semolina protein content. They observed genetic gain between 10.7 and 1.78 by using GS for quality traits than conventional phenotypic selection in durum wheat. The forward prediction accuracy can be increased by augmenting the size of training population for quality attributes in durum wheat.

## *GS for Baking Quality*

Wheat breeding faces tremendous challenges particularly for genetic improvement of quantitative traits such as baking quality due to exhaustive labour demand, time devouring and high cost testing procedure for its associated traits. Bread wheat quality is directly and positively associated with its baking quality ranking. Thus, traits associated with baking quality are being tested at the end of multi-location and multiyear yield trials. Wheat genotypes, about exceeding the number of 400, were phenotyped (Michel et al. 2018) for dough elasticity, sum of protein content and all related mixing properties in across the environments for eight years since 2009 to 2016 for estimation of baking quality. The accuracy of prediction was acceptable for all dough-related characters when they used modelling major QTLs as fixed effect and applying multi-trait prediction models. GS can be used more than 2–3 years ahead of direct phenotypic selection, and selection efficiency was prognosticated about two times higher than indirect selection for protein content. Wheat storage protein composition is as much essential as protein content for quality in wheat breeding program, milling and food processing. Forward prediction accuracies can be increased by using multi-trait prediction model over standard RR-BLUP method by increasing the size of training population of protein content associated traits and extensive phenotyping of genotypes. Thus accuracy of prediction was increased for water intake, farino quality intake, dough extensibility and development, and this might be good strategy for early generation genomic selection. GS is a better approach than conventional approach like direct phenotypic selection as well as marker-driven selection and might be increased by using the available public domain information about genetic constitution of traits for prediction of GEBV. Grain yield, protein composition and total protein yield of more than 650 inbred lines were independently evaluated at five different breeding cycles and estimated the biasness impact within cycle during cross-validation (Michel et al. 2016). Protein composition ranked first for heritability, and protein yield showed least heritability. Protein yield showed considerable biasness of prediction accuracy followed by grain yield and protein content by using fivefold cross-validation using populations from individual cycles. The prediction accuracy was increased by using breeding cycles in cross-validation and reached a maximum prediction for protein component and least for protein yield. Haile et al. (2018), used two separate breeding panels, one panel comprising 170 genotypes from released varieties and elite breeding material while 154 genotypes derived from doubled haploid breeding in second panel of durum wheat, and analysed then prediction accuracy by using genomic prediction models such as multi- and single-trait method for complex traits such as yield per se and other quality attribute, viz. gluten index, alveograph measures and protein content. Besides BayesB and BayesA GS models, they found significantly improved single-trait prediction accuracy for almost all traits. BayesB and BayesA model predicted grater for tenacity, gluten strength and index in doubled haploid population only. The multi-trait models were much better than single-trait models for only end crop grain yield. Albeit, all six GS models, viz. RKHS, Bayesian LASSO, BayesB, BayesA, GBLUP and RRBLUP, were applicable for prediction accuracy for gluten strength with other economic traits in durum wheat, but two GS models, viz. G BLUP and RR BLUP, being high prediction accuracy, were endorsed for their simple computation programme and MT-SI model for simultaneous improvement of protein and yield.

GS is widely accepted and a pivotal tool to increase the genetic gain of crops including wheat by avoiding labour cumbersome, time consuming and relatively high cost of phenotyping of large breeding mapping population in each and early generation of breeding program. It is a very important tool to enhance genetic gain through effective selection method for low heritable traits such as grain and flour quality.

## 6 Genomic Selection for Drought/Heat Stress Tolerance

### *Need of Genomic Selection to Breed for Drought/Heat Tolerance in Wheat*

Based on population growth and warming predictions, between one to three billion people are likely to be left outside the climate conditions in the next 50 years which have served mankind well over the past 6000 years (Xu et al. 2020). Specifically, 3.5 billion people would be exposed to mean annual temperature ≥29.0 °C, a condition found only in 0.8% area but expected to cover 19% of the world's land in 2070. Similarly, climate change is a major threat to most crops grown in tropical and subtropical countries worldwide. As a consequence of climate change, abiotic stresses cause significant yield losses in plants as much as 50% (Qin et al. 2011). Among these drought and heat stress majorly affects the crop growth and yield. Wheat being the staple food for many regions worldwide is significantly affected by drought and heat stresses with up to 86% and 69% yield loss, respectively (Fischer and Maurer 1978; Prasad et al. 2011). With the same pace of climate change, scientists believed that by this century, up to 60% of existing wheat-growing regions around the world would be under simultaneous, severe and prolonged droughts (Trnka et al. 2019), and by 2050, 51 per cent of the Indo-Gangetic plains are projected to be reclassified as a heat-stressed mega-environment (Ortiz et al. 2008). Furthermore, a study predicted that with every 1 °C rise in global temperature, wheat yield will be decreased by 4.1 per cent to 6.4 per cent (Liu et al. 2016), while on contrary wheat consumption will be increased by more than 30 per cent over the next 40 years (Weigand 2011). By seeing this scenario and prediction, it is extremely important to use various statistical models to predict breeding value of the genotypes to cope up with the climate change in the form of drought and heat stress (Sallam et al. 2019).

Developing the resistant variety is the ideal approach to manage biotic and abiotic stresses. Creation of variability and selection has been the fundamental of plant breeding. First step being the selection of potential germplasm with wide genotypic differences for drought and heat tolerance (Baenziger 2016) followed by development of mapping populations using tolerant genotypes as donor parents. Traditionally, phenotypic selection is the sole tool for breeders; however, drought and heat tolerance being quantitative trait are highly influenced by the environment and have low heritability (Yang et al. 2002; Bernardo 2008). To overcome these issues, molecular markers played an important role from genetic diversity studies to mapping of minor as well as major effects QTLs associated with drought and heat tolerance in wheat (Malik and Malik 2015; Gahlaut et al. 2017; Batool et al. 2018; Devi et al. 2019). The advent and evolution of next-generation sequencing (NGS) technology particularly methods such as GBS (Poland and Rife 2012) have led to a rapid generation of genome-wide markers with low cost. Genomic selection (GS) is such a tool which has power to combine low-cost genotyping, HTP and robust statistical model to enhance selection for quantitative traits so that genetic gain can be increased over time and could further speed up varietal development for tolerance to abiotic stresses like drought and heat.

## *Previous Work Done in Genomic Selection for Drought/Heat Tolerance in Wheat*

GS has been used for abiotic stress tolerance on major crops like rice (Bhandari et al. 2019) and maize (Shikha et al. 2017; Xu et al. 2018; Wang et al. 2019). Similarly, experiments were also carried out to check the effectiveness of different GS models for drought and heat stress in wheat. Rutkoski et al. (2016), evaluated 557 wheat lines for canopy temperature and NDVI through HTP in optimal, early heat, late heat, drought and severe drought environments and observed that secondary traits enhanced prediction accuracies for grain yield by 56 percent through pedigree and by 70 percent through genomic prediction models. Meanwhile, Haghighattalab et al. (2017), used UAS for HTP of advanced wheat breeding lines in irrigated and drought-stressed environments. By analysing the data obtained from UAS imagery and its relationship with grain yield, they evaluated potential of UAS imagery for predicting GY at plot level and found high correlation between imagery-derived phenotypic traits and grain yield. Further, Dunckel et al. (2017), hypothesized that "to increase the speed of introgression of exotic germplasm, genomic selection approaches could be applied to enable rapid cycles of selection". In order to validate this hypothesis, some selected lines from DH and RILs population were evaluated for grain yield and other agronomic traits under irrigated, heat and drought-stressed environments using five distinct genomic prediction models and observed that these models had moderate prediction ability and were slightly lower than expected. In addition to this, Sun et al. (2017), compared three statistical models, namely, simple repeatability (SR), multi-trait (MT) and random regression (RR) for their predictive abilities for grain yield under five different drought and heat stress environments and found that predictive abilities were enhanced by an average of ~70 percent, by including secondary trait BLUPs from SR, MT or RR models in multivariate pedigree and GS models. Crain et al. (2018), evaluated 1170 advanced wheat lines in drought (2014, 2015) and heat (2015) environment. For precise HTP a portable phenotyping platform known as "Phenocart" and for marker discovery and genotyping GBS was used. They assessed several GS models using genotypic and phenotypic data obtained from these platforms. They finally concluded that ongoing advances in yield prediction models and huge data generation through genomics and phonemics will make GS feasible for plant breeders to achieve enhanced genetic gain in every possible environment. A Durum panel was evaluated by Sukumaran et al. (2018) in well-watered, heat-stress and drought stress environment, and reaction norm model for genomic prediction was applied. They found that in all cross-validation schemes, prediction accuracy was improved by addition of G × E interaction terms to the model. Their result showed that integration of G × E interaction to various GS models could further enhance genetic gain in durum wheat breeding for drought and heat stress environment. An experiment was conducted to assess the efficacy of GS through prediction accuracy and response to selection. Breeding material was evaluated across field season from 2014 to 2016 in drought and normal condition. As a result, they found that Reproducing Kernel Hilbert Space (RKHS) and Random Forest (RF) (non-parametric algorithms) provided higher accuracies for line selection in same year cross-validation as well as in cross year prediction (Hu et al. 2019). Juliana et al. (2019a), combined genomic-enabled prediction and HTP to evaluate bread wheat to predict grain yield (GY) in late sown heat-stressed and drought-stressed environments and found average accuracy of the genomic prediction as 0.50 and 0.51 in drought-stressed and heat-stressed environments, respectively, utilizing fivefold cross-validation. Genomic predictions gave better result than pedigree-based predictions across nurseries. Finally, they concluded that across-year prediction of GY is challenging, but combination of HTP, screening of large population and evaluation of un-phenotyped large nurseries could improve the efficiency of GS for this purpose. Integration of GWAS and GS was done by Merida-Garcia et al. (2019), to evaluate 179 durum wheat lines in rainfed condition of Southern Spain (Andalusia) for various agronomic and quality traits of interest. Based on the promising results of GS prediction ability values, they concluded that the GS could be effectively used for improvement of various agronomic and quality traits targeted for rainfed condition.

## 7 Conclusion and Future Prospects

In the era of climate change and population explosion, there is an urgent need to adopt modern tools and techniques of plant breeding which would increase the selection efficiency, shorten the breeding cycle and ultimately fasten the development of varieties tolerant to biotic and abiotic stresses. Previous work done by the researchers suggests that the different models of GS used for predicting the performance of the lines have high prediction accuracy and hence will be a useful tool to breed wheat for multiple traits. Different GS models could be combined with HTP; enabling high prediction accuracy will speed up the selection of line consequently varietal development for various traits will be fast. The GS prediction ability values showed promising results for quality traits in drought and heat stress environment. Further integration of technologies like GWAS and GS with marker-assisted breeding, will definitely accelerate the breeding cycles. There remains immense scope for integrating GS with speed breeding and gene editing to fasten the breeding programme and hence varietal development of wheat for multiple traits.

## References


Araus JL, Cairns JE (2014) Field high-throughput phenotyping: the new crop breeding frontier. Trends Plant Sci 19:52–61

Araus JL, Kefauver SC, Zaman-Allah M, Olsen MS, Cairns JE (2018) Translating high-throughput phenotyping into genetic gain. Trends Plant Sci 23:451–466

Arruda MP, Brown PJ, Lipka AE, Krill AM, Thurber C, Kolb FL (2015) Genomic selection for predicting head blight resistance in a wheat breeding program. Plant Genome 8:1–12. https://doi.org/10.3835/plantgenome2015.01.0003

Arruda MP, Lipka AE, Brown PJ, Krill AM, Thurber C, Brown-Guedira G, Dong Y, Foresman BJ, Kolb FL (2016) Comparing genomic selection and marker-assisted selection for Fusarium head blight resistance in wheat (*Triticum aestivum* L.). Mol Breed 36:1–11

Asad MA, Bai B, Lan C, Yan J, Xia X, Zhang Y, He Z (2014) Identification of QTL for adult-plant resistance to powdery mildew in Chinese wheat landrace Pingyuan 50. Crop J 2:308–314

Baenziger PS (2016) Wheat breeding and genetics. Ref Modul Food Sci 2016

Barmeier G, Schmidhalter U (2016) High-throughput phenotyping of wheat and barley plants grown in single or few rows in small plots using active and passive spectral proximal sensing. Sensors 16(11):1860–1873

Batool N, Ilyas N, Shahzad A, Hauser BA, Arshad M (2018) Quantitative trait loci (QTLs) mapping for salt stress tolerance in wheat at germination stage. Pak J Agric Sci 55(1):47–55

Battenfield SD, Guzmán C, Gaynor RC, Singh RP, Peña RJ, Dreisigacker S, Fritz AK, Poland JA (2016) Genomic selection for processing and endues quality traits in the CIMMYT spring bread wheat breeding program. Plant Genome 9(2):1–12. https://doi.org/10.3835/plantgenome2016.01.0005

Bekele D, Tesfaye K, Fikre A (2019) Recent developments in genomic selection for minor gene quantitative disease resistance plant breeding. J Plant Pathol Microbiol 10:1–8

Bernardo R (2008) Molecular markers and selection for complex traits in plants: learning from the last 20 years. Crop Sci 48:1649–1664

Bhandari A, Bartholomé J, Cao-Hamadoun TV, Kumari N, Frouin J, Kumar A, Ahmadi N (2019) Selection of trait-specific markers and multi-environment models improve genomic predictive ability in rice. PLoS One 14(5):1–21. https://doi.org/10.1371/journal.pone.0208871

Bokore FE, Cuthbert RD, Knox RE, Singh A, Campbell HL, Pozniak CJ, N'Diaye A, Sharpe AG, Ruan Y (2019) Mapping quantitative trait loci associated with common bunt resistance in a spring wheat (*Triticum aestivum* L.) variety Lillian. Theor Appl Genet 132:3023–3033

Bonnett D, Rebetzke GJ, Spielmeyer W (2005) Strategies for efficient implementation of molecular markers in wheat breeding. Mol Breed 15:75–85

Breiman L (2001) Random forests. Mach Learn 45:5–32

Burgueno J, de los Campos G, Weigel K, Crossa J (2012) Genomic prediction of breeding values when modeling genotype × environment interaction using pedigree and dense molecular markers. Crop Sci 52:707–719

CGIAR Research Program on Wheat (CRP) (2018) Wheat in the world. https://wheat.org/wheat-in-the-world/

Chakradhar T, Hindu V, Reddy PS (2017) Genomic-based-breeding tools for tropical maize improvement. Genetica 145(6):525–539

Chantret N, Mingeot D, Sourdille P, Bernard M, Jacquemin JM, Doussinault G (2001) A major QTL for powdery mildew resistance is stable over time and at two development stages in winter wheat. Theor Appl Genet 103:962–971

Charmet G, Tran LG, Auzanneau J, Rincent R, Bouchet S (2020) BWGS: AR package for genomic selection and its application to a wheat breeding programme. PLoS One 15(4):e0222733

Cooper M, Messina CD, Podlich D, Totir LR, Baumgarten A, Hausmann NJ, Wright D, Graham G (2014) Predicting the future of plant breeding: complementing empirical evaluation with genetic prediction. Crop Pasture Sci 65:311–336

Crain JL, Wei Y, Barker J, Thompson SM, Alderman PD, Reynolds M, Zhang N, Poland J (2016) Development and deployment of a portable field phenotyping platform. Crop Sci 56:965–975

Crain J, Mondal S, Rutkoski J, Singh RP, Poland J (2018) Combining high-throughput phenotyping and genomic information to increase prediction and selection accuracy in wheat breeding. Plant Genome 11(1):1–14. https://doi.org/10.3835/plantgenome2017.05.0043

Crossa J, de los Campos G, Perez P, Gianola D, Burgueño J, Araus JL, Makumbi D, Singh RP, Dreisigacker S, Yan JB, Arief V, Banziger M, Braun HJ (2010) Prediction of genetic values of quantitative traits in plant breeding using pedigree and molecular markers. Genetics 186:713–724

Crossa J, Perez P, Campos GDL, Mahuku G, Dreisigacker S, Magorokosho C (2011) Genomic selection and prediction in plant breeding. J Crop Improv 25:239–261

Crossa J, Pérez P, Hickey J, Burgueño J, Ornella L, Cerón-Rojas J, Zhang X, Dreisigacker S, Babu R, Li Y, Bonnett D (2014) Genomic prediction in CIMMYT maize and wheat breeding programs. Heredity 112:48–60

Crossa J, Pérez-Rodríguez P, Cuevas J, Montesinos-López O, Jarquín D, de los Campos G, Burgueño J, González-Camacho JM, Pérez-Elizalde S, Beyene Y, Dreisigacker S (2017) Genomic selection in plant breeding: methods, models, and perspectives. Trends Plant Sci 22:961–975

Daetwyler HD, Bansal UK, Bariana HS, Hayden MJ, Hayes BJ (2014) Genomic prediction for rust resistance in diverse wheat landraces. Theor Appl Genet 127:1795–1803

De Souza CHW, Lamparelli RAC, Rocha JV, Magalhães PSG (2017) Height estimation of sugarcane using an unmanned aerial system (UAS) based on structure from motion (SfM) point clouds. Int J Remote Sens 38:2218–2230

Deery D, Jimenez-Berni J, Rebetzke GJ, James R, Bovill WD, Furbank R, Condon AG (2016) Methodology for high-throughput field phenotyping of canopy temperature using airborne and ground based thermography. Front Plant Sci 7:1808–1817

Deery DM, Rebetzke G, Jimenez-Berni JA, Bovill B, James RA, Condon AG, Furbank R, Chapman S, Fischer R (2019) Evaluation of the phenotypic repeatability of canopy temperature in wheat using continuous-terrestrial and airborne measurements. Front Plant Sci 10:875–893

Deng SM, Wu XR, Wu YY, Zhou RH, Wang HG, Jia JZ, Liu SB (2011) Characterization and precise mapping of a QTL increasing spike number with pleiotropic effects in wheat. Theor Appl Genet 122:281–289

Desta ZA, Ortiz R (2014) Genomic selection: genome-wide prediction in plant improvement. Trends Plant Sci 19:592–601

Devi R, Ram S, Rana V, Malik VK, Pande VSingh GP (2019) QTL mapping for salt tolerance associated traits in wheat (*Triticum aestivum* L.). Euphytica 215(12):210–232. https://doi.org/10.1007/s10681-019-2533-z

Dexter J, Clear R, Preston K (1996) Fusarium head blight: effect on the milling and baking of some Canadian wheats. Cereal Chem 73:695–701

Dunckel S, Crossa J, Wu S, Bonnett D, Poland J (2017) Genomic selection for increased yield in synthetic-derived wheat. Crop Sci 57(2):713–725

FAO (2018) The state of the world's land and water resources for food and agriculture: managing systems at risk. Food and Agriculture Organization of the United Nations, Rome, Italy, 2011 FAOSTAT. http://www.fao.org/faostat/

Fiedler JD, Salsman E, Liu Y, Michalak de Jiménez M, Hegstad JB, Chen B, Manthey FA, Chao S, Xu S, Elias EM, Li X (2017) Genome-wide association and prediction of grain and semolina quality traits in durum wheat breeding populations. Plant Genome 10(3):1–12. https://doi.org/10.3835/plantgenome2017.05.0038

Fischer RA, Maurer R (1978) Drought resistance in spring wheat cultivars. I. Grain yield responses. Aust J Agric Res 29:897–912

Frisch M, Melchinger AE (2005) Selection theory for marker-assisted backcrossing. Genetics 170:909–917

Furbank RT, Tester M (2011) Phenomics–technologies to relieve the phenotyping bottleneck. Trends Plant Sci 16:635–644

Gahlaut V, Jaiswal V, Tyagi BS, Singh G, Sareen S, Balyan HS, Gupta PK (2017) QTL mapping for nine drought-responsive agronomic traits in bread wheat under irrigated and rain-fed environments. PLoS One 12(8):e0182857. https://doi.org/10.1371/journal.pone.0182857

Gebrewahid TW, Zhang P, Zhou Y, Yan X, Xia X, He Z, Liu D, Li Z (2020) QTL mapping of adult plant resistance to stripe rust and leaf rust in a Fuyu 3/Zhengzhou 5389 wheat population. Crop J. https://doi.org/10.1016/j.cj.2019.09.013

Gianola D, Van Kaam JB (2008) Reproducing Kernel Hilbert spaces regression methods for genomic assisted prediction of quantitative traits. Genetics 178:2289–2303

Habier D, Fernando RL, Kizilkaya K, Garrick DJ (2011) Extension of the Bayesian alphabet for genomic selection. BMC Bioinformatics 12(1):186–197

Haghighattalab A, Pérez LG, Mondal S, Singh D, Schinstock D, Rutkoski J, Ortiz-Monasterio I, Singh RP, Goodin D, Poland J (2016) Application of unmanned aerial systems for high throughput phenotyping of large wheat breeding nurseries. Plant Methods 12(1):35–49

Haghighattalab A, Crain J, Mondal S, Rutkoski J, Singh RP, Poland J (2017) Application of geographically weighted regression to improve grain yield prediction from unmanned aerial system imagery. Crop Sci 57(5):2478–2489

Haile JK, Nachit MM, Hammer K, Badebo A, Röder MS (2012) QTL mapping of resistance to race Ug99 of *Puccinia graminis* f. sp. *tritici* in durum wheat (*Triticum durum* Desf.). Mol Breed 30:1479–1493

Haile JK, N'Diaye A, Clarke F, Clarke J, Knox R, Rutkoski J, Bassi FM, Pozniak CJ (2018) Genomic selection for grain yield and quality traits in durum wheat. Mol Breed 38(6):75–92. https://doi.org/10.1007/s11032-018-0818-x

HarvestPlus (2017) Crops. http://www.harvestplus.org/what-we-do/crops

Hassan MA, Yang M, Rasheed A, Yang G, Reynolds M, Xia X, Xiao Y, He Z (2019) A rapid monitoring of NDVI across the wheat growth cycle for grain yield prediction using a multi-spectral UAV platform. Plant Sci 282:95–103

Hayes B, Goddard ME (2001) The distribution of the effects of genes affecting quantitative traits in livestock. Genet Sel Evol 33:209–229

Hayes BJ, Panozzo J, Walker CK, Choy AL, Kant S, Wong D, Tibbits J, Daetwyler HD, Rochfort S, Hayden MJ, Spangenberg GC (2017) Accelerating wheat breeding for end-use quality with multi-trait genomic predictions incorporating near infrared and nuclear magnetic resonance-derived phenotypes. Theor Appl Genet 130:2505–2519

He S, Schulthess AW, Mirdita V, Zhao Y, Korzun V, Bothe R, Ebmeyer E, Reif JC, Jiang Y (2016) Genomic selection in a commercial winter wheat population. Theor Appl Genet 129:641–651

Heffner EL, Sorrells ME, Jannink JL (2009) Genomic selection for crop improvement. Crop Sci 49:1–12

Heffner EL, Lorenz AJ, Jannink JL, Sorrells ME (2010) Plant breeding with genomic selection: gain per unit time and cost. Crop Sci 50:1681–1690

Heffner EL, Jannink JL, Iwata H, Souza E, Sorrells ME (2011a) Genomic selection accuracy for grain quality traits in biparental wheat populations. Crop Sci 51(6):2597–2606

Heffner EL, Jannink JL, Sorrells ME (2011b) Genomic selection accuracy using multifamily prediction models in a wheat breeding program. Plant Genome 4:65–75

Heidari B, Sayed-Tabatabaei BE, Saeidi G, Kearsey M, Suenaga K (2011) Mapping QTL for grain yield, yield components, and spike features in a doubled haploid population of bread wheat. Genome 54(6):517–527

Henderson CR (1963) Selection index and expected genetic advance. Stat Genet Plant Breed 982:141–163

Herrera-Foessel SA, Lagudah ES, Huerta-Espino J, Hayden MJ, Bariana HS, Singh D, Singh RP (2011) New slow rusting leaf rust and stripe rust resistance genes *Lr67* and *Yr46* in wheat are pleiotropic or closely linked. Theor Appl Genet 122:239–249

Herter CP, Ebmeyer E, Kollers S, Korzun V, Miedaner T (2019) An experimental approach for estimating the genomic selection advantage for Fusarium head blight and *Septoria tritici* blotch in winter wheat. Theor Appl Genet 132:2425–2437

Hickey LT, Hafeez AN, Robinson H, Jackson SA, Leal-Bertioli SCM, Tester M, Gao C, Godwin ID, Hayes BJ, Wulff BBH (2019) Breeding crops to feed 10 billion. Nat Biotechnol 37:744–754

Hoffstetter A, Cabrera A, Huang M, Sneller C (2016) Optimizing training population data and validation of genomic selection for economic traits in soft winter wheat. G3: Genes Genomes Genetics 6(9):2919–2928

Holman FH, Riche AB, Michalski A, Castle M, Wooster MJ, Hawkesford MJ (2016) High throughput field phenotyping of wheat plant height and growth rate in field plot trials using UAV based remote sensing. Remote Sens 8(12):1031–1053

Howes NK, Woods SM (1998) Simulations and practical problems of applying multiple marker assisted selection and doubled haploids to wheat breeding programs. Eur J Plant Pathol 100:225–230

Hu X, Carver BF, Powers C, Yan L, Zhu L, Chen C (2019) Effectiveness of genomic selection by response to selection for winter wheat variety improvement. Plant Genome 12(3):1–15. https://doi.org/10.3835/plantgenome2018.11.0090

Hu Y, Knapp S, Schmidhalter U (2020) Advancing high-throughput phenotyping of wheat in early selection cycles. Remote Sens 12(3):574–583

IWGSC (International Wheat Genome Sequencing Consortium) (2018) Shifting the limits in wheat research and breeding using a fully annotated reference genome. Science 361:eaar7191. https://doi.org/10.1126/science.aar7191

Jannink J, Lorenz AJ, Iwata H (2010) Genomic selection in plant breeding: from theory to practice. Brief Funct Genomics 9:166–177

Jiang Y, Schulthess AW, Rodemann B, Ling J, Plieske J, Kollers S, Ebmeyer E, Korzun V, Argillier O, Stiewe G, Ganal MW (2017) Validating the prediction accuracies of marker-assisted and genomic selection of Fusarium head blight resistance in wheat using an independent sample. Theor Appl Genet 130:471–482

Jiang Y, Li C, Robertson JS, Sun S, Xu R, Paterson AH (2018) Gphenovision: a ground mobile system with multi-modal imaging for field-based high throughput phenotyping of cotton. Sci Rep 8:1213–1228

Jimenez-Berni JA, Deery DM, Rozas-Larraondo P, Condon AT, Rebetzke GJ, James RA, Bovill WD, Furbank RT, Sirault XR (2018) High throughput determination of plant height, ground cover, and above-ground biomass in wheat with LiDAR. Front Plant Sci 9:237–255. https://doi.org/10.3389/fpls.2018.00237

Jin F, Zhang D, Bockus W, Baenziger PS, Carver B, Bai G (2013) Fusarium head blight resistance in US winter wheat cultivars and elite breeding lines. Crop Sci 53:2006–2013

Juliana P, Singh RP, Singh PK, Crossa J, Huerta-Espino J, Lan C, Bhavani S, Rutkoski JE, Poland JA, Bergstrom GC, Sorrells ME (2017a) Genomic and pedigree-based prediction for leaf, stem, and stripe rust resistance in wheat. Theor Appl Genet 130:1415–1430

Juliana P, Singh RP, Singh PK, Crossa J, Rutkoski JE, Poland JA, Bergstrom GC, Sorrells ME (2017b) Comparison of models and whole-genome profiling approaches for genomic-enabled prediction of *Septoria tritici* blotch, *Stagonospora nodorum* blotch, and tan spot resistance in wheat. Plant Genome 10(2):1–16. https://doi.org/10.3835/plantgenome2016.08.0082

Juliana P, Singh RP, Poland J, Mondal S, Crossa J, Montesinos-López OA, Dreisigacker S, Pérez-Rodríguez P, Huerta-Espino J, Crespo-Herrera L, Govindan V (2018) Prospects and challenges of applied genomic selection – a new paradigm in breeding for grain yield in bread wheat. Plant Genome 11(3):180017

Juliana P, Montesinos-López OA, Crossa J, Mondal S, Pérez LG, Poland J, Huerta-Espino J, Crespo-Herrera L, Govindan V, Dreisigacker S, Shrestha S, Shrestha S (2019a) Integrating genomic-enabled prediction and high-throughput phenotyping in breeding for climate-resilient bread wheat. Theor Appl Genet 132(1):177–194

Juliana P, Poland J, Huerta-Espino J, Shrestha S, Crossa J, Crespo-Herrera L, Toledo FH, Govindan V, Mondal S, Kumar U, Bhavani S (2019b) Improving grain yield, stress resilience and quality of bread wheat using large-scale genomics. Nat Genet 51(10):1530–1539

Kelly AM, Smith AB, Eccleston JA, Cullis BR (2007) The accuracy of varietal selection using factor analytic models for multi-environment plant breeding trials. Crop Sci 47:1063–1070

Kipp S, Mistele B, Baresel P, Schmidhalter U (2014) High-throughput phenotyping early plant vigour of winter wheat. Eur J Agron 52:271–278

Krattinger SG, Lagudah ES, Spielmeyer W, Singh RP, Huerta-Espino J, McFadden H, Bossolini E, Selter LL, Keller B (2009) A putative ABC transporter confers durable resistance to multiple fungal pathogens in wheat. Science 323:1360–1363

Krause MR, González-Pérez L, Crossa J, Pérez-Rodríguez P, Montesinos-López O, Singh RP, Dreisigacker S, Poland J, Rutkoskiy J, Sorrells M, Gore MA (2018) Use of hyperspectral reflectance-derived relationship matrices for genomic prediction of grain yield in wheat. BioRxiv:e389825. https://doi.org/10.1101/389825

Kthiri D, Loladze A, N'Diaye A, Nilsen KT, Walkowiak S, Dreisigacker S, Ammar K, Pozniak CJ (2019) Mapping of genetic loci conferring resistance to leaf rust from three globally resistant durum wheat sources. Front Plant Sci 10:1247–1262. https://doi.org/10.3389/fpls.2019.01247

Lee U, Chang S, Putra GA, Kim H, Kim DH (2018) An automated, high-throughput plant phenotyping system using machine learning-based plant segmentation and image analysis. PLoS One 13(4). https://doi.org/10.1371/journal.pone.0196615

Lehermeier C, Krämer N, Bauer E, Bauland C, Camisan C, Campo L, Flament P, Melchinger AE, Menz M, Meyer N, Moreau L (2014) Usefulness of multiparental populations of maize (*Zea mays* L.) for genome-based prediction. Genetics 198(1):3–16

Lelong C, Burger P, Jubelin G, Roux B, Labbé S, Baret F (2008) Assessment of unmanned aerial vehicles imagery for quantitative monitoring of wheat crop in small plots. Sensors 8:3557–3585

Liu G, Jia L, Lu L, Qin D, Zhang J, Guan P (2014) Mapping QTLs of yield-related traits using RIL population derived from common wheat and Tibetan semi-wild wheat. Theor Appl Genet 127:2415–2432

Liu B, Asseng S, Müller C, Ewert F, Elliott J, Lobell DB, Martre P, Ruane AC, Wallach D, Jones JW, Rosenzweig C (2016) Similar estimates of temperature impacts on global wheat yield by three independent methods. Nat Clim Chang 6:1130–1136

Lopez-Cruz M, Crossa J, Bonnett D, Dreisigacker S, Poland J, Jannink JL, Singh RP, Autrique E, de los Campos G (2015) Increased prediction accuracy in wheat breeding trials using a marker× environment interaction genomic selection model. G3: Genes Genomes Genetics 5(4):569–582

Lorenz AJ, Smith KP, Jannink JL (2012) Potential and optimization of genomic selection for Fusarium head blight resistance in six-row barley. Crop Sci 52:1609–1621

Lozada DN, Mason RE, Sarinelli JM, Brown-Guedira G (2019) Accuracy of genomic selection for grain yield and agronomic traits in soft red winter wheat. BMC Genet 20(1):82–94

Lu J, Hu J, Zhao G, Mei F, Zhang C (2017) An in-field automatic wheat disease diagnosis system. Comput Electron Agric 142:369–379

Malik S, Malik TA (2015) Genetic mapping of potential QTLs associated with drought tolerance in wheat. J Anim Plant Sci 25(4):1032–1040

Manickavelu A, Hattori T, Yamaoka S, Yoshimura K, Kondou Y, Onogi A, Matsui M, Iwata H, Ban T (2017) Genetic nature of elemental contents in wheat grains and its genomic prediction: toward the effective use of wheat landraces from Afghanistan. PLoS One 12(1):e0169416

Martin LBB, Fei Z, Giovannoni JJ, Rose JKC (2013) Catalyzing plant science research with RNA-seq. Front Plant Sci 4:66–75

Merida-Garcia R, Liu G, He S, Gonzalez-Dugo V, Dorado G, Galvez S, Solis I, Zarco-Tejada PJ, Reif JC, Hernandez P (2019) Genetic dissection of agronomic and quality traits based on association mapping and genomic selection approaches in durum wheat grown in Southern Spain. PLoS One 14(2):e0211718

Meuwissen THE, Hayes BJ, Goddard ME (2001) Prediction of total genetic value using genome-wide dense marker maps. Genetics 157:1819–1829

Michel S, Ametz C, Gungor H, Epure D, Grausgruber H, Löschenberger F, Buerstmayr H (2016) Genomic selection across multiple breeding cycles in applied bread wheat breeding. Theor Appl Genet 129(6):1179–1189

Michel S, Kummer C, Gallee M, Hellinger J, Ametz C, Akgöl B, Epure D, Löschenberger F, Buerstmayr H (2018) Improving the baking quality of bread wheat by genomic selection in early generations. Theor Appl Genet 131(2):477–493

Michel S, Löschenberger F, Ametz C, Pachler B, Sparry E, Bürstmayr H (2019) Simultaneous selection for grain yield and protein content in genomics-assisted wheat breeding. Theor Appl Genet 132(6):1745–1760

Miedaner T, Korzun V (2012) Marker-assisted selection for disease resistance in wheat and barley breeding. Phytopathology 102:560–566

Mirdita V, He S, Zhao Y, Korzun V, Bothe R, Ebmeyer E, Reif JC, Jiang Y (2015) Potential and limits of whole genome prediction of resistance to Fusarium head blight and *Septoria tritici* blotch in a vast central European elite winter wheat population. Theor Appl Genet 128:2471–2481

Moore JK, Manmathan HK, Anderson VA, Poland JA, Morris CF, Haley SD (2017) Improving genomic prediction for pre-harvest sprouting tolerance in wheat by weighting large-effect quantitative trait loci. Crop Sci 57(3):1315–1324

Moose SP, Mumm RH (2008) Molecular plant breeding as the foundation for 21st century crop improvement. Plant Physiol 147:969–977

Muleta KT, Bulli P, Zhang Z, Chen X, Pumphrey M (2017) Unlocking diversity in germplasm collections via genomic selection: a case study based on quantitative adult plant resistance to stripe rust in spring wheat. Plant Genome 10(3):1–15. https://doi.org/10.3835/plantgenome2016.12.0124

Ornella L, Singh S, Perez P, Burgueño J, Singh R, Tapia E, Bhavani S, Dreisigacker S, Braun HJ, Mathews K, Crossa J (2012) Genomic prediction of genetic values for resistance to wheat rusts. Plant Genome 5:136–148

Ortiz R, Sayre KD, Govaerts B, Gupta R, Subbarao GV, Ban T, Hodson D, Dixon JM, Ortiz-Monasterio JI, Reynolds M (2008) Climate change: can wheat beat the heat? Agric Ecosyst Environ 126:46–58

Ortiz-Monasterio JI, Palacios-Rojas N, Meng E, Pixley K, Trethowan R, Pena RJ (2007) Enhancing the mineral and vitamin content of wheat and maize through plant breeding. J Cereal Sci 46(3):293–307

Park T, Casella G (2008) The Bayesian lasso. J Am Stat Assoc 103:681–686

Parlevliet JE (2002) Durability of resistance against fungal, bacterial and viral pathogens; present situation. Euphytica 124:147–156

Perez-Rodriguez P, Gianola D, González-Camacho JM, Crossa J, Manes Y, Dreisigacker S (2012) Comparison between linear and non-parametric regression models for genome-enabled prediction in wheat. G3: Genes Genomes Genetics 2:1595–1605

Poland JA, Rife TW (2012) Genotyping-by-sequencing for plant breeding and genetics. Plant Genome J 5(3):92–102

Poland J, Rutkoski J (2016) Advances and challenges in genomic selection for disease resistance. Annu Rev Phytopathol 54:79–98

Poland J, Endelman J, Dawson J, Rutkoski J, Wu SY, Manes Y, Dreisigacker S, Crossa J, Sanchez-Villeda H, Sorrells M, Jannink JL (2012) Genomic selection in wheat breeding using genotyping-by-sequencing. Plant Genome 5:103–113

Prasad PVV, Pisipati SR, Momčilović I, Ristic Z (2011) Independent and combined effects of high temperature and drought stress during grain filling on plant yield and chloroplast EF-Tu expression in spring wheat. J Agron Crop Sci 197:430–441

Qin F, Shinozaki K, Yamaguchi-Shinozaki K (2011) Achievements and challenges in understanding plant abiotic stress responses and tolerance. Plant Cell Physiol 52(9):1569–1582

Rapp M, Lein V, Lacoudre F, Lafferty J, Müller E, Vida G, Bozhanova V, Ibraliu A, Thorwarth P, Piepho HP, Leiser WL (2018) Simultaneous improvement of grain yield and protein content in durum wheat by different phenotypic indices and genomic selection. Theor Appl Genet 131(6):1315–1329

Rasmusson DC, Phillips RL (1997) Plant breeding progress and genetic diversity from de novo variation and elevated epistasis. Crop Sci 37:303–310

Ray DK, Ramankutty N, Mueller ND, West PC, Foley JA (2012) Recent patterns of crop yield growth and stagnation. Nat Commun 3:1293–1299

Ray DK, Mueller ND, West PC, Foley JA (2013) Yield trends are insufficient to double global crop production by 2050. PLoS One 8(6):e66428

Reynolds D, Baret F, Welcker C, Bostrom A, Ball J, Cellini F, Lorence A, Chawade A, Khafif M, Noshita K, Mueller-Linow M (2019) What is cost-efficient phenotyping? Optimizing costs for different scenarios. Plant Sci 282:14–22

Rincent R, Charpentier JP, Faivre-Rampant P, Paux E, Le Gouis J, Bastien C, Segura V (2018) Phenomic selection is a low-cost and high-throughput method based on indirect predictions: proof of concept on wheat and poplar. G3: Genes Genomes Genetics 8(12):3961–3972

Rutkoski J, Benson J, Jia Y, Brown-Guedira G, Jannink JL, Sorrells M (2012) Evaluation of genomic prediction methods for Fusarium head blight resistance in wheat. Plant Genome 5:51–61

Rutkoski JE, Poland JA, Singh RP, Huerta-Espino J, Bhavani S, Barbier H, Rouse MN, Jannink JL, Sorrells ME (2014) Genomic selection for quantitative adult plant stem rust resistance in wheat. Plant Genome 7(3):1–10. https://doi.org/10.3835/plantgenome2014.02.0006

Rutkoski J, Singh RP, Huerta-Espino J, Bhavani S, Poland J, Jannink JL, Sorrells ME (2015) Genetic gain from phenotypic and genomic selection for quantitative resistance to stem rust of wheat. Plant Genome 8(2):1–10. https://doi.org/10.3835/plantgenome2014.10.0074

Rutkoski J, Poland J, Mondal S, Autrique E, Pérez LG, Crossa J, Reynolds M, Singh R (2016) Canopy temperature and vegetation indices from high-throughput phenotyping improve accuracy of pedigree and genomic selection for grain yield in wheat. G3: Genes Genomes Genetics 6:2799–2808

Saccomanno A, Matny O, Marone D, Laidò G, Petruzzino G, Mazzucotelli E, Desiderio F, Blanco A, Gadaleta A, Pecchioni N, De Vita P (2018) Genetic mapping of loci for resistance to stem rust in a tetraploid wheat collection. Int J Mol Sci 19:3907–3922. https://doi.org/10.3390/ijms19123907

Saintenac C, Zhang W, Salcedo A, Rouse MN, Trick HN, Akhunov E, Dubcovsky J (2013) Identification of wheat gene *Sr35* that confers resistance to *Ug99* stem rust race group. Science 341:783–786

Sallam AH, Endelman JB, Jannink JL, Smith KP (2015) Assessing genomic selection prediction accuracy in a dynamic barley breeding population. Plant Genome 8:1):1–1)15. https://doi.org/10.3835/plantgenome2014.05.0020

Sallam A, Alqudah AM, Dawood MF, Baenziger PS, Börner A (2019) Drought stress tolerance in wheat and barley: advances in physiology, breeding and genetics research. Int J Mol Sci 20(13):3137–3172. https://doi.org/10.3390/ijms20133137

Sarinelli JM, Murphy JP, Tyagi P, Holland JB, Johnson JW, Mergoum M, Mason RE, Babar A, Harrison S, Sutton R, Griffey CA (2019) Training population selection and use of fixed effects to optimize genomic predictions in a historical USA winter wheat panel. Theor Appl Genet 132:1247–1261

Shikha M, Kanika A, Rao AR, Mallikarjuna MG, Gupta HS, Nepolean T (2017) Genomic selection for drought tolerance using genome-wide SNPs in maize. Front Plant Sci 8:550–561. https://doi.org/10.3389/fpls.2017.00550

Singh RP, Singh PK, Rutkoski J, Hodson DP, He X, Jørgensen LN, Hovmøller MS, Huerta-Espino J (2016) Disease impact on wheat yield potential and prospects of genetic control. Annu Rev Phytopathol 54:303–322

Singh PK, He X, Sansaloni CP, Juliana P, Dreisigacker S, Duveiller E, Kumar U, Joshi AK, Singh RP (2018) Resistance to spot blotch in two mapping populations of common wheat is controlled by multiple QTL of minor effects. Int J Mol Sci 19(4054):4069. https://doi.org/10.3390/ijms19124054

Singh D, Wang X, Kumar U, Gao L, Noor M, Imtiaz M, Singh RP, Poland J (2019) High-throughput phenotyping enabled genetic dissection of crop lodging in wheat. Front Plant Sci 10:394–405

Sukumaran S, Jarquin D, Crossa J, Reynolds M (2018) Genomic-enabled prediction accuracies increased by modeling genotype × environment interaction in durum wheat. Plant Genome 11(2):1–11. https://doi.org/10.3835/plantgenome2017.12.0112

Sun J, Rutkoski JE, Poland JA, Crossa J, Jannink JL, Sorrells ME (2017) Multitrait, random regression, or simple repeatability model in high-throughput phenotyping data improve genomic prediction for wheat grain yield. Plant Genome 10(2):1–12. https://doi.org/10.3835/plantgenome2016.11.0111

Tattaris M, Reynolds MP, Chapman SC (2016) A direct comparison of remote sensing approaches for high-throughput phenotyping in plant breeding. Front Plant Sci 7:1131–1140

Thompson AL, Thorp KR, Conley M, Andrade-Sanchez P, Heun JT, Dyer JM, White JW (2018) Deploying a proximal sensing cart to identify drought-adaptive traits in upland cotton for high-throughput phenotyping. Front Plant Sci 9:507–520. https://doi.org/10.3389/fpls.2018.00507

Trnka M, Feng S, Semenov MA, Olesen JE, Kersebaum KC, Rötter RP, Semerádová D, Klem K, Huang W, Ruiz-Ramos M, Hlavinka P (2019) Mitigation efforts will not fully alleviate the increase in water scarcity occurrence probability in wheat-producing areas. Sci Adv 5(9):eaau2406. https://doi.org/10.1126/sciadv.aau2406

van Eeuwijk FA, Bustos-Korts D, Millet EJ, Boer MP, Kruijer W, Thompson A, Malosetti M, Iwata H, Quiroz R, Kuppe C, Muller O, Blazakis KN, Yu K, Tardieu F, Chapman SC (2019) Modelling strategies for assessing and increasing the effectiveness of new phenotyping techniques in plant breeding. Plant Sci 282:23–39. https://doi.org/10.1016/j.plantsci.2018.06.018

Vanderplank JE (2012) Disease resistance in plants. Elsevier, Philadelphia

Vega FA, Ramirez FC, Saiz MP, Rosua FO (2015) Multi-temporal imaging using an unmanned aerial vehicle for monitoring a sunflower crop. Biosyst Eng 132:19–27

Velu G, Ortiz-Monasterio I, Cakmak I, Hao Y, Singh RP (2014) Biofortification strategies to increase grain zinc and iron concentrations in wheat. J Cereal Sci 59(3):365–372

Velu G, Crossa J, Singh RP, Hao Y, Dreisigacker S, Perez-Rodriguez P, Joshi AK, Chatrath R, Gupta V, Balasubramaniam A, Tiwari C (2016) Genomic prediction for grain zinc and iron concentrations in spring wheat. Theor Appl Genet 129(8):1595–1605

Wang N, Liu B, Liang X, Zhou Y, Song J, Yang J, Yong H, Weng J, Zhang D, Li M, Nair S (2019) Genome-wide association study and genomic prediction analyses of drought stress tolerance in China in a collection of off-PVP maize inbred lines. Mol Breed 39(8):113–132

Watanabe K, Guo W, Arai K, Takanashi H, Kajiya-Kanegae H, Kobayashi M, Yano K, Tokunaga T, Fujiwara T, Tsutsumi N, Iwata H (2017) High-throughput phenotyping of sorghum plant height using an unmanned aerial vehicle and its application to genomic prediction modeling. Front Plant Sci 8:421–431. https://doi.org/10.3389/fpls.2017.00421

Watt M, Fiorani F, Usadel B, Rascher U, Muller O, Schurr U (2020) Phenotyping: new windows into the plant for breeders. Annu Rev Plant Biol 71:689–712. https://doi.org/10.1146/annurev-arplant-042916-041124

Weckwerth W, Ghatak A, Bellaire A, Chaturvedi P, Varshney RK (2020) PANOMICS meets germplasm. Plant Biotechnol J 18(7):1507–1525. https://doi.org/10.1111/pbi.13372

Weigand C (2011) Wheat import projections towards 2050. US Wheat Associates, Arlington

White JW, Conley MM (2013) A flexible, low-cost cart for proximal sensing. Crop Sci 53:1646–1649

White JW, Andrade-Sanchez P, Gore MA, Bronson KF, Coffelt TA, Conley MM, Feldmann KA, French AN, Heun JT, Hunsaker DJ, Jenks MA (2012) Field-based phenomics for plant genetics research. Field Crops Res 133:101–112. https://doi.org/10.1016/j.fcr.2012.04.003

William M, Singh R, Huerta-Espino J, Islas SO, Hoisington D (2003) Molecular marker mapping of leaf rust resistance gene *Lr46* and its association with stripe rust resistance gene *Yr29* in wheat. Phytopathology 93:153–159

Würschum T, Leiser WL, Kazman E, Longin CFH (2016) Genetic control of protein content and sedimentation volume in European winter wheat cultivars. Theor Appl Genet 129:1685–1696. https://doi.org/10.1007/s00122-016-2732-0

Xu C, Zhang H, Sun J, Guo Z, Zou C, Li WX, Xie C, Huang C, Xu R, Liao H, Wang J (2018) Genome-wide association study dissects yield components associated with low-phosphorus stress tolerance in maize. Theor Appl Genet 131(8):1699–1714

Xu C, Kohler TA, Lentonf TM, Svenning J, Schefferc M (2020) Future of the human climate niche. Proc Natl Acad Sci 117(21):11350–11355. https://doi.org/10.1073/pnas.1910114117

Yang J, Sears R, Gill B, Paulsen G (2002) Growth and senescence characteristics associated with tolerance of wheat-alien amphiploids to high temperature under controlled conditions. Euphytica 126(2):185–193

York LM (2019) Functional phenomics: an emerging field integrating high-throughput phenotyping, physiology, and bioinformatics. J Exp Bot 70(2):379–386

Yu X, Ren S, Zhao L, Guo J, Bao Y, Ma Y, Wang H, Ohm HW, Yu D, Li H, Kong L (2018) Molecular mapping of a novel wheat powdery mildew resistance gene Ml92145E8-9 and its application in wheat breeding by marker-assisted selection. Crop J 6:621–627

Yuan L, Zhang H, Zhang Y, Xing C, Bao Z (2017) Feasibility assessment of multi-spectral satellite sensors in monitoring and discriminating wheat diseases and insects. Optik 131:598–608

Yuan FP, Zeng QD, Wu JH, Wang QL, Yang ZJ, Liang BP et al. (2018) QTL mapping and validation of adult plant resistance to stripe rust in Chinese wheat landrace Humai 15. Front Plant Sci 9:968

Zampieri M, Ceglar A, Dentener F, Toreti A (2017) Wheat yield loss attributable to heat waves, drought and water excess at the global, national and subnational scales. Environ Res Lett 12(6):064008. https://doi.org/10.1088/1748-9326/aa723b

Zhang P, Li X, Gebrewahid TW, Liu H, Xia X, He Z, Li Z, Liu D (2019) QTL mapping of adult-plant resistance to leaf and stripe rust in wheat cross SW 8588/Thatcher using the wheat 55K SNP array. Plant Dis 103:3041–3049

Zhao Y, Gowda M, Longin FH, Würschum T, Ranc N, Reif JC (2012) Impact of selective genotyping in the training population on accuracy and bias of genomic selection. Theor Appl Genet 125:707–713. https://doi.org/10.1007/s00122-012-1862-2

Zhao YS, Zeng J, Fernando R, Reif JC (2013) Genomic prediction of hybrid wheat performance. Crop Sci 53:802–810

Zhao Y, Mette MF, Reif JC (2015) Genomic selection in hybrid breeding. Plant Breed 134(1):1–10

# Genetic Dissection for Yield and Yield-Related Traits in Bread Wheat (*Triticum aestivum* L.)

**Reyazul Rouf Mir, Sachin Kumar, and Safoora Shafi**

**Abstract** Wheat is one of the most important cereal crops of global importance. Wheat crop provides one-fifth of the daily calories and dietary proteins for human consumption. Improving grain yield (GY) and yield-contributing traits is considered important for increasing wheat production and therefore for food security. The yield and traits related to yield in wheat are mostly quantitative traits controlled by several small effect/minor genes/QTLs. Hundreds of studies have been conducted in wheat for the discovery of genes/QTLs for yield and related traits using different approaches. Among different approaches, linkage-based QTL mapping and association mapping are most commonly used approaches. Traditional QTL mapping involves the use of biparental mapping populations derived from crossing two contrasting parental genotypes. The other recently emerged mapping approach "association mapping" also known as genome-wide association studies (GWAS) that involves the use of diverse germplasm is considered the method of choice nowadays to unravel and understand genetics of yield and yield-related traits. Using these different mapping approaches, several genes/QTLs have been already identified for GY and yield-related traits in wheat. The QTLs/genes identified belong to all the 21 bread wheat chromosomes. In addition, QTL × Environment, QTL × QTL, and QTL × QTL × Environment interactions have been also worked out in detail. The important stable and major QTLs identified will prove useful in wheat molecular breeding programs aimed at enhancing GY for food security.

**Keywords** Wheat · Yield · Yield contributing traits · QTLs · Genes · Epistatic interactions

R. R. Mir (✉) · S. Shafi
Division of Genetics and Plant Breeding, Faculty of Agriculture, SKUAST-Kashmir, Wadura Sopore, India
e-mail: rrmir@skuastkashmir.ac.in

S. Kumar
Department of Genetics and Plant Breeding, Chaudhary Charan Singh University, Meerut, Uttar Pradesh, India

© The Editor(s) (if applicable) and The Author(s), under exclusive license to Springer Nature Switzerland AG 2021

S. H. Wani et al. (eds.), *Physiological, Molecular, and Genetic Perspectives of Wheat Improvement*, https://doi.org/10.1007/978-3-030-59577-7_10

# 1 Introduction

Bread wheat is the third most important staple food crop followed by rice and maize. Its popularity is not only because of its wide adaptability but also for its global food demand and nutritional security. It has been estimated that wheat accounts for 20% the dietary protein and 21% food calories consumed by the human population (International Wheat Yield Partnership; IWYP). However, in order to meet the demands of rapidly increasing population, wheat grain production must exceed by 2% rate annually on the same cultivated land available at present (Gill et al. 2004) or should increase by at least 50% by 2030 (Tshikunde et al. 2019). Therefore, it is a need of the hour to increase wheat production despite the fact ~700 million metric tons of wheat grains are annually produced. In addition, there is urgent need for the increase in worldwide average wheat yields from 3–5 t $ha^{-1}$ (Rosegrant and Agcaoili 2010; Tshikunde et al. 2019). Furthermore, it is important to increase the wheat production/productivity under extreme weather conditions like water stress, drought stress, and heat stress (Curtis and Halford 2014). During the past decade, the continuous plant breeding efforts have led to increase in the GY in wheat (Laidig et al. 2017; Piepho et al. 2014; Würschum et al. 2018). Further, to increase the wheat yield, IWYP was formed with a goal to increase wheat yield by 50% in the next 20 years.

The grain yield (GY) in bread wheat is considered one of the most complex quantitative traits inherited quantitatively and significantly influenced by the environment (Gupta et al. 2007). The GY trait can be dissected into several component traits with higher heritability (Kato et al. 2000; Hai et al. 2008). In addition, it has been reported that individual traits showing correlation with GY are most often controlled by the same set of QTLs/genes (Li et al. 2007; Hai et al. 2008; Moeller et al. 2014; Nasseer et al. 2016; Xu et al. 2017; Zhang et al. 2018). The GY in bread wheat usually constitutes three component traits including number of spikelet's/spike, number of grains per spike, and 1000 grain/kernel weight. Among the three yield components, highest heritability (59–80%) is shown by grain weight (GW) (Xiao and He 2003), and this indicates that selection for grain weight in early breeding/segregating generations will prove effective. However, grain weight as a trait also constitutes other traits including grain length and grain width (Tyagi et al. 2015). Several efforts have been made to genetically dissect GW component traits to know about the genetics of GW in wheat (Breseghello and Sorrells 2006a; Sun et al. 2009; Cui et al. 2011; Mir et al. 2012a, b; Tyagi et al. 2015). The component traits of grain weight (length, width, and area) are also considered very stable with high heritability (Mir et al. 2012a, b; Distelfeld et al. 2014; Tyagi et al. 2015).

The important reproductive organ harboring grains in wheat is spike and therefore traits related to wheat spike are considered very important for manipulating GY. A variety of published studies have indicated a very strong and positive correlation between different spike traits like spike length with GY (Kumar et al. 2007) and yield-related traits including 1000 grain/kernel weight (Wu et al. 2012; Mir et al. 2012a, b; Gao et al. 2015). Therefore, from a breeding point of view, genes/QTLs already identified or to be identified in future for traits related to wheat spike are important for wheat molecular breeding programs aimed at enhancing GY. The three genomic regions/loci identified for the domestication of wheat spike traits include *Q*, *C*, and *S* (Faris et al. 2014). The domestication locus "Q" also known as super domestication gene in wheat has pleiotropic effects on several traits like plant height, rachis fragility, and spike length (Simons et al. 2006). On the other hand, other important domestication locus "C-locus" also known as "compactum locus" affects spike compactness, grain morphology traits, and grain number, and locus "S" is responsible for round seeds and glumes of a wheat spike (Salina et al. 2000; Johnson et al. 2008).

Genetic dissection for quantitative traits has been done using two important approaches: QTL mapping and more recent association mapping. Each approach has its own advantages and disadvantages (Mir et al. 2012a). Both approaches have been now used extensively for genetic dissection leading to gene/QTL discovery for variety of traits in wheat (Gupta et al. 2008; Mohan et al. 2009; Kulwal et al. 2010; Gupta et al. 2011; Jaiswal et al. 2012; Mir et al. 2012b; Jaiswal et al. 2016). The QTL/genes ones identified have been deployed in wheat molecular breeding programs through modern breeding approaches including marker-assisted selection (Gupta et al. 2010a, b; Gupta et al. 2011; Kumar et al. 2010, 2011; Fig. 1). QTL mapping and association mapping studies in wheat crop have shown that yield and yield-related traits are complex quantitative traits controlled by several QTLs/genes having small (minor) effects on the GY or yield-related traits (for a recent

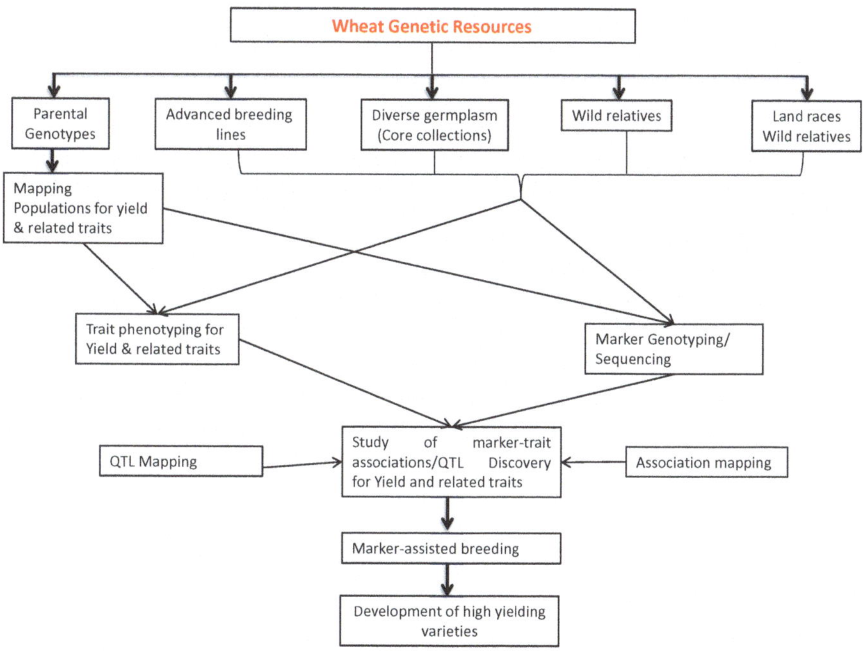


**Fig. 1** Different steps involved in genetic dissection for yield and related traits in bread wheat

publication and references therein, see Mir et al. 2012b; Guo et al. 2018; Hu et al. 2020). In this manuscript the results of genetic dissection leading to gene/QTL discovery for yield and yield-related traits have been compiled and discussed in detail.

## 2 QTL Analyses for Yield-Contributing Traits

As already mentioned, in wheat, GY and yield-contributing traits are quantitative traits controlled by several small effect minor genes/QTL. In the past, a number of studies were undertaken in wheat for the detection, chromosomal localization, and the study of the effects of genes/QTL for the above traits (for details see Table 1 and references cited there). These studies involved classical cytogenetic analyses as well as molecular approaches of QTL analysis. The QTL analyses were conducted using either the genetic/linkage maps of single or few chromosomes or genetic/linkage maps of whole bread wheat genome. For instance, Kumar et al. (2006) used individual wheat chromosome (only three chromosomes; 1A, 2B, and 7A) to map genes/QTLs for grain weight, and while using the same mapping population and whole genome genetic maps, different genes/QTLs for grain weight were reported (Mir et al. 2012b).

For GY and yield-related traits in wheat, genes/QTLs have been identified on 8–11 different chromosomes through cytogenetic studies. For instance, genes/QTLs were discovered on 11 wheat chromosomes for GY, on 8 chromosomes for tillers per plant, on 10 chromosomes for spike length, on 10 chromosomes for spikelets per spike, and on 11 chromosomes for grains per spike (see Mir (2012) for more details; Table 1). Dozens of published reports are available where QTLs have been detected and their associated markers identified for GY and yield-related traits in bread wheat. The number of chromosomes reported to carry genes/QTLs following QTL analyses is higher than those reported through cytogenetic studies (Kulwal et al. 2004; Huang et al. 2003, 2004; Quarrie et al. 2005; Tyagi et al. 2015; Bhusal et al. 2017). For the component traits of GY, QTLs were reported on 8 to 21 different chromosomes of bread wheat (for details see Table 1 for a summary).

For instance, QTLs have been discovered and validated for GY (14 QTLs), grain number/spike (1 QTL), number of spikelet's/spike (15 QTLs), grain weight (11 QTLs), and for water status (9 QTLs) in wheat (Zhang et al. 2018). It is important to mention that these QTLs showed no association with plant height and days to heading. As expected, significant correlations were found between the studied traits and identified colocalized. Similarly, genes that are responsible for spike morphology have also been identified using recombinant inbred lines (RILs) mapping population developed from crossing between two winter wheat cultivars. A set of four novel QTLs/genes for spike morphological traits were identified which provided insights into the genetics that shaped spike morphology in wheat (Zhai et al. 2016).

**Table 1** A summary of QTL mapping studies for yield and yield-contributing traits in bread wheat

| Trait/ no. of QTL | Range of PVE (%) | Chromosome (arm) | Population | Linked markers | Method of QTL mapping | Software used | Reference |
|---|---|---|---|---|---|---|---|
| 1000 grain weight | | | | | | | |
| 1 | 17.0 | 4AS | CRSL | RFLP | SIM, sCIM | MQTL | Araki et al. (1999) |
| 2 | 7.5–17.4 | 3AS | RICL | RFLP | MR | PROC REG | Shah et al. (1999a, b) |
| 2 | 11.0–19.0 | 5AL | RIL | RFLP | SIM, sCIM | MQTL | Kato et al. (2000) |
| 10 | 4.7–19.7 | 1D, 2BS, 2DS, 3AL, 5BL, 6AL, 6DS, 7AS, 7DL | RIL | RFLP, SSR | MR | Splus | Groos et al. (2003) |
| 8 | 14.3–25.9 | 2AS, 2DL, 4DS, 5BS, 7AS, 7BL, 7DS | $BC_2F_2$ | SSR | SIM | QGene | Huang et al. (2003) |
| 14 | 8.4–25.2 | 1BS, 1DL, 2AS, 2DL, 3AL, 3BS, 3BL, 3DS, 4BL, 6AS, 6AL, 7AS, 7AL, 7DS | $BC_2F_1$ | SSR | MR, SIM | QGene | Huang et al. (2004) |
| 6 | 3.0–31.8 | 2AS, 3DL, 4AL, 4BL, 4DS, 6DL | DH | SSR | CIM | QTL Cartographer | McCartney et al. (2005) |
| 6 | 3.7–26.3 | 2BS, 2DS, 3BL, 4BL, 4DS, 6AL | DH | SSR | CIM | QTL Cartographer | Huang et al. (2006) |
| 3 | 9.1–19.9 | 1AS, 2BS, 7AS | RIL | SSR, AFLP | CIM | QTL Cartographer | Kumar et al. (2006) |
| 9 | 6.1–44.1 | 1DS, 3BL, 5DL, 6AL, 7DL | RIL | SSR, EST-SSR, ISSR, SRAP | MCIM | QTL Mapper | Li et al. (2007) |
| 2 | – | 6AL, 7DL | DH | SSR | CIM | Map Manager QTX | Kuchel et al. (2007) |
| 6 | 3.5–10.2 | 2DL, 3BS, 5AS, 5AL, 7AL | DH | SSR | CIM | QTL Cartographer | Cuthbert et al. (2008) |
| 10 | 4.3–16.80 | 1A, 1B, 2A, 2D, 3B, 4A, 4D, 5A, 6D, 7D | RIL | SSR | CIM | QTL Cartographer | Wang et al. (2009) |

(continued)

**Table 1** (continued)

| Trait/ no. of QTL | Range of PVE (%) | Chromosome (arm) | Population | Linked markers | Method of QTL mapping | Software used | Reference |
|---|---|---|---|---|---|---|---|
| 4 | 5.9–20.1 | 1DS, 2AL, 5DS, 6AL | RIL | SSR, EST-SSR, ISSR, SRAP | MCIM | QTLNetwork | Sun et al. (2009) |
| 10 | 5.0–15.0 | 1A, 1D, 2B, 2D, 4B, 5B, 6B | RIL | SSR | CIM, MCIM | QTL Cartographer | Ramya et al. (2010) |
| 20 | 5.0–19.7 | 1B, 2A, 2B, 2D, 3A, 3D, 4A, 5B | | DArT, ISSR, SSR, STS, SRAP, RAPD | ICIM | | Cui et al. (2014) |
| 10 | 4.37–23.27% | 1A, 1B, 2B,5A, 6A, 6B, 7A, 7D | RIL | SSR, AFLP | ICIM | IciMapping 3.0 | Mir et al. (2012b) |
| 6 | 4.7–8.3 | 1D, 2B, 3B, 4A, 5B, 7A | RIL | SSR | CIM | IciMapping 3.0 | Wang et al. (2012) |
| 69 | 1.9–48.8 | 2A, 2D, 4A, 4B, 5A, 6A, 7A | RIL | SNP | – | – | Guan et al. (2019) |
| Other grain traits (excluding GW) | | | | | | | |
| 17 | 5.6–21.9 | 1A, 2B, 2D, 3B, 7A, 7B | RIL | RFLP, SSR | One-way ANOVA | QGENE | Campbell et al. (1999) |
| 11 | 3.3–16.6 | 2B, 2D, 5B, 6B, 7B | RIL | RAPD, SSR, ISSR | MR | QGENE | Dholakia et al. (2003) |
| 28 | a = −3.09–1.80 | 1AS, 1B, 2A,2B,2D,3A,3B,4A,4B,5B,5D,6A,7B | RIL, DH | SSR, RFLP, AFLP | Multitrait CIM | QTL Cartographer | Breseghello and Sorrells (2007) |
| 20 | 5.9–26.4 | 1A, 1B, 1D, 2A, 2B, 3B, 4A, 4B, 5D, 6A, 6B, 7B | RIL | SSR, EST-SSR,ISSR,SRAP, TRAP | MCIM | QTLNetwork | Sun et al. (2009) |
| Grain yield | | | | | | | |
| 1 | 17.0–27.0 | 4AS | CRSL | RFLP | SIM, sCIM | MQTL | Araki et al. (1999) |
| 4 | 8.0–27.0 | 5AL, 5AS | RIL | RFLP | SIM, sCIM | MQTL | Kato et al. (2000) |
| 9 | 3.9–15.7 | 2B, 3BS, 4AL, 4BL, 5AL, 5BL, 7DL | RIL | RFLP, SSR | MR | Splus | Groos et al. (2003) |
| 11 | 9.6–21.6 | 1AL, 1BL, 2AS, 2BL, 2DS, 2DL, 3BS, 4DS, 4DL, 5BS | $BC_2F_2$ | SSR | SIM | QGene | Huang et al. (2003) |

| | | | | | | | |
|---|---|---|---|---|---|---|---|
| 9 | 10.0–23.0 | 1AS, 3DS, 4DL, 5AS, 5AL, 5BL, 6BS, 6DS, 6DL | $BC_2F_1$ | SSR | MR, SIM | QGene | Huang et al. (2004) |
| 5 | 4.0–13.3 | 2AS, 2BS, 3DL, 4AL, 4DS | DH | SSR | CIM | QTL Cartographer | McCartney et al. (2005) |
| 3 | 8.1–11.0 | 5AL, 7AS, 7BS | DH | SSR | CIM | QTL Cartographer | Huang et al. (2006) |
| 10 | 7.3–21.1 | 1AL, 1B, 2BL, 4AL, 4B, 5A, 5B, 6B, 7A, 7DL | RIL | AFLP, SSR | CIM | QTL Cartographer | Marza et al. (2006) |
| 2 | 8.8–13.3 | 2DS, 7DL | $BC_2F_{2:4}$ | SSR | MR, SIM | QGene | Narasimhamoorthy et al. (2006) |
| 1 | 15.0–20.0 | 4AL | RIL | SSR, EST | CIM | QTL Cartographer | Kirigwi et al. (2007) |
| 9 | – | 1BS, 2DL, 3DL, 4AL, 4DL, 5BL, 6AL, 6DL, 7B | DH | SSR | CIM | Map Manager QTX | Kuchel et al. (2007) |
| 8 | 6.5–47.3 | 1DL, 2DL, 3BL, 4AS, 4DL, 7AS, 7AL | RIL (PI) | SSR | CIM, MCIM | QTL Cartographer, QTLNetwork | Kumar et al. (2007) |
| 6 (7)[a] | 9.9–37.2 | 1AL, 2AS, 2DS, 4BL, 6DL | RIL (PII) | RFLP, SSR | CIM, MCIM | QTL Cartographer, QTLNetwork | Kumar et al. (2007) |
| 5 (8)[a] | 10.3–22.9 | 2AS, 2DL, 3BS, 6AL | RIL | SSR, EST-SSR, ISSR | MCIM | QTL Mapper | Li et al. (2007) |
| 5 | 3.7–12.6 | 1AS, 2DL, 3BS, 5AL | DH | SSR | CIM | QTL Cartographer | Cuthbert et al. (2008) |
| Tiller per plant (TPP) | | | | | | | |
| 1 | 10.0–16.0 | 4AS | CRSL | RFLP | SIM, sCIM | MQTL | Araki et al. (1999) |
| 3 | – | 2BS | RIL | RFLP | SMA | QGene | Ahmed et al. (2000) |
| 3 | 7.0–37.0 | 5AS, 5AL | RIL | RFLP | SIM, sCIM | MQTL | Kato et al. (2000) |

(continued)

**Table 1** (continued)

| Trait/ no. of QTL | Range of PVE (%) | Chromosome (arm) | Population | Linked markers | Method of QTL mapping | Software used | Reference |
|---|---|---|---|---|---|---|---|
| 3 | 11.0–31.0 | 1DS, 2DS, 6AS | RIL | RFLP, SSR | SIM | QGene | Li et al. (2002) |
| 8 | 9.2–12.6 | 1BL, 2AS, 2DL, 3BS, 4DS, 5DL, 6DL, 7AS | $BC_2F_2$ | SSR | SIM | QGene | Huang et al. (2003) |
| 2 | 7.0–13.9 | 1BS, 7AS | $BC_2F_1$ | SSR | MR, SIM | QGene | Huang et al. (2004) |
| 1 | 3.4–6.2 | 3BL | $BC_2F_{2:4}$ | RFLP | MR, SIM | QGene | Narasimhamoorthy et al. (2006) |
| 3 (3)[a] | 6.1–12.5 | 3AL, 7AL, 7BL | RIL (PI) | SSR | CIM, MCIM | QTL Cartographer, QTLNetwork | Kumar et al. (2007) |
| 8 (9)[a] | 7.4–31.0 | 1AL, 1BS, 3BL, 3DL, 4AL, 6DL, 7AS | RIL (PII) | RFLP, SSR | CIM, MCIM | QTL Cartographer, QTLNetwork | Kumar et al. (2007) |
| Spike length (SL) | | | | | | | |
| 2 | 10.0–86.0 | 5AL | RIL | RFLP | SIM, sCIM | MQTL | Kato et al. (1999) |
| 5 | 6.9–11.6 | 1AL, 2BS, 2DS, 4AS, 5AL | DH | RFLP, SSR | One-way ANOVA, MR | – | Sourdille et al. (2003) |
| 4 | 15.0–22.0 | 1AL, 1BS, 4AL, 7AL | RIL | RFLP, SSR | SIM | QGene | Li et al. (2002) |
| 4 | 11.0–23.0 | 1BS, 4AL, 4DL, 7AS | RIL | RFLP | CIM | QTL Cartographer | Jantasuriyarat et al. (2004) |
| 10 | 7.4–18.0 | 1AL, 1AS, 1B, 2BS, 2BL, 3BL, 4B, 5B, 7A, 7BS | RIL | AFLP, SSR | CIM | QTL Cartographer | Marza et al. (2006) |
| 2 (23)[a] | 9.3–18.0 | 2BL, 2DL | RIL (PI) | SSR | CIM, MCIM | QTL Cartographer, QTLNetwork | Kumar et al. (2007) |

| | | | | | | | |
|---|---|---|---|---|---|---|---|
| 9 (5)[a] | 9.1–30.1 | 1AS, 1BL, 1DL, 2DS, 4AL, 5AL, 5DL | RIL (PII) | RFLP, SSR | CIM, MCIM | QTL Cartographer, QTLNetwork | Kumar et al. (2007) |
| 3 | 9.0–19.0 | 3DS, 4AL, 5AL | DH | SSR, TRAP | CIM | QTL Cartographer | Chu et al. (2008) |
| 6 | 4.0–10.0 | 1A, 2B, 2D, 5A, 5B, 5D | RIL | SSR | CIM, MCIM | QTL Cartographer | Ramya et al. 2010 |
| Spikelets per spike (SPS) | | | | | | | |
| 1 | 46.0–52.0 | 4AS | CRSL | RFLP | SIM, sCIM | MQTL | Araki et al. (1999) |
| 4 | 14.0–49.0 | 5AS, 5AL | RIL | RFLP | SIM, sCIM | MQTL | Kato et al. (2000) |
| 3 | 8.8–15.6 | 2AS, 2BS, 5AL | DH | RFLP, AFLP | One-way ANOVA | – | Sourdille et al. (2003) |
| 2 | 16.0–22.0 | 2DS, 7AL | RIL | RFLP, SSR | SIM | QGene | Li et al. (2002) |
| 5 | 8.0–27.0 | 3AS, 3DL, 4AL, 7AL | RIL | SSR, RFLP | CIM | QTL Cartographer | Jantasuriyarat et al. (2004) |
| 3 (4)[a] | 7.1–15.8 | 2BS, 4AL, 6AL | RIL (PI) | SSR | CIM, MCIM | QTL Cartographer, QTLNetwork | Kumar et al. (2007) |
| 5 (6)[a] | 5.7–29.6 | 2DS, 4AL, 4DS, 5AL, 6AS | RIL (PII) | RFLP, SSR | CIM, MCIM | QTL Cartographer, QTLNetwork | Kumar et al. (2007) |
| 1 | 51.7 | 5DL | RIL | SSR, EST-SSR | MCIM | QTL Mapper | Li et al. (2007) |
| 1 | 15.0–29.0 | 4DL | DH | SSR, TRAP | CIM | QTL Cartographer | Chu et al. (2008) |
| Number of grains/spike (GS) | | | | | | | |
| 3 | 8.8–18.3 | 3AS, 3AL | RICL | RFLP | MR | PROC REG | Shah et al. (1999a, b) |
| 8 | 8.2–15.0 | 1DL, 2AS, 3DS, 6AS, 6AL, 7AS, 7AL, 7DS | $BC_2F_1$ | SSR | MR, SIM | QGene | Huang et al. (2004) |

(continued)

**Table 1** (continued)

| Trait/ no. of QTL | Range of PVE (%) | Chromosome (arm) | Population | Linked markers | Method of QTL mapping | Software used | Reference |
|---|---|---|---|---|---|---|---|
| 8 | 8.7–21.0 | 1AL, 1B, 2BS, 2DL, 3BS, 4B, 6A, 7BS | RIL | AFLP, SSR | CIM | QTL Cartographer | Marza et al. (2006) |
| 1 | 12.3–13.2 | 3DS | $BC_2F_{2:4}$ | SSR | MR, SIM | QGene | Narasimhamoorthy et al. (2006) |
| 4 (2)[a] | 8.9–17.6 | 2AS, 4BS, 7AS, 7AL | RIL (PI) | SSR | CIM, MCIM | QTL Cartographer, QTLNetwork | Kumar et al. (2007) |
| 10 (4)[a] | 6.4–43.3 | 1AL, 1BL, 2BS, 2DS, 2DL, 3BS, 3DL, 7AL | RIL (PII) | RFLP, SSR | CIM, MCIM | QTL Cartographer, QTLNetwork | Kumar et al. (2007) |
| 7 | 8.6–20.7 | 1DS, 2AS, 3BS, 6AL,6BS | RIL | SSR, EST-SSR, ISSR, TRAP | MCIM | QTL Mapper | Li et al. (2007) |
| 5 | 4.3–8.6 | 1AS, 2DL, 3BS, 5AL, 7AL | DH | SSR | CIM | QTL Cartographer | Cuthbert et al. (2008) |
| 6 | 6.8–15.8 | 1D, 2AS, 2D, 3A, 4D, 6D | RIL | SSR | CIM | QTL Cartographer | Wang et al. (2009) |
| 17 | 4.1–26.3 | – | | DArT, ISSR, SSR, STS, SRAP, RAPD | ICIM | – | Cui et al. (2014) |

*BIL* backcross inbred line, *CRSL* chromosome recombinant substitution line, *RIL* recombinant inbred line, *DH* doubled haploid, *BC* backcross, *RICL* recombinant inbred chromosome line, *IL* introgression line, *MR* marker regression, *SMA* single marker association, *IM* interval mapping, *SIM* simple interval mapping, *CIM* composite interval mapping, *a* additive effect, *sCIM* simplified composite interval mapping, *MCIM* mixed model-based composite interval mapping

[a]Figure in parenthesis indicates the number of QTL detected by two-locus QTL analysis

## 3 QTL Analyses for Grain Weight (GW)

Grain weight in bread wheat is one of the most important GY-contributing traits. The trait possess high phenotypic stability/heritability and is having favorable effect on flour yield (Kumar et al. 2006; Mir et al. 2012b; Mir 2012). Although grain weight as a trait is considered very easy to select for plant breeders due to its high heritability, still the need of the hour is to know about genetics of GWand related traits and identify genes/QTLs for grain weight and some related traits like grain length and grain width. Different approaches like single marker analysis (SMA), simple interval mapping (SIM), composite interval mapping (CIM), and multi-trait composite interval mapping (MCIM) along with modern genomics approaches including genome-wide association studies (GWAS) were used to dissect grain weight (Mir 2012 for details). In some of the studies, only individual chromosomes have been used, and in some other studies, whole genome genetic maps have been constructed and used in QTL discovery approaches (Kumar et al. 2006; Mir et al. 2012b; Mir 2012). The use of genomics tools and techniques has led to the identification of dozens of major/stable genes/QTLs for grain weight (Table 1).

Like QTL mapping, genome-wide association studies (GWAS) have also been conducted in detail for the identification of GW trait-linked markers in wheat (Breseghello and Sorrells, 2006b, 2007; Neumann et al. 2011; Wang et al. 2012; Mir et al. 2012b; Rasheed et al. 2014; Chen et al. 2016; Arora et al. 2017; Sukumaran et al. 2018). One of the most important advantages of association mapping used for genetic dissection of grain weight and related traits is that several QTLs/genes that have been discovered through QTL mapping have been validated and fine-mapped through GWAS. For instance, in our earlier study on grain weight, we have identified ten QTLs including four major QTLs for grain weight through biparental QTL mapping (Mir et al. 2012a, b). Through association mapping, 11 QTLs/genes/associated markers were identified that were significantly associated with GW. Of these, eight QTLs were validated, and six QTLs were fine-mapped with closely linked markers (Mir et al. 2012b). Further, in our another study, we have also validated some of the QTLs that have been identified in Mir et al. (2012b) using near-isogenic line (NILs) developed through marker-assisted selection (Kumari et al. 2019). High-throughput genotyping platforms have been also used to scan and discover GW genes/QTLs in wheat. For instance, genes/SNPs have been identified for grain length, width, and weight using genotyping-by-sequencing-based single nucleotide polymorphic (GBS-SNP) (Arora et al. 2017). More details about gene discovery for GW and the list of QTLs/genes are available elsewhere (https://shigen.nig.ac.jp/wheat/komugi/genes/symbolClassList.jsp) and also in Table 1. The GW is a polygenic trait and has been associated with ~332 QTLs and other related grain traits covering all the 21 chromosomes in bread wheat (Tyagi et al. 2015). Attempts have also been made to clone GW-associated gene like recent cloning of candidate gene *TaTGW-7A* present on the short arm of homoeologous chromosome 7A (Hu et al. 2016).

## 4 QTL Analysis for Other Grain Traits (Grain Size and Shape)

As discussed earlier, GW is affected/associated with grain size and grain shape. Both of these traits have been found to influence milling and baking quality of bread wheat (Breseghello and Sorrells 2006a) and show a positive correlation with GW. Several dozen QTLs have been identified for both grain size and shape in bread wheat (Gegas et al. 2010; Mir 2012; Valluru et al. 2014; Williams and Sorrells, 2014; Tyagi et al. 2014, 2015; Kumari et al. 2019). QTLs have been identified on 16 chromosomes for grain shape, on 8 chromosomes for 1000 grain weight, and on 5 chromosomes for flour yield (Cabral et al. 2018). In another study, a total of 20 QTLs have been identified for grain length, grain width, thousand-grain weight, and test weight (Sun et al. (2009). The QTLs are contributing 5.9–26.4% of the phenotypic variations in different environments for these traits.

It is interesting to compare ancestral wheat species (showing huge variability for grain size) with modern wheat varieties (showing higher grain width and lower grain length and shape), and the comparison provides us indication that genes/QTLs for grain size and shape could provide better insight into domestication process of wheat crop (Simons et al. 2006; Gegas et al. 2010; Williams et al. 2013). Several other QTLs have been identified for grain shape and size, and correlations/relationships have been worked out between plant height and grain morphology traits including grain weight using different mapping populations (Kumar et al. 2016; Cabral et al. 2018).

The QTLs identified for the above traits are now subject for fine-mapping followed by cloning. The genes once cloned will prove useful in wheat molecular breeding programs aimed at enhancing grain weight and related traits in bread wheat. It is pertinent to mention here that several genes for grain weight and size have been characterized and cloned in rice. Since rice is very close relative of wheat genome, most often scientists tried to isolate the orthologous genes of rice in wheat followed by their cloning. For instance, gene "*Grain Size 3*" also known as *GS3* in rice associated with grain length has been characterized and cloned in wheat (*TaGS3*). The gene *TaGS3* in wheat has also influenced other traits including grain weight and grain size (Yang et al. 2019). Similarly, several other rice genes responsible for grain size and weight have been now cloned in wheat using translational genomics approaches (see Yang et al. (2019) for more details and reference therein).

## 5 QTL × QTL Epistatic Interactions for GY and Yield-Contributing Traits

Epistasis is a complex phenomenon of interaction of alleles or genes or QTLs at two or more than two loci. The phenomenon (qualitative epistasis) was early observed by Batson in 1909, and later the quantitative epistasis was suggested by Fisher in

1919 by coining the term "epistasy." Although the phenomenon of epistasis was observed earlier in diploid crops, later it was found to be an important contributor for genetic variation and evolution in polyploid crops like wheat (Santantonio et al. 2019). With the recent advances in genomics tools and technologies, hundreds and thousands of markers have become available in wheat genome, and the whole genome scans with dense markers have led to more precise detection of epistatic genes/QTLs. The accuracy and prediction of epistatic interaction increases with the increase in marker density and population size. The value of allele or genotype during epistatic interactions at a locus depends on the value of the alleles or genotype at other epistatically interacting loci. Our current knowledge of biochemical and physiological genetics, as well as that of the regulation of gene expression, strongly suggests the ubiquity of epistatic interactions involving QTL. It is important to mention that precise estimation of epistasis is crucial because epistasis complicates the genotype × phenotype relationships. It is now well-known that almost all-important traits in crop plants are complex quantitative traits, and their genetic architecture is not typically the result of variation attributed to a single locus but is rather the result of a number of individual QTL, the interactions among them (epistasis), and the interactions among QTL and environments (Wade 2001).

As mentioned above, GY and its component traits discussed here are quantitative, i.e., controlled by several polygenes with small effects involved in epistatic interactions (Mir et al. 2012b; Mir 2012). In order to evaluate GY and to gain insight into the genetic control and relationship between the traits, it is therefore useful to dissect these individual components by using molecular maps. The phenomenon of gene interactions for GY and related traits has been worked in detail in a variety of crop plants including maize rice, barley, and wheat (Kuchel et al. 2007; Maccaferri et al. 2008; Huang et al. 2009; Xue et al. 2010; Mir 2012; Li et al. 2016).

## 6 Main Effect QTLs for GY and Contributing Traits

The genes/QTLs that have been identified for GY and yield-related traits in wheat are of different types including main effect QTLs (M-QTLs) and epistatic QTLs (E-QTLs). The M-QTLs are the QTLs having their main effect, and their effect is not influenced by any other QTL/gene or by environment. Hundreds of M-QTLs have been identified for GY and yield-related traits in wheat, and the number is continuously increasing (Mir et al. 2012b). In our own study, we have identified a set of ten QTLs including four major QTLs on chromosome IA, 5A, 6A, and 6B and three stable QTLs (Gupta et al. 2011; Mir et al. 2012b). Some of the major QTLs identified by us including "*QGw.ccsu-1A.3*" have been repeatedly used in marker-assisted wheat breeding programs for enhancing grain weight in bread wheat (Kumari et al. 2019). In another study, we have also identified a set of 45 QTLs for 6 grain traits including grain length, grain width, grain surface area, grain volume, horizontal axis proportion, and vertical perimeter. The QTLs of these traits are present on 19 of the 21 bread wheat chromosomes (Tyagi et al. 2015). The meta-analysis of more than 300 QTLs selected from ~35 publications by us earlier led to the identification of 23 meta-QTLs including 17 most reliable M-QTLs. The examination of results revealed that 17 M-QTLs were due to clustering of QTLs for grain weight only, 15 M-QTLs from clustering of QTLs for grain weight and other grain traits, and one M-QTL due to clustering of QTLs for grain traits other than grain weight (Tyagi et al. 2015). Several M-QTLs under different water regimes for yield in durum wheat have also been identified (Maccaferri et al. 2008, 2011). Similarly, 33 M-QTLs were identified for seven yield and yield-related traits (Patil et al. 2013).

**Acknowledgements** Thanks are due to Dean, faculty of Agriculture, Wadura for proving necessary facilities to the corresponding author of the manuscript.

**Author's Contributions** All the authors contributed toward the writing of this manuscript.

## References

Ahmed TA, Tsujimoto H, Sasakuma T (2000) QTLs associated with plant height and related characters in haploid wheat. Breed Sci 50:267–273
Araki E, Miura H, Sawada S (1999) Identification of genetic loci affecting amylose content and agronomic traits on chromosome 4A of wheat. Theor Appl Genet 98:977–984
Arora S, Singh N, Kaur S, Bains NS, Uauy C, Poland J, Chhuneja P (2017) Genome-wide association study of grain architecture in wild wheat *Aegilops tauschii*. Front Plant Sci 8:886
Bhusal N, Sarial AK, Sharma P, Sareen S (2017) Mapping QTLs for grain yield components in wheat under heat stress. PLoS One 12:e0189594
Breseghello F, Sorrells ME (2006a) Association mapping of kernel size and milling quality in wheat (*Triticum aestivum* L.) cultivars. Genetics 172:1165–1177
Breseghello F, Sorrells ME (2006b) Association analysis as a strategy for improvement of quantitative traits in plants. Crop Sci 46:1323–1330
Breseghello F, Sorrells ME (2007) QTL analysis of kernel size and shape in two hexaploid wheat mapping populations. Field Crop Res 101(2):172–179
Cabral AL, Jordan MC, Larson G, Somers DJ, Humphreys DG, McCartney CA (2018) Relationship between QTL for grain shape, grain weight, test weight, milling yield, and plant height in the spring wheat cross RL4452/'AC Domain'. PLoS One 13:e0190681
Campbell KG, Bergman CJ, Gualberto DG, Anderson JA, Giroux MJ, Hareland G, Fulcher RG, Sorrells ME, Finney PL (1999) Quantitative trait loci associated with kernel traits in soft × hard wheat cross. Crop Sci 39:1184–1195
Chen G, Zhang H, Deng Z, Wu R, Li D, Wang M, Tian J (2016) Genome-wide association study for kernel weight-related traits using SNPs in a Chinese winter wheat population. Euphytica 212(2):173–185
Chu C-G, Xu SS, Friesen TL, Faris JD (2008) Whole genome mapping in a wheat doubled haploid population using SSRs and TRAPs and the identification of QTL for agronomic traits. Mol Breed 22:251–266
Cui F, Ding AM, Li J, Zhao CH, Li XF, Feng DS, Wang XQ, Wang L, Gao JR, Wang HG (2011) Wheat kernel dimensions: how do they contribute to kernel weight at an individual QTL? J Genet 90:409–425
Cui F, Zhao CH, Ding AM, Li J, Wang L, Li XF, Bao YG, Li JM, Wang HG (2014) Construction of an integrative linkage map and QTL mapping of grain yield-related traits using three related wheat RIL populations. Theor Appl Genet 127:659–675

Curtis T, Halford NG (2014) Food security: the challenge of increasing wheat yield and the importance of not compromising food safety. Ann Appl Biol 164:354–372

Cuthbert JL, Somers DJ, Brule-Babel AL, Brown PD, Crow GH (2008) Molecular mapping of quantitative trait loci for yield and yield components in spring wheat (*Triticum aestivum* L.). Theor Appl Genet 117(4):595–608

Dholakia BB, Ammiraju JSS, Singh H, Lagu MD, Roder MS, Rao VS, Dhaliwal HS, Ranjekar PK, Gupta VS (2003) Molecular marker analysis of kernel size and shape in bread wheat. Plant Breed 122:392–395

Distelfeld A, Avni R, Fischer AM (2014) Senescence, nutrient remobilization, and yield in wheat and barley. J Exp Bot 65(14):3783–3798

Faris JD, Zhang Z, Garvin DF, Xu SS (2014) Molecular and comparative mapping of genes governing spike compactness from wild emmer wheat. Mol Gen Genomics 289:641–651

Gao FM, Wen WE, Liu JD, Rasheed A, Yin GH, Xia XC et al (2015) Genome-wide linkage mapping of QTL for yield components, plant height and yield-related physiological traits in the Chinese wheat cross Zhou 8425B/Chinese Spring. Front Plant Sci 6:1099

Gegas VC, Nazari A, Griffiths S, Simmonds J, Fish L, Orford S et al (2010) A genetic framework for grain size and shape variation in wheat. Plant Cell 22(4):1046–1056

Gill BS, Appels R, Botha-Oberholster AM, Buell CR, Bennetzen JL, Chalhoub B, Li W (2004) A workshop report on wheat genome sequencing: International Genome Research on Wheat Consortium. Genetics 168(2):1087–1096

Groos C, Robert N, Bervas E, Charmet G (2003) Genetic analysis of grain protein-content, grain yield and thousand-kernel weight in bread wheat. Theor Appl Genet 106(6):1032–1040

Guan P, Di N, Mu Q, Shen X, Wang Y, Wang X et al (2019) Use of near-isogenic lines to precisely map and validate a major QTL for grain weight on chromosome 4AL in bread wheat (*Triticum aestivum* L.). Theor Appl Genet 132(8):2367–2379

Guo Z, Zhao Y, Roder MS, Reif JC, Ganal MW, Chen D, Schnurbusch T (2018) Manipulation and prediction of spike morphology traits for the improvement of grain yield in wheat. Sci Rep 8(1):1–10

Gupta PK, Balyan HS, Kulwal PL, Kumar N, Kumar A, Mir RR, Mohan A, Kumar J (2007) QTL analysis for some quantitative traits in bread wheat. J Zhejiang Univ Sci B 8:807–814

Gupta PK, Mir RR, Mohan A, Kumar J (2008) Wheat genomics: present status and future prospectus. Int J Plant Genomics 896451:1–36

Gupta PK, Kumar J, Mir RR, Kumar A (2010a) Marker-assisted selection as a component of conventional plant breeding. Plant Breed Rev 33:145–217

Gupta PK, Langridge P, Mir RR (2010b) Marker-assisted wheat breeding: present status and future possibilities. Mol Breed 26:145–161

Gupta PK, Balyan HS, Mir RR, Kumar J, Kumar A, Kumar S, Jaiswal V, Tyagi S, Kumari S (2011) QTL analysis, association mapping and marker-assisted selection for some quality traits in bread wheat – an overview of the work lone at CCS University, Meerut. J Wheat Res 3:1–11

Hai L, Guo H, Wagner C, Xiao S, Friedt W (2008) Genomic regions for yield and yield parameters in Chinese winter wheat (*Triticum aestivum* L.) genotypes tested under varying environments correspond to QTL in widely different wheat materials. Plant Sci 175(3):226–232

Hu M-J, Zhang H-P, Cao J-J, Zhu X-F, Wang S-X, Jiang H, Wu ZY, Lu J, Chang C, Sun G-L, Ma C-X (2016) Characterization of an IAA-glucose hydrolase gene *TaTGW6* associated with grain weight in common wheat (*Triticum aestivum* L.). Mol Breed 36:25

Hu J, Wang X, Zhang G et al (2020) QTL mapping for yield-related traits in wheat based on four RIL populations. Theor Appl Genet 133:917–933

Huang XQ, Coster H, Ganal MW, Roder MS (2003) Advanced backcross QTL analysis for the identification of quantitative trait loci alleles from wild relatives of wheat (*Triticum aestivum* L.). Theor Appl Genet 106:1379–1389

Huang XQ, Kempf H, Ganal MW, Roder MS (2004) Advanced backcross QTL analysis in progenies derived from a cross between a German elite winter wheat variety and a synthetic wheat (*Triticum aestivum* L.). Theor Appl Genet 109:933–943

Huang XQ, Cloutier S, Lycar L, Radovanovic N, Humphreys DG, Noll JS, Somers DJ, Brown PD (2006) Molecular detection of QTL for agronomic and quality traits in a doubled haploid population derived from two Canadian wheats (*Triticum aestivum* L.). Theor Appl Genet 113:753–766

Huang X, Qian Q, Liu Z, Sun H, He S, Luo D et al (2009) Natural variation at the DEP1 locus enhances GY in rice. Nat Genet 41(4):494

Jaiswal V, Mir RR, Mohan A, Balyan HS, Gupta PK (2012) Association mapping for pre-harvest sprouting tolerance in bread wheat (*Triticum aestivum* L.). Euphytica 188(1):89–102

Jaiswal V, Gahlaut V, Meher PK, Mir RR, Jaiswal JP et al (2016) Genome wide single locus single trait, multi-locus and multi-trait association mapping for some important agronomic traits in common wheat (*T. aestivum* L.). PLoS One 11(7):e0159343

Jantasuriyarat C, Vales MI, Watson CJ, Riera-Lizarazu O (2004) Identification and mapping of genetic loci affecting the free-threshing habit and spike compactness in wheat (*Triticum aestivum* L.). Theor Appl Genet 108:261–273

Johnson EB, Nalam VJ, Zemetra RS, Riera-Lizarazu O (2008) Mapping the compactum locus in wheat (*Triticum aestivum* L.) and its relationship to other spike morphology genes of the Triticeae. Euphytica 163:193–201

Kato K, Miura H, Sawada S (1999) QTL mapping of genes controlling ear emergence time and plant height on chromosome 5A of wheat. Theor Appl Genet 98:472–477

Kato K, Miura H, Sawada S (2000) Mapping QTLs controlling GY and its components on chromosome 5A of wheat. Theor Appl Genet 101:1114–1121

Kirigwi FM, Van Ginkel M, Brown-Guedira G, Gill BS, Paulsen GM, Fritz AK (2007) Markers associated with a QTL for grain yield in wheat under drought. Mol Breed 20(4):401–413

Kuchel H, Williams KJ, Langridge P, Eagles HA, Jefferies SP (2007) Genetic dissection of grain yield in bread wheat. Theor Appl Genet 115:1029–1041

Kulwal PL, Singh R, Balyan HS, Gupta PK (2004) Genetic basis of pre-harvest sprouting tolerance using single-locus and two-locus QTL analyses in bread wheat. Funct Integr Genomics 4:94–101

Kulwal PL, Mir RR, Kumar S, Gupta PK (2010). QTL analysis and molecular breeding for seed dormany and pre-harvest sprouting tolerance in bread wheat. J Plant Biol 37: 59–74

Kumar N, Kulwal PL, Gaur A et al (2006) QTL analysis for grain weight in common wheat. Euphytica 151(2):135–144

Kumar N, Kulwal PL, Balyan HS, Gupta PK (2007) QTL mapping for yield and yield contributing traits in two mapping populations of bread wheat. Mol Breed 19:163–177

Kumar J, Mir RR, Kumar N, Kumar A, Mohan A, Prabhu KV, Bȧlyan HS, Gupta PK (2010) Marker assisted selection for pre-harvest sprouting tolerance and leaf rust resistance in bread wheat. Plant Breed 129:617–621

Kumar J, Kumar N, Kumar A, Mir RR, Kumar S, Kumar R, Jaiswal V, Tyagi S, Prabhu KV, Balyan HS, Gupta PK (2011) Introgression of a major gene for high grain protein content in some Indian bread wheat cultivars. Field Crop Res 123:226–233

Kumar A, Mantovani EE, Seetan R, Soltani A, Echeverry-Solarte M, Jain S et al (2016) Dissection of genetic factors underlying wheat kernel shape and size in an elite × non-adapted cross using a high density SNP linkage map. Plant Genome 9(1):27898771

Kumari S, Mir RR, Tyagi S, Balyan HS, Gupta PK (2019) Validation of QTL for grain weight using MAS-derived pairs of NILs in bread wheat (*Triticum aestivum* L.). J Plant Biochem Biotech 28(3):336–344

Laidig F, Piepho HP, Rentel D, Drobek T, Meyer U, Huesken A (2017) Breeding progress, environmental variation and correlation of winter wheat yield and quality traits in German official variety trials and on-farm during 1983–2014. Theor Appl Genet 130(1):223–245

Li WL, Nelson JC, Chu CY et al (2002) Chromosomal locations and genetic relationships of tiller and spike characters in wheat. Euphytica 125:357–366

Li S, Jia J, Wei X, Zhang X, Li L, Chen H et al (2007) A intervarietal genetic map and QTL analysis for yield traits in wheat. Mol Breed 20(2):167–178

Li X, Li YX, Chen L, Wu X, Qin W, Song Y et al (2016) Fine mapping of qkw7, a major QTL for kernel weight and kernel width in maize, confirmed by the combined analytic approaches of linkage and association analysis. Euphytica 210:221–232

Maccaferri M, Sanguineti MC, Corneti S, Ortega JLA, Salem MB, Bort J, DeAmbrogio E, Garcia del Moral LF, Demontis A, El-Ahmed A, Maalouf F, Machlab H, Martos V, Moragues M, Motawaj J, Nachit M, Nserallah N, Ouabbou H, Royo C, Slama A, Tuberosa R (2008) Quantitative trait loci for grain yield and adaptation of durum wheat (*Triticum durum* Desf.) across a wide range of water availability. Genetics 178:489–511

Maccaferri M, Sanguineti MC, Demontis A, El-Ahmed A, Garcia del Moral L, Maalouf F, Nachit M, Nserallah N, Ouabbou H, Rhouma S, Royo C, Villegas D, Tuberosa R (2011) Association mapping in durum wheat grown across a broad range of water regimes. J Exp Bot 62:409–438

Marza F, Bai GH, Carver BF, Zhou WC (2006) Quantitative trait loci for yield and related traits in the wheat population. Ning7840 × Clark. Theor Appl Genet 112:688–698

McCartney CA, Somers DJ, Humphreys DG, Lukow O, Ames N, Noll J, Cloutier S, McCallum BD (2005) Mapping quantitative trait loci controlling agronomic traits in the spring wheat cross RL44529 'AC Domain'. Genome 48:870–883

Mir RR (2012) Genetic dissection of grain weight & related traits in bread wheat. LAP: Lambert Academic Publishing, p 188

Mir RR, Zaman-Allah M, Sreenivasulu N, Trethowan R, Varshney RK (2012a) Integrated genomics, physiology and breeding approaches for improving drought tolerance in crops. Theor Appl Genet 125:625–645

Mir RR, Kumar N, Jaiswal V, Girdharwal N, Prasad M, Balyan HS, Gupta PK (2012b) Genetic dissection of grain weight in bread wheat through quantitative trait locus interval and association mapping. Mol Breed 29(4):963–972

Moeller C, Evers JB, Rebetzke G (2014) Canopy architectural and physiological characterization of near-isogenic wheat lines differing in the *tiller inhibition* gene *tin*. Front Plant Sci 5:617

Mohan A, Kulwal PL, Singh R, Kumar V, Mir RR, Kumar J, Prasad M, Balyan HS, Gupta PK (2009) Genome wide QTL analysis for pre-harvest sprouting tolerance in bread wheat. Euphytica 168:319–329

Narasimhamoorthy B, Gill BS, Fritz AK, Nelson JC, Brown-Guedira GL (2006) Advanced backcross QTL analysis of a hard winter wheat × synthetic wheat population. Theor Appl Genetics 112(5):787–796

Nasseer AM, Martin JM, Heo HY, Blake NK, Sherman JD, Pumphrey M et al (2016) Impact of a quantitative trait locus for tiller number on plasticity of agronomic traits in spring wheat. Crop Sci 56(2):595–602

Neumann K, Kobiljski B, Dencic S, Varshney RK, Borner A (2011) Genome-wide association mapping: a case study in bread wheat (*Triticum aestivum* L.). Mol Breed 27:37–58

Patil RM, Tamhankar SA, Oak MD et al (2013) Mapping of QTL for agronomic traits and kernel characters in durum wheat (*Triticum durum* Desf.). Euphytica 190:117–129

Piepho HP, Laidig F, Drobek T, Meyer U (2014) Dissecting genetic and non-genetic sources of long-term yield trend in German official variety trials. Theor Appl Genet 127(5):1009–1018

Quarrie SA, Steed A, Calestani C, Semikhodskii A, Lebreton C et al (2005) A high-density genetic map of hexaploid wheat (Triticum aestivum L.) from the cross Chinese Spring × SQ1 and its use to compare QTLs for grain yield across a range of environments. Theor Appl Genet 110:865–880

Ramya P, Chaubal A, Kulkarni K, Gupta L, Kadoo N, Dhaliwal HS et al (2010) QTL mapping of 1000-kernel weight, kernel length, and kernel width in bread wheat (*Triticum aestivum* L.). J Appl Genet 51:421–429

Rasheed A, Xia X, Ogbonnaya F, Mahmood T, Zhang Z, Mujeeb-Kazi A, He Z (2014) Genome-wide association for grain morphology in synthetic hexaploid wheats using digital imaging analysis. BMC Plant Biol 14(1):128

Rosegrant MW, Agcaoili M (2010) Global food demand, supply, and price prospects to 2010. International Food Policy Research Institute, Washington, DC

Salina E, Borner A, Leonova I, Korzun V, Laikova L, Maystrenko O et al (2000) Microsatellite mapping of the induced sphaerococcoid mutation genes in *Triticum aestivum*. Theor Appl Genet 100:686–689

Santantonio N, Jannink JL, Sorrells M (2019) A low resolution epistasis mapping approach to identify chromosome arm interactions in allohexaploid wheat. G3: Genes Genomes Genetics 9:675–684

Shah MM, Baenziger PS, Yen Y, Gill KS, Moreno-Sevilla B, Haliloglu K (1999a) Genetic analyses of agronomic traits controlled by wheat chromosome 3A. Crop Sci 39:1016–1021

Shah MM, Gill KS, Baenziger PS, Yen Y, Kaeppler SM, Ariyarathne HM (1999b) Molecular mapping of loci for agronomic traits on chromosome 3A of bread wheat. Crop Sci 39:1728–1732

Simons KJ, Fellers JP, Trick HN, Zhang ZC, Tai YS, Gill BS et al (2006) Molecular characterization of the major wheat domestication gene Q. Genetics 172:547–555

Sourdille P, Cadalen T, Guyomarc'h H, Snape J, Perretant M, Charmet G, Boeuf C, Bernard S, Bernard M (2003) An update of the Courtot × Chinese Spring intervarietal molecular marker linkage map for the QTL detection of agronomic traits in wheat. Theor Appl Genet 106:530–538

Sukumaran S, Lopes M, Dreisigacker S, Reynolds M (2018) Genetic analysis of multi-environmental spring wheat trials identifies genomic regions for locus-specific trade-offs for grain weight and grain number. Theor Appl Genet 131(4):985–998

Sun X-Y, Wu K, Zhao Y, Kong F-M, Han G-Z, Jiang H-M, Huang X-J, Li R-J, Wang H-G, Li S-S (2009) QTL analysis of kernel shape and weight using recombinant inbred lines in wheat. Euphytica 165:615–624

Tshikunde NM, Mashilo J, Shimelis H, Odindo A (2019) Agronomic and physiological traits, and associated quantitative trait loci (QTL) affecting yield response in wheat (*Triticum aestivum* L.): a review. Front Plant Sci 10:1428

Tyagi S, Mir RR, Kaur H, Chhuneja P, Ramesh B, Balyan HS, Gupta PK (2014) Marker-assisted pyramiding of eight QTLs/genes for seven different traits in common wheat (*Triticum aestivum* L.). Mol Breed 34:167–175

Tyagi S, Mir RR, Balyan HS, Gupta PK (2015) Interval mapping and meta-QTL analysis of grain traits in common wheat (*Triticum aestivum* L.). Euphytica 201(3):367–380

Valluru R, Reynolds MP, Salse J (2014) Genetic and molecular bases of yield-associated traits: a translational biology approach between rice and wheat. Theor Appl Genet 127(7):1463–1489

Wade MJ (2001) Epistasis, complex traits, and mapping genes. Genetica 112:59–69

Wang RX, Hai L, Zhang XY, You GX, Yan CS, Xiao SH (2009) QTL mapping for grain filling rate and yield-related traits in RILs of the Chinese winter wheat population Heshangmai 3 Yu8679. Theor Appl Genet 118:313–325

Wang L, Ge H, Hao C, Dong Y, Zhang X (2012) Identifying loci influencing 1,000-kernel weight in wheat by microsatellite screening for evidence of selection during breeding. PLoS One 7(2):e29432

Williams K, Sorrells ME (2014) Three-dimensional seed size and shape QTL in hexaploid wheat (*Triticum aestivum* L.) populations. Crop Sci 54(1):98–110

Williams K, Munkvold J, Sorrells M (2013) Comparison of digital image analysis using elliptic fourier descriptors and major dimensions to phenotype seed shape in hexaploid wheat (*Triticum aestivum* L.). Euphytica 190:99–116

Wu X, Chang X, Jing R (2012) Genetic insight into yield-associated traits of wheat grown in multiple rain-fed environments. PLoS One 7:e31249

Würschum T, Leiser WL, Langer SM, Tucker MR, Longin CFH (2018) Phenotypic and genetic analysis of spike and kernel characteristics in wheat reveals long-term genetic trends of grain yield components. Theor Appl Genet 131(10):2071–2084

Xiao SH, He ZH (2003) Wheat yield and end use quality improvement in China. In: Zhuang QS (ed) Chinese wheat improvement and pedigree analysis. China Agricultural Publish Press, China

Xu T, Bian N, Wen M et al (2017) Characterization of a common wheat (*Triticum aestivum* L) high-tillering dwarf mutant. Theor Appl Genet 130(3):483–494

Xue DW, Zhou MX, Zhang XQ, Chen S, Wei K, Zeng FR et al (2010) Identification of QTLs for yield and yield components of barley under different growth conditions. J Zhejiang Univ Sci B 11(3):169–176

Yang J, Zhou Y, Zhang Y et al (2019) Cloning, characterization of *TaGS3* and identification of allelic variation associated with kernel traits in wheat (*Triticum aestivum* L.). BMC Genet 20:98

Zhai H, Feng Z, Li JX, Xiao S, Ni Z, Sun Q (2016) QTL analysis of spike morphological traits and plant height in winter wheat (*Triticum aestivum* L.) using a high-density SNP and SSR-based linkage map. Front Plant Sci 7:1617

Zhang J, Gizaw SA, Bossolini E, Hegarty J, Howell T, Carter AH et al (2018) Identification and validation of QTL for grain yield and plant water status under contrasting water treatments in fall-sown spring wheats. Theor Appl Genet 131(8):1741–1759

# Marker-Assisted Breeding for Resistance Against Wheat Rusts

**Maneet Rana, Rahul Kaldate, Sajad Un Nabi, Shabir H. Wani, and Hanif Khan**

**Abstract** Leaf, stem and stripe rust diseases seriously threaten wheat production worldwide. The obligate biotrophic rust pathogens are highly capable of producing new virulent races that can overcome resistance. To defy the emerging threat of rusts in wheat, attempts are going on at global level for identification of pathotypes, new sources of resistance and deployment of resistant varieties for management of these rusts. Several rust resistance genes (>200) and their associated molecular markers are available to breeders for their use in rust resistance breeding programme. Molecular markers have been extensively used in wheat breeding programmes for various reasons, of which linkage and QTL map being on the top list. Linkage maps depict the presence of major genes and QTLs on chromosomal regions. Molecular markers associated with many effective resistance genes are now increasingly available and have been successfully utilized in many wheat breeding programmes. Marker systems like SSR, STS, SCAR, DArT and SNPs are now popularly being used because of their robustness. MAS and more traditional screening methods like seedling resistance test (SRT) against the rust pathotypes are often complementary, and both are utilized depending on the generation. MAS has important application when two or more resistance genes against particular rust need to be pyramided in a single wheat genotype. In such cases where screening with the pathotypes is not able to confirm the presence of two or more resistance

M. Rana
ICAR-Indian Grassland and Fodder Research Institute, Jhansi, Uttar Pradesh, India

R. Kaldate
Department of Agricultural Biotechnology, Assam Agricultural University, Jorhat, Assam, India

S. U. Nabi
ICAR- Central Institute of Temperate Horticulture, Srinagar, Jammu and Kashmir, India

S. H. Wani
Mountain Research Centre for Field Crops, Khudwani, Anantnag, Jammu and Kashmir, India

Sher-e-Kashmir University of Agricultural Sciences and Technology of Kashmir, Shalimar, Badgam, Jammu and Kashmir, India

H. Khan (✉)
ICAR- Indian Institute of Wheat and Barley Research, Karnal, Haryana, India
e-mail: Hanif.Khan@icar.gov.in

© The Editor(s) (if applicable) and The Author(s), under exclusive license to Springer Nature Switzerland AG 2021

S. H. Wani et al. (eds.), *Physiological, Molecular, and Genetic Perspectives of Wheat Improvement*, https://doi.org/10.1007/978-3-030-59577-7_11

genes conferring resistance against all the pathotypes of the disease, MAS helps in the selection. The various approaches presently being used for MAS has been discussed in this chapter along with the successful examples of mapping various rust resistance genes and their introgression into elite wheat cultivars using various marker-assisted breeding strategies.

**Keywords** Wheat · *Puccinia* · Host-pathogen interactions · Genetics · Rust resistance · Durable resistance · Molecular markers · MAS

## 1 Introduction

Wheat (*Triticum aestivum* L.) being the third major food crop worldwide is a leading source of calories and protein for humans as well as livestock (Mondal et al. 2016). Limited genetic diversity at the farmers' field (Wang et al. 2017) along with changing climate scenarios posed multiple threats (biotic as well as abiotic) to wheat crop resulting in significant yield losses worldwide. Among various biotic stresses, wheat is susceptible to nearly about 30 viral, 45 fungal and 80 bacterial diseases and among the various fungal diseases; rusts are reasonably most important due to the huge economic losses with up to 7–30% in case of leaf rust and 100% in case of stem rust (Leonard and Szabo 2005; Bolton et al. 2008; Singh et al. 2011a). To defy the emerging threat of rusts in wheat, attempts are going on at the global level for identification of races and pathotypes, new sources for resistance and deployment of resistant varieties for management of these rusts (Figueroa et al. 2018). Several rust resistance genes (>200) and associated molecular markers are available to breeders for their use in rust resistance breeding programme. However, continuous emergence of novel races and pathotypes has posed a challenge to breeders to overcome this and develop rust-resistant cultivars (Bhardwaj et al. 2019).

Rust is governed by both qualitative and quantitative inheritance. Qualitative disease resistance is controlled by a single resistance gene with large effects and follows the gene-for-gene mechanism of resistance against a particular race of a known pathogen species (race specificity). In contrast, quantitative resistance is controlled by many genes with small effects, and does not involve race specificity. Qualitative resistance is often overcome by pathogen through rapid evolution of new race virulent over the deployed resistance gene, while quantitative resistance provides durable resistance since the overcoming of multiple modes of resistance is very difficult for the pathogen until superrace is evolved (Parlevliet 2002). For incorporating durable rust resistance among elite cultivars, marker-assisted approach proves to be the tool of choice to the breeders. For marker-assisted breeding programme, the first and foremost requirement is the availability of tightly linked molecular marker with the trait of interest. Rust is being controlled by single gene in case of qualitative trait and many genes with small effects in case of quantitative resistance. Mapping of these genes/loci requires high-throughput molecular markers. Technological advances have provided a range of molecular markers to breeders from conventional RFLP (restriction fragment length polymorphism), RAPD (random amplified polymorphic DNA), AFLP (amplified fragment length polymorphism), ISSR (inter-simple sequence repeats) to advanced SSR (simple sequence repeats), S/TRAP (sequence-/target-related amplified polymorphism), DArT (diversity arrays technology), STS (sequence-tagged sites), SNPs (single-nucleotide polymorphisms), etc. However, SNPs are the marker of choice in next-generation sequencing era. Molecular markers offer an alternative approach to plant breeders for improving cultivars for resistance to biotic stresses including rust, very rapidly and precisely in addition to conventional selection schemes. Markers linked to the targeted trait/QTLs can be used for marker-assisted breeding (MAB) for crop improvement endeavours. There are various molecular breeding strategies for introgression of trait of interest; these include marker-assisted selection (MAS), backcrossing (MABC), gene pyramiding (MAGP), recurrent selection (MARS), genome-wide selection (GWS) and genomic selection (GS). In this chapter, we address recent advances in MAB for the development of resistance against wheat rusts. We begin with the history and various types of wheat rust, pathotype surveillance, virulence/avirulence variations, different screening methodologies for rust resistance, its genetic basis, breeding for durable rust resistance and finally successful examples of mapping various rust resistance genes and their introgression into elite wheat lines using various MAB strategies.

## 2 Wheat Rusts

Wheat is susceptible to various abiotic and biotic stresses, among them rust is the most important pathogens causing devastating damage and continues to be a threat to crop production (Zhao et al. 2016). According to the Biblical records, rusts are the earliest known diseases and have coexisted with wheat since the times immemorial. Many centuries ago, these were documented as grave pests by the ancients. The festival 'Robigalia' was celebrated by Romans on 25 April every year during which the priest prayed the Robigus to save the crops from these pests. Most of the other earlier and ancient records, however, have also dealt with sacrifices and festivals to please the God to keep away their crops from damaging rusts (Gupta et al. 2017). As evidenced by urediniospore excavations, in Israel dated back 1300 BC, the wheat rusts are oldest plant pathogens (Bhardwaj et al. 2016). Globally wheat crop is infected with three rusts, i.e. stem rust incited by *Puccinia graminis* f. sp. *tritici*, leaf rust by *Puccinia triticina* and stripe rust by *Puccinia striiformis* (also called as black, brown and yellow rust, respectively) (Berlin et al. 2015). All these rusts are major threats to wheat production causing significant losses in different regions of the world and environments suitable for disease development (Gessese 2019). The significant yield losses of millions of tons have been reported from several countries in a single crop season (Figueroa et al. 2018; Zhao et al. 2016). The rust fungus belongs to *Basidiomycota* (phylum), *Urediniomycetes* (class), *Uredinales* (order), *Pucciniaceae* (family) and genus *Puccinia* (Bolton et al. 2008). The *Puccinia* species are further classified as formae speciales, which invade different hosts, but are morphologically similar. Formae speciales are further divided into races (physiological races, infection types or pathotypes) which can parasitize certain cultivars of host species and are identified by differential host cultivars (Gessese 2019). Being obligate in nature, rust fungi infect and reproduce only in living host tissues, even though some axenic cultures were successfully obtained in the 1960s (Zhao et al. 2016). Being biotrophic in nature, it takes several days for symptom development due to intimate association between fungus and host (Duplessis et al. 2012). Teliospores are developed to survive during unfavourable weather conditions, until the onset of favourable conditions for infection. The rust fungi are heteroecious in nature that requires botanically two different hosts to complete its life cycle (Schumann and D'Arcy 2006). The rust fungi is having macrocyclic life cycle, with five stages of spore. Out of five spore stages, three (uredinial, telial and basidial) occurred on the primary host while other two (pycnial and aecial) on alternate host. Alternate hosts played an important role in pathogen variation and disease epiphytotics (Beddow et al. 2015; Singh et al. 2015). At the beginning of the twentieth century, vast studies in deciphering the genetics of disease resistance and host-pathogen interaction and life cycle of rust pathogen were initiated due to the widespread occurrence of rust epidemics in wheat (Berlin et al. 2012). The genetic and molecular basis of pathogenicity is not well characterized in rusts, due to the inability of the generation of rusts in vitro and also the unavailability of robust methods of genetic transformation in rusts. Globally several research centres developed their own systems of race designation and analysis. Researchers are continuously monitoring the race frequencies, virulence frequencies and its combinations, evolution and *Puccinia* diversity to know the effective resistance genes for their use in downstream breeding programmes (Figueroa et al. 2018).

## Stem Rust

Stem or black rust is one of the most devastating rust diseases of wheat throughout the world. It caused huge crises at both political and economic fronts globally in general and in South Asia in particular, which laid the foundation of Green Revolution in the 1960s. It mainly parasitizes stem and leaf surfaces and sometimes infects leaf sheaths, glume awns, spikes and grains (Figueroa et al. 2016). The above-ground parts are mostly damaged, and infected plants are characterized by production of small number of tillers with few kernels per spike. The kernels are usually shrunken and small in size with huge reduction in milling and quality (Figueroa et al. 2018). The disease was being incited by *Puccinia graminis* f. sp.

*tritici* Ericks and Henn. (Pgt) and is widely distributed throughout the world. It is a heteroecious rust with telial stage on wheat and an aecial stage on the *Berberis* spp. The rust is macrocyclic having five stages of spore (Singh et al. 2015). The Pgt is predominant in warm regions with moist conditions, and typical symptoms as masses of brick-red urediniospores are observed. The optimum and maximum temperatures at which spores can germinate are 15–24 °C and 30 °C, respectively (Chen et al. 2014). An estimated global annual yield loss of wheat due to this rust is up to 6.12 million tons, which is equivalent to 1.10 billion dollars (Singh et al. 2015). Recently, stem rust has gained importance due to the emergence of new virulence traits in Pgt populations, signifying the susceptibility of wheat cultivars broadly being used throughout the world (Tomar et al. 2014). The emergence of a new virulent race in 1998 in Uganda, viz. *Ug99*, and its spread within Africa and up to the Middle East subsequently alarmed the return of this dreaded disease which has about 40 years of successful control (Singh et al. 2015). A number of commercial cultivars (90%) succumbed to this race considered as precarious to wheat production throughout the world. Several other unrelated races, like Digalu, also appeared in Germany, Ethiopia and other parts of the world and considerably reduced the effectiveness of the resistant cultivars worldwide (Olivera Firpo et al. 2017).

## *Leaf Rust*

Leaf rust also called brown rust is common and widely distributed throughout the world. Generally, it infects leaf blades, but under severe conditions, glumes and leaf sheaths can also get infected. This rust is pervasive in almost all major wheat-growing regions of the world and more frequent than other two rusts. It is a main concern in Asia, North Africa, Europe, North and South America, Australia and New Zealand (Gessese 2019). Yield losses of the disease are substantial, even though losses caused by brown rust exhibit spatial and temporal variation (Figueroa et al. 2018). The disease is caused by an obligate parasite *P. triticina* prevalent in areas having moist conditions with mild temperatures. It is a heteroecious and macrocyclic rust. The primary hosts are durum, bread, wild and cultivated emmer wheat, whereas the alternate hosts are *Thalictrum speciosissimum* and *Isopyrum fumaroides*; however, in most of the wheat-growing regions, alternate hosts are absent (Zhao et al. 2016; Singh et al. 2015). The optimum temperature at which spore germinates on leaf surfaces is 10–25 °C along with the availability of moisture in the form of water. Due to the high adaptability of the pathogen to a broad range of climates, there is continuous emergence of new virulent pathotypes (McCallum et al. 2016). Urediniospore drives the devastating asexual reproductive phase, which mediates infection via multiple developmental stages, such as haustoria (Zhao et al. 2016). The emergence of new and virulent races with use of resistance genes allows the existing variants or mutants to be selected and perpetuated at low frequency.

### *Stripe Rust*

The stripe also called as yellow rust being equally destructive as stem rust is ubiquitous in regions having cool and wet weather (temperate) with varied cropping systems (Bux et al. 2012). Presently, it is economically the most important rust disease reaching yield losses up to 100%, which results in monetary losses nearly US 1 billion dollars annually worldwide. More than 50 major wheat-growing countries have been reported to be affected from this rust across the world, viz. East Asia, the USA, South Asia, Western Europe, Oceania and East Africa and Arab Peninsula (Beddow et al. 2015). The disease is incited by *P. striiformis* Westend. f. sp. *tritici* (Pst). It is characterized by production of yellow-coloured uredinia in the form of stripes on the lower surface of leaf and leaf sheaths. In severe conditions, awns, glumes and immature green kernels are also getting infected (Chen et al. 2014). The uredinial spores can germinate on the leaf surface at 10–12 °C and a suitable amount of water in the form of dew (Bux et al. 2012). Till 2010, it was thought that Pst has no alternate host; however, various *Berberis* spp., viz. *chinensis*, *koreana*, *holstii* and *vulgaris*, were identified as potential alternate hosts of rust (Zhao et al. 2016). From 2000 onwards, several aggressive races of Pst has been spread to different and less affected areas of wheat-growing regions, due to their adaptability to higher temperatures (Ali et al. 2014). Although the same Pst populations was reported in countries like Australia, Europe and North America, a substantial level of genetic diversity was existing among themselves (Chen et al. 2014). The centre of diversity, the place where recombination occurs normally, is obvious in pathogen populations of Central Asia and Himalayan and nearby regions. Recently, new races have emerged and spread to Europe and other temperate regions, and their genetic analysis confirmed the Himalayan region as their origin (Hubbard et al. 2015). Attempts have been made to explore the genetic structure of Pst population at a global level, elucidating sources of invasions, universal population subdivisions and the existence of the centre of diversity in the Himalayan and nearby regions (Thach et al. 2016; Walter et al. 2016).

## 3 Pathotype Surveillance

Many countries like the USA, Canada, Australia, France and India established an annual programme for pathotype surveillance in rust pathogens, after physiological races in *P. graminis* f. sp. *tritici* were discovered by Stakman. The countries which have actual knowledge of the deployed resistance genes in the commercial cultivars and surveillance of pathotypes form the basis for information on the pathogenic variations or virulence in those countries or regions (Park et al. 2011; Gessese 2019). However, the main aim of surveillance is to trace the emergence of new and virulent pathotypes and their control measures before the same causes damage. Pathotype variation studies are conducted on definite groups of host genotypes under controlled environments using inoculum of isolates of interest (Bhardwaj et al. 2019). Each race is identified as virulent or avirulent based on the qualitative infection type score on a set of differential host genotypes (Kolmer et al. 2011).

## 4 Virulence/Avirulence Variations

To interpret the nature of pathogenicity of rust fungi and deciphering the nature of host resistance, genetic understanding of avirulence/virulence is necessary. The inheritance pattern of traits is the same for rust fungi and higher plants with similar pattern of Mendelian inheritance. Thus, using segregating populations, the genetic analysis of pathogenicity can be studied for all three wheat rusts (Zhao et al. 2016). Avirulence genes can be either dominant or recessive but mostly controlled by single dominant gene. For example, the avirulences of Pst to stripe rust resistance genes (*Yr6*, *Yr7*, *Yr8*, *Yr19*, *Exp2* and *Tye*) against PST-127 race are governed by single dominant gene, while for *YrExp1* and *Yr17*, it is governed by single recessive gene (Wang et al. 2016). In Pt, the various leaf rust resistance genes (*Lr3*, *Lr11*, *Lr16*, *Lr21*, *Lr26* and *Lr30*) segregate in 15:1 ratio and gives indication of two independent dominant genes. Studies have also shown interaction of two complementary genes, with one being dominant for avirulence loci and another being dominant for their suppression, for example, in *Pgt* and *Pt* (Tian et al. 2016).

## 5 Screening Methodologies for Rust Resistance

To avoid the huge yield losses wreaked by the wheat rusts, wheat breeders are continuously adding new effective rust resistance genes in their breeding material. To discover and decipher the new rust resistance gene/genes from available germplasm, screening or evaluation is being done in various nurseries from national and international breeding programmes of wheat (Riaz et al. 2016). The germplasm is being continuously screened with the predominant races of rust pathogen in the region/country. The most common stages to screen wheat germplasm are at seedling and adult plant stages (Draz et al. 2015).

### *Seedling Screening*

In this method, the 8-day-old seedlings are inoculated with identified and most predominant races/pathotypes of the rust pathogen in a region/country. The inoculated seedlings are further incubated at 18 °C overnight in a dark dew chamber. The plants on the next day are moved to greenhouse where these are maintained at optimum temperature and humidity necessary for rust development. Seedlings are then kept under surveillance till the symptoms of rust are developed and response from each germplasm line is scored based on the infection types expressed according to the scale available for each rust (Draz et al. 2015).

### *Adult Plant Screening*

Adult plant resistance also called slow rusting tends to slow the progression of the rust pathogen and is activated generally at third leaf stage onwards (Gupta et al. 2017). The screening method involves inoculation of plants mostly at booting stage of the crop. The plants are first moisturized with water using sprinkler type of spray and then dusted with powder of urediniospore from most prevailing pathotypes/races in region/country (1:20 –urediniospores/talcum powder) generally at sunset before dew onset. Mostly at early dough stage, full development of rust symptoms occurs. The reactions of adult plants are scored as plant response based on rust severity according to the scale available for each type of rust (Draz et al. 2015; Riaz et al. 2016).

## 6 Genetic Basis of Resistance to Rust Pathogen

Physiological specialization does occur in case of obligate parasites like rusts in wheat. The races differ in their infection on differential hosts. While working on flax and *Melampsora lini* system, Flor (1995) showed that the inheritance of both resistance in the host and the ability pathogenicity of the parasite are controlled by pairs of matching genes. Gene-for-gene relationships are widespread and very important aspect of plant disease resistance and evolved through a series of steps in the evolution of each. Therefore, if a host is resistant to pathogen, a virulent mutant would have an advantage over avirulent. Likewise, if a host is susceptible to a pathogen, a resistant mutant in host would be at advantage. Physiological races of the rusts were originally identified on the basis of their infection types on the wheat varieties with a combination of genes. However, with the understanding of the genetic basis, they are presently identified in relation to the known genes for the resistance and are called 'pathotypes'. A vital outcome of this work is the assemblage of a collection of well-characterized pathogen isolates which over a period of time can provide a basis for predicting the presence of genes and gene combinations in wheat stocks of unknown resistance genotype – the classic 'gene postulation studies'. Biffen (1905) working with yellow rust was first to show that resistance to a pathogen could be governed by a single gene which inherits in a Mendelian way. He demonstrated that resistance to yellow rust in Rivet wheat was governed by one recessive gene. Further this monogenic inheritance was reported in several studies; however, reports on duplicate complementary additive and other interactions were also there. It is now recognized that disease resistance may show the following three modes of inheritance: (1) oligogenic, (2) polygenic and (3) cytoplasmic. The typical steps taken in characterizing new sources of rust resistance are the following: (i) Determine the number of genes and mode of inheritance. (ii) Generate single-gene lines. (iii) Determine the chromosomal location and genetic map position. (iv) Perform tests of allelism with known rust resistance genes found in the same chromosomal region.

Resistance and avirulence result from an active interaction between active gene products from resistant host and avirulent pathogen. Genetic analyses have indicated that resistance to diseases in wheat is controlled generally by dominant genes (and few recessive), while virulence in pathogen is generally due to recessive genes (or avirulence is controlled by dominant genes). Wheat rust resistance genes can be divided into two categories: seedling resistance genes or all-stage resistance (ASR) gene that confers resistance during the life of plant and adult plant resistance (APR) genes that normally not expressed in seedlings but become active as the plant reaches the adult stage. Seedling resistance genes are characterized by a hypersensitive response (HR) that includes chlorosis or necrosis surrounding the site of infection and a reduction in uredinium size. Rust resistance gene of ASR class generally confirms to Flor's gene-for-gene hypothesis (Flor 1971). Some ASR genes are of 'broad spectrum' type, as they provide resistance to all types of established pathotypes of a single pathogen species. Since seedling genes are pathogen race-specific, the cell death phenomenon is activated due to plant hypersensitive response preventing pathogen spread (Ellis et al. 2014; Mondal et al. 2016). Some of the known seedling or ASR genes can be assumed by testing different genotypes of host with wide array of pathotypes that have different virulence pattern (Browder 1973). In this method, the genotype/cultivar under gene postulation is planted along with isogenic lines having known resistance genes and inoculated separately at seedling stage with an array of rust pathotypes differing in virulences. Of late rust resistance genes are characterized through genetic linkage, molecular mapping techniques specifically and genome-wide association studies (GWAS) which exploit historical recombination's events in nonrandom natural populations (Hall et al. 2010). Resistance gene cloning using wild crop relative provides insight into the evolutionary forces that have shaped the mechanism of disease resistance. Arora et al. (2019) have recently reported cloning of four *Sr* genes using AgRenSeq (association genetics with R gene enrichment sequencing) approach which identifies NLR region without any reference genome.

There are two classes of APR genes: (i) those that produce a hypersensitive response like that found in seedling genes and which may or may not be race-specific and (ii) those that confer quantitative resistance that is presumed to be non-race-specific. ASR genes may confer a differential response based on the particular virulence of the race or may confer resistance that is broad spectrum and does not discriminate between races of the fungus. This is also true of the hypersensitive-type APR genes. Partial or quantitative APR is characterized by reduced receptivity, smaller uredinia and increased latent period.

In order to accommodate newly identified rust resistance genes in wheat, a standard nomenclature both for ASR and APR classes has been defined as *Lr*, *Sr* and *Yr* where *Lr* denotes for leaf rust resistance, *Sr* denotes for stem rust resistance and Yr denotes for stripe rust resistance. In general, genes for APR confer a partial often slow-rusting phenotype (Singh et al. 2011b). Landraces of wheat, wild relatives and related species, viz. *Aegilops tauschii* Coss. *Ae. squarrosa* and *Triticum tauschii*, are major sources of many genes including rust resistance. A summary of all the *Lr*, *Sr* and *Yr* genes was given by McIntosh et al. (1995). Rust resistance genes have been identified progressively in wheat, and there are currently 78, 77 and 59 genes for yellow, brown and black rust resistance, respectively (McIntosh et al. 2017). Two recently identified stripe rust resistance (*Yr*) genes, namely, *Yr79* and *Yr80*, have been added to the rust-resistant gene library (Feng et al. 2018; Nsabiyera et al. 2018). In case of leaf rust resistance, *Lr78* and *Lr79* has been reported recently (Kolmer et al. 2018a, b, c; Qureshi et al. 2018a). Australian scientists have done characterizing and naming of qualitatively inherited rust resistance loci *Yr47*, *Yr51*, *Yr55*, *Yr56*, *Yr58*, *Yr63*, *Yr66*, *Yr67*, *Sr48*, *Sr49*, *Lr71* and *Lr73* and APR genes *Yr46/Lr67/Sr55*, *Yr49*, *Sr56* and *Lr74* using landraces collected by English Botanist Arthur Watkins from 32 nations in the 1920s and modern cultivars from different geographical regions followed by molecular studies (Bariana and Bansal 2017). In addition to 59 *Sr* genes, a novel *Sr60* gene has been reported recently which confers resistance to stem rust by encoding for protein having 2 putative kinase domains (Chen et al. 2018, 2020).

## 7 Breeding for Durable Rust Resistance

The durability of genetic resistance against rust diseases in wheat still remains a major challenge and is of special concern to wheat breeders and farmers. Despite arguments among researchers on strategies and genetic mechanisms to achieve durable resistance a reflection of the different host-pathogen systems they all share the common objective of its utilization for the protection of crops. The association of durable resistance with both major and minor genes depending on the different host-pathogen systems and the parasitic behaviour of pathogens and their degree of host specialization has been much discussed (Parlevliet 1993). However, there seems to be a general agreement on the utilization of quantitative resistance controlled by minor genes for achieving durable resistance particularly with heterocyclic fungi that are biotrophic such as rust on cereals.

Johnson and Law (1973) at first proposed the term of durable resistance in the context of the generalized idea that resistance expressed as a low but positive apparent infection rate 'r' was an attribute of horizontal resistance effective against all pathotypes and controlled by polygenes (VanderPlank 1975, 2012). Durable resistance was more specifically redefined as 'the resistance that remains effective in a cultivar that is widely grown for a long period of time in an environment favourable to the disease' (Johnson 1983). In wheat, *Sr2* and *Lr34 are* best known durable resistance genes *which* provide resistance to stem rust and leaf rust, stripe rust and powdery mildew, respectively. *Sr2* and *Lr34/Yr18/Sr57/Pm38* gene complex has widely been used in breeding programme at CIMMYT and major wheat breeding organizations. In order to impart durable resistance, genes should be used in combinations of three or more, as APR genes individually provide low levels of resistance (Bariana et al. 2007). Some wheat breeders and pathologists have a common view that more emphasis should be given for the use of APR genes than ASR genes. This is due to the fact that ASR genes lack durability. Polygenic resistance of APR nature is governed by multiple genes and quantitatively gets less influenced by race-specific pathogens. The involved genes provide non race-specific partial resistance to all the pathotypes of a given pathogen species, thus making it more durable (Lagudah 2011; Burdon et al. 2014). Despite the fact that incorporating APR into new cultivars can be difficult as compared to ASR, it was found that many wheat cultivars possessing APR showed slow rusting contributing to durable resistance. Primary gene pool including indigenous collections comprising of landraces, old cultivars and breeding lines are considered as a valuable genetic resource for providing new and durable resistance that can be exploited for the development of current-day high-yielding varieties (Mujeeb-Kazi et al. 2013). To widen the spectrum of rust resistance and durability in modern wheat varieties, crop scientists keep searching lines having new sources of resistance along with newer alleles for known resistance genes. The pleiotropic genes/QTL can confer slow-rusting resistance against the three rusts of wheat. Biparental/multiparental populations developed using resistant landraces and modern varieties shall enable towards this endeavour. Further for achieving durable rust resistance, multiple resistance genes should be pyramided in the same elite cultivar using marker-assisted pyramiding approach. This will not only prevents the pathogens to overcome it but also prolongs the life of the individual resistance genes.

## 8 Marker-Assisted Selection (MAS)

Marker-assisted selection (MAS) refers to the use of tightly linked markers for the selection of individual (s) with desired gene/trait of interest. MAS is preferred over conventional approach for those traits where low heritability, recessive nature and destructive phenotyping hamper the varietal development process. There are various forms of MAS used for the introgression of desired gene (s) for imparting resistance against the desired trait of interest; these include marker-assisted backcrossing, gene pyramiding or gene stacking, gene pyramiding through multiple-parent crossing, marker-assisted backcrossing gene pyramiding, marker-assisted recurrent selection and genomic selection. The readers can find more details on their procedure in an excellent review by Rana et al. (2019) on gene pyramiding and multiple character breeding. For any introgression programme, the first and foremost step is the mapping of gene of interest with tightly linked molecular markers. Wheat, being an important and major cereal crop, has excellent availability of genomic resources say in terms of molecular markers, linkage maps, QTL maps, physical maps as well as genome sequence information. Several approaches have been suggested and applied in wheat breeding using effective use of MAS (Gupta et al. 2010; Randhawa et al. 2019).

## 9 MAS for Leaf Rust Resistance

Rust being a devastating disease results in huge economic losses to wheat production worldwide. In order to control or reduce the effect of rust on wheat production, several rust resistance genes (>200) and associated molecular markers are available to breeders for their use in rust resistance breeding programme. Molecular markers play an important role for imparting leaf rust resistance in wheat specifically molecular marker-assisted breeding programmes in several ways using marker-assisted selection (Singh et al. 2004, 2018; Gupta et al. 2005; Nocente et al. 2007; Vida et al. 2009; Kuraparthy et al. 2011; Riar et al. 2012; Yadawad et al. 2015), marker-assisted backcrossing (Chhuneja et al. 2008; Kumar et al. 2010; Pietrusińska et al. 2011; Tiwari et al. 2014; Savitha et al. 2016; Singla et al. 2017; Yadawad et al. 2017; Koujalagi et al. 2019) and marker-assisted gene pyramiding (Singh et al. 2004, 2017; Samsampour et al. 2009; Bhawar et al. 2011; Chhuneja et al. 2011; Charpe et al. 2012; Tiwari et al. 2014) which in turn speed up the process of recurrent parent recovery using background selection and identification of gene of interest using foreground selection. Thus, it becomes the tool of choice to breeders for introgression of resistance genes to the elite cultivars.

Further, in order to map resistance genes with tightly linked markers, various workers employed different strategies for mapping of leaf rust resistance genes using various genotyping platforms, viz. SSR (Wang et al. 2015; Qureshi et al. 2017b, 2018a; Sadeghabad et al. 2017), DArT (Calvo-Salazar et al. 2015; Kolmer 2015; Lan et al. 2015; Chhetri et al. 2016; Lan et al. 2017a; Ren et al. 2017; Kolmer et al. 2018a; Ponce-Molina et al. 2018), KASP assay (Kassa et al. 2017; Qureshi et al. 2018a; Gill et al. 2019), SNP assays using different chips (*9 K* (Li et al. 2017; Aoun et al. 2019), *55 K* (Zhang et al. 2019a, b; Gebrewahid et al. 2020), *90 K* (Nsabiyera et al. 2016; Lu et al. 2017; Zhang et al. 2017; Kolmer et al. 2018b, c; Kolmer et al. 2019; Kthiri et al. 2019; Nsabiyera et al. 2020)) and more recently genotyping-by-sequencing (GBS) approach (Yuan et al. 2020), and were able to

**Table 1** Molecular markers closely associated with wheat leaf rust resistance for their utilization in MAB

| S. no. | Donor (its description)/Cross Type/mapping population | QTL/gene | Associated markers | Genotyping technology | Reference |
|---|---|---|---|---|---|
| 1. | Kenya Kongoni (Kenyan wheat), $F_5$ RILs | *QLr.cim-1DS* | *wPt-3738* and *rPt-4471* (DArT markers) | DArT and SSR | Calvo-Salazar et al. (2015) |
| 2. | Americano 25e, $F_6$ RILs | *Lr46* | *csLV46* (STS) | DArT | Kolmer (2015) |
| | | *QLr.cdl-5BL* | wPt-*4091* wPt-*2373(*DArT*)*; 4.4 cM | | |
| 3. | Sujata (bread wheat cultivar), $F_{4:5}$ RILs | *QLr.cim-1AS/QYr.cim-1AS* | *wPt-9752* and *Xgdm33* | DArT and SSR | Lan et al. (2015) |
| | | *Lr46/Yr29* PAPR gene | *csLV46G22 wPt-8168 Xwmc216* | | |
| | | *Lr67/Yr46* | *Xgwm192* | | |
| | | *QLr.cim-2BS* | *wPt-8058 and Xwmc661* | | |
| | | *QLr.cim-7BL/YrSuj* | *Xcfa2040 and Xwmc526,* 6.0 cM | | |
| 4. | Weimai 8 (Chinese wheat cultivar), $F_{2:3}$ | *QLr.cim-5DS/QYr.cim-5DS* | 100002510 and 3948152 (DArT markers) | SSR | Wang et al. (2015) |
| | | *QLr.hbau-2AS* | *Xcfd36* and *Xbarc1138* (SSR); 2.58 cM | | |
| 5. | W195 × BTSS; $F_7$ RILs | *Lr46/Yr29* | *csLV46G22* (tightly linked)<br>*wPt-8168* and *Xwmc216* (flanking), 3.4 cM | DArTseq | Chhetri et al. (2016) |
| | | *QLr.sun-2BS* | *11058385* | | |
| 6. | CSP44/WL711, RILs | *Lr48* | *IWB70147*; 0.3 cM proximal<br>*IWB31002 IWB39832 IWB34324 IWB72894* and *IWB36920* (co-segregating SNP markers) | iSelect 90 K Infinium SNP assay | Nsabiyera et al. (2016) |
| 7. | Four mapping populations, RILs/DHs | *Lr16* | BS00108724_kwm461 2BS-5192454_kwm677 2BS-5203447_kwm742 2BS-5194460_kwm747 2BS-5175914_kwm847 and 2BS-5175914_kwm849 (SNP markers) | KASP assay SSR SCAR | Kassa et al. (2017) |

(continued)

**Table 1** (continued)

| S. no. | Donor (its description)/Cross Type/mapping population | QTL/gene | Associated markers | Genotyping technology | Reference |
|---|---|---|---|---|---|
| 8. | Bairds (durum wheat cultivar); $F_{4:5}$ RILs | *Lr17/Yr37* | VENTRIUP-LN2 | 50 K DArT and SSR | Lan et al. (2017a) |
| | | *QLr/Yr.hebau-7BL* | Xcfa2040 | | |
| | | *QLr.hebau-2DS* | AX-111031419 and AX-109507893 | | |
| | | *QLr.hebau-6DS* | AX-86175025 and AX-108780681 | | |
| | | *Lr46* | *csLV46G22* | | |
| 9. | UC1110 × PI610750, $F_{7:8}$ RILs | *QLr.hebau-2DS2* | *AX-108903243* and *AX-110055353*, 23 (Mb)[a] | SSRs DArTs and ESTs | Lan et al. (2017b) |
| | | *QLr.hebau-3AL* | *3/AX-109036576* and *AX-111464284*, 9.4 (Mb)[a] | | |
| | | *QLr.hebau-3AL* | *AX-109394676 AX-109994125 AX-110023742* (developed KASP markers) | | |
| | | *QLr.hebau-3DS* | *AX-110909845* and *AX-109395143*, 12.4 (Mb)[a] | | |
| | | *QLr.hebau-3DS* | *AX-109971456* (developed KASP marker) | | |
| | | *QLr.hebau-4DL* | *AX-110476142-AX-111092299* | | |
| | | *QLr.hebau-5AL* | *AX-110414471-AX-109408919*, 19.8 (Mb)[a] | | |
| | | *QLr.hebau-5DL* | *AX-1 10374586 and AX-108770574, 5.2 (Mb)[a]* | | |
| | | *QLr.hebau-7DL* | *AX-109932021 and AX-108763519*, 39.3 (Mb)[a] | | |
| | | *QLr.cim-2DS* | *cfd51* and *gwm455* (SSR markers), 7.5 cM | | |
| 10. | Ning7840 × Clark, $F_{12}$ RILs | *QLr.hwwg-7DS/Lr34* | *Lr34exon11-KASP* | 9 K Illumina chips and SSR | Li et al. (2017) |
| | | *QLr.hwwg-3BS.1* | *IWA4654- IWA1702*, 0.5 cM | | |
| | | *QLr.hwwg-6AS* | *Xbarc23- IWA3321*, 14.2 cM | | |
| | | *QLr.hwwg-5AS* | IWA2145- IWA5053, 7.4 cM | | |

| | | | | | |
|---|---|---|---|---|---|
| 11. | CI13227 (US winter wheat line), DHs | *QLr.cim-1BL.2* | 100041322\|F\|0-38:A>G-38:A>G- ---100084210\|F\|0-16:A>G-16:A> | 90 K SNP assays SSR | Lu et al. (2017) |
| | | *QLr.cim-3DS* | 1233406----3941237 | | |
| | | *QLr.hwwg-2DS/ QLrlp.osu-2DS* | IWB34642 and IWB8545 (SNP markers), 16.4 cM | | |
| 12. | Aus28183 (Australian landrace), RILs | *Yr47* and *Lr52* | *sun180* (SSR), 0.4 cM | Set of SSR STS SNP | Qureshi et al. (2017a) |
| 13. | Aus26582 (Australian landrace), RILs | *LrAW2* (locus) | *sun684* (SSR) *sunKASP_60* (KASP) | SSR STS | Qureshi et al. (2017b) |
| 14. | Kundan, $F_{5:6}$ RILs | *Lr46/Yr29* | *1092272* and *102414*, 1.2 Cm | DArT-Seq and SSR | Ren et al. (2017) |
| | | *QLr.cim-2BL* | *1237388* and *1081780_35:C>T* | | |
| | | *QLr.cim-2DS* | *1020115* and *1242814* | | |
| 15. | Thatcher *Lr18*, $F_3$ | *QLr.hbau-2AS* | Lr37/Yr17 – AX-111661031, 1.1 cM | SSR | Sadeghabad et al. (2017) |
| | | *QLr.hbau-2DS* | AX-109486828 – AX-109403444, 1.1 cM | | |
| | | *QLr.hbau-7BL* | AX-108959083 – Lr68; 0.4 cM | | |
| | | *Lr18* | *Xgpw7425* and *Xwmc75*, 1.5 cM | | |
| 16. | Forno (Swiss winter wheat cultivar), $BC_3F_2$ | *QLr.sfr-1BS/Lr75* | *gwm604-swm271* (SSR), 4.3 cM | 1BS-specific SSR | Singla et al. (2017) |
| 17. | Zhou 8425B × Chinese Spring, $F_8$ RILs | *QLr.hebau-2AL* | *wmc181* and BS00057060_51 | 90 K iSelect SNP array and SSR | Zhang et al. (2017) |
| 18. | Toropi, RILs | *Lr78* | *IWA6289* (KASP) | DArT | Kolmer et al. (2018a) |
| 19. | Caldwell (US soft red winter wheat), $F_6$ RILs | *QLr.cdl-3BS/Lr74* | cfb5006 (SSR) | iSelect 90 K Infinium SNP assay | Kolmer et al. (2018b) |

(continued)

**Table 1** (continued)

| S. no. | Donor (its description)/Cross Type/mapping population | QTL/gene | Associated markers | Genotyping technology | Reference |
|---|---|---|---|---|---|
| 20. | Santa (Fe winter wheat), $F_6$ RILs | *QLr.cdl-3BL/Lr77* | IWB32805 IWB73555 (SNP), 4.4 cM | iSelect 90 K Infinium SNP assay | Kolmer et al. (2018c) |
| | | | *IWB10344* (SNP) | | |
| 21. | Chilero (Spring wheat), $F_{4:5}$ RILs | *QLr.sun-3BS* | *1373089* | DArT, GBS and SSR | Ponce-Molina et al. (2018) |
| | | *QLr.sun-4DL* | *1052292* | | |
| | | *QLr/Yr.cim-1BL* | 1164928 and 2289154 | | |
| 22. | Aus26582 (Australian Landrace), RIL | *Lr79* | *sun786* (SSR), 1.8 cM | SSR and KASP | Qureshi et al. (2018a) |
| 23. | PI 192051 (landrace), RILs | *Lr.ace-4A* | *IWA232 and IWA1793*, 4.0 cM | Illumina iSelect 9 K SNP array. | Aoun et al. (2019) |
| 24. | KS93U50 × Morocco, $F_5$ RILs | *QLr.cim-2BC* | *1151114 and 3222199* | SSR EST KASP | Gill et al. (2019) |
| | | *QLr.cim-5BL* | *1092064_17:G>A and 1090228* | | |
| | | *QLr.cim-6B* | *2277143_7:T>C and 1234305_34:G>A* | | |
| | | *Lr42* | *TC387992* and *WMC432*<br>*SNP113325* and *TC387992* (KASP markers), 3.7 cM | | |
| 25. | (Thatcher*2/Duster) $F_4$ line × Thatcher, $F_6$ RILs | *QLr/Yr.cim-3DC* | *gwm341* and *barc1119* (SSR markers), 2.7 cM | iSelect 90 K Infinium SNP assay, KASP and STS | Kolmer et al. (2019) |
| | | *Lr46* | *csLV46G22* (STS marker) | | |

| | | | | | |
|---|---|---|---|---|---|
| 26. | Gaza (Middle East), $F_8$ RILs | *Lr77* | *IWB10344* (SNP marker) | Illumina iSelect 90 K SNP array | Kthiri et al. (2019) |
| | | *QLr.usw-6BS* | *CAP7_c10772_156* | | |
| | Arnacoris (France); $F_8$ RILs | *QLr.usw-6BL* | *GENE-3689_293* (SNP marker) | | |
| | | *QLr.usw-7BL* (major) | *BS00010355_51* | | |
| | | *QLr.usw-1BL.1* (minor) | *BS00060686_51* and *Kukri_c46030_412* (SNP markers), 9.8 cM | | |
| | | *QLr.usw-2BS* | *Tdurum_contig76118_145* and *wsnp_Ex_c18354_27181086* (SNP markers), 3.9 cM | | |
| | Saragolla (Italy); $F_8$ RILs | *QLr.usw-3B* | *Tdurum_contig33168_461 and RAC875_rep_c82061_78*, 1.0 cM | | |
| | | *QLr.usw-1BL.2* | *wsnp_Ex_c4436_7981188* and *BS00000010_51* | | |
| 27. | Pingyuan 50 × Mingxian 169 (Chinese landraces), DHs | *QLr.hebau-2BS* | JD_c767_567 and RFL_Contig1483_1765; 0.9 cM | Wheat Affymetrix 55 K SNP array and additional SSR markers | Zhang et al. (2019a) |
| | | *QLr.hebau-3A* | wsnp_Ex_c1660_3159173 and Tdurum_contig5096_193 | | |
| | | *QLr.hebau-3BS* | BobWhite_c9711_71 and Excalibur_c6330_1158 | | |
| | | *QLr.hebau-4AL* | BobWhite_c15697_675 and Excalibur_c2827_580 | | |
| | | *QLr.hebau-4B/Lr12* | BS00022181_51 and Excalibur_c37565_709 | | |
| | | *QLr.hebau-5BL* | wsnp_Ex_c3175_5864335 and BobWhite_c16916_658 | | |
| | | *QLr.hebau-7DS/Lr34* | Kukri_c92151_216 and csLV34 | | |
| | | *QLr.hebau-1BL* | *AX-109949200* and *AX-110968810*, 11.8 (Mb)[a] | | |
| 28. | SW 8588 × Thatcher, RILs | *QLr.hebau-1BL/QYr.hebau-1BL* (*Lr46/Yr29* gene) | csLV46G22 | Wheat 55 K SNP array and SSR | Zhang et al. (2019b) |

(continued)

**Table 1** (continued)

<table>
<tr><th>S. no.</th><th>Donor (its description)/Cross Type/mapping population</th><th>QTL/gene</th><th>Associated markers</th><th>Genotyping technology</th><th>Reference</th></tr>
<tr><td>29.</td><td>Duster (Oklahoma hard red winter wheat cultivar PI 644016), DHs</td><td>Lr34 gene</td><td>Lr34E11-KASP and Lr34-E22-KASP</td><td>SNP</td><td>Fang et al. (2020)</td></tr>
<tr><td>30.</td><td>Mianyang 351-15, $F_6$ RILs</td><td>QLr.hwwg-7BL<br>QLr.hwwg-7AL</td><td>IWB9496 and IWB24039 (SNP markers), 1.8 cM<br>IWB42182 and IWB73053 (SNP markers), 10.2 cM</td><td>Wheat 55 K SNP array and SSR</td><td>Gebrewahid et al. (2020)</td></tr>
<tr><td></td><td></td><td>QLr.hwwg-3B</td><td>IWB35536 and IWB5899 (SNP markers), 0.4 cM</td><td></td><td></td></tr>
<tr><td></td><td></td><td>QLr.hbau-1BL</td><td>AX-110946149 – Lr46/Yr29, 1.1 cM</td><td></td><td></td></tr>
<tr><td>31.</td><td>VL404 (Indian cultivar), $F_{6:8}$ RILs</td><td>Lr49</td><td>sunKASP_21 and sunKASP_24, 1.0 cM</td><td>Infinium iSelect 90 K SNP array</td><td>Nsabiyera et al. (2020)</td></tr>
<tr><td>32.</td><td>Arableu#1 (CIMMYT spring wheat line), $F_5$ RILs</td><td>QLr.cim-2BL</td><td>wPt-6174 and wPt-8548 (DArT markers)</td><td>GBS</td><td>Yuan et al. (2020)</td></tr>
<tr><td></td><td></td><td>QLr.cim-3BS</td><td>wPt-5209 and csSr2 (DArT markers)</td><td></td><td></td></tr>
<tr><td></td><td></td><td>QLr.cim-5AC</td><td>wPt-3187 and wPt-7769 (DArT markers)</td><td></td><td></td></tr>
<tr><td></td><td></td><td>QLr.cim-1BL.1 L, Lr46</td><td>5411162|F|0-35:A>C-35:A>C---- 1132278|F|0-20:C>T-20:C>T</td><td></td><td></td></tr>
</table>

achieve mapping of *Lr* gene to a distance of nearly less than 0.1 cM. Although good process has been made in identification of closely linked markers for pathotype-specific genes, progress in terms of identification of slow-rusting genes is slow and limited. Table 1 summarized the salient examples of molecular markers closely associated with wheat *Lr* resistance genes for their utilization using marker-assisted breeding (MAB). It is because of this fine mapping which not only reduces the linkage drag but also helped breeders to successfully introgressed leaf rust resistance in wheat, using marker-assisted gene pyramiding and backcrossing strategy. Table 4 summarizes the successful MAS events mentioning source of resistance gene, information about improved cultivar and the approach used for imparting resistance to the elite cultivars.

## 10 MAS for Stem Rust Resistance

Similarly, for achieving durable *Sr* resistance, molecular markers have been used in different breeding programmes using MAS (Sivasamy et al. 2009) and MABC (Prasad et al. 2014; Yadav et al. 2015). For mapping of *Sr* resistance genes in wheat, various workers employed different strategies using various genotyping platforms, viz. SSR (Bansal et al. 2015; Briggs et al. 2015; Chen et al. 2015; Yu et al. 2015; Sharma et al. 2019), DArT (Basnet et al. 2015; Chhetri et al. 2016), SNP assays using different chips (*9 K* (Aoun et al. 2019), *90 K* (Nirmala et al. 2016; Hiebert et al. 2017; Chen et al. 2018)) and more recently GBS approach (Qureshi et al. 2018b), and were able to achieve mapping of stem rust resistance to a distance of nearly less than 0.1 cM. Table 2 summarized the prominent examples of molecular markers closely associated with wheat *Sr* resistance genes for their utilization using MAB. With this fine mapping, Prasad et al. (2014) and Yadav et al. (2015) could be able to achieve stem rust resistance in elite cultivars of wheat using marker-assisted backcrossing strategy. Table 4 summarizes these successful MAS strategies for stem rust resistance in wheat mentioning source of resistance gene, information about improved cultivar and the approach used for imparting resistance to the elite cultivars.

## 11 MAS for Stripe Rust Resistance

In a similar way like for leaf and stem rust resistance, molecular markers have been used in different breeding programmes, viz. MAS (Kuraparthy et al. 2009), MABC (Randhawa et al. 2009) and MAGP (Revathi et al. 2010; Qie et al. 2018; Liu et al. 2020), for imparting stripe rust resistance to wheat. Further, for mapping of *Yr* resistance genes, various workers employed different strategies in wheat using various genotyping platforms, viz. SSR (Hou et al. 2015; Yaniv et al. 2015), DArT (Calvo-Salazar et al. 2015; Lan et al. 2015; Chhetri et al. 2016; Ren et al. 2017; Ponce-Molina et al. 2018), KASP assay (Wu et al. 2018b), SNP assays using different chips (*55 K* (Wu et al. 2018a; Ma et al. 2019; Gebrewahid et al. 2020), *90 K* (Liu et al. 2015; Pakeerathan et al. 2019; Zhang et al. 2019a, b, c)) and more recently GBS approach (Yuan et al. 2020), and were able to achieve mapping of stripe rust resistance to a distance of nearly less than 0.1 cM. Table 3 summarized the successful examples of molecular markers closely associated with wheat *Yr* resistance genes for their utilization using MAB. However, in spite of all this, there is only few reports of MAB for stripe rust resistance in wheat by Qie et al. (2018) and Liu et al. (2020) where they used molecular markers, marker-assisted selection, gene pyramiding and backcrossing strategy to identify individuals which impart durable stripe rust resistance. Table 4 summarizes these successful events mentioning source of resistance gene, information about the improved cultivar and the approach used.

Meanwhile, since pathogen races are also evolving and continuously breaking the resistance, so in order to develop durable resistance, breeders need to focus only on those genes which are quantitatively inherited and that too in combinations with resistance genes from other sources. This could be possible by pyramiding of different seedling as well as adult plant resistance genes. Kumar et al. (2010) successfully demonstrated the pyramiding of QTL/genes for more than one trait in elite wheat cultivar. Further, Tyagi et al. (2014) combined three rust resistance genes (*Lr24*, *Sr24* and *Yr36*) in the background of a PBW343 along with four grain quality traits, while Mallick et al. (2015) pyramided *Lr19*, *Sr26* and *Yr10* genes in the genetic background of HD2932 with an aim to develop combined resistance against all three types of rust, which is the need of the hour for reducing the risk of development of new rust pathotypes.

## 12 Conclusion and Future Prospects

Wheat rust being responsible for huge economic losses has been genetically mapped very extensively being compared to other important trait of interests. Now since several rust resistance genes, associated molecular markers and information about its genetics are available to breeders for their use in rust resistance breeding programme, emphasis should be on their introgression to the elite cultivars and that too using several gene combinations via marker-assisted approach specially gene pyramiding so that no new race could break the resistance very easily and the introgressed wheat remains to be durable for long. Further, both qualitative and quantitative resistance need to be utilized for developing resistance varieties, and the breeding programmes in the future should have central focus on combined selection of both types of resistance.

**Table 2** Molecular markers closely associated with wheat stem rust resistance for their utilization in MAB

| S. no. | Donor (its description)/ cross-type/mapping population | QTL/gene | Associated markers | Genotyping technology | Reference |
|---|---|---|---|---|---|
| 1. | AUS28011 (Mahmoudi landrace collected from Ghardimaou Tunisia), $F_6$ RILs | *Sr49* | *sun209* and *sun479*, 2.4 cM | SSR | Bansal et al. (2015) |
| 2. | ND643/2*Weebill1, $F_{4:5}$ RILs | *SrND643* | Xgwm350, 0.5 cM Xwmc219, 4.1 cM and Xwmc776, 2.9 cM | DArT SSR SNP | Basnet et al. (2015) |
| 3. | G3116 (PI 427992 wild *T. monococcum* ssp. *aegilopoides* and PI 306540 cultivated spring *T. monococcum* ssp. *monococcum* accession), two $F_2$ populations | *SrTm4* a recessive stem rust resistance gene | *BQ461276* and *DR732348* (STS), 2.1 cM | SSR | Briggs et al. (2015) |
| 4. | DV92 (spring growth habit) G3116 (wild winter *T. monococcum* subsp. *Aegilopoides*: $F_{2:3}$ families) | *Sr21* | *FD527726* and *EX594406*, 0.2 cM | SSR | Chen et al. (2015) |
| 5. | Clae 25 (*Aegilops tauschii* accession), $F_2$ population | *Sr46* | *Xgwm210* and *Xwmc111* | SSR and STS | Yu et al. (2015) |
| 6. | W195 × BTSS; F7 RILs | *QSr. sun-2BL* | *1125978* | DArTseq | Chhetri et al. (2016) |
| | | *QSr. sun-6AS* | *1213304* | | |
| | | *QSr. sun-4DL* | *1052292* | | |
| 7. | Triumph 64 (winter wheat cultivar), DHs | *SrTmp*, *Sr* gene conferring resistance to TTKSK | *gpw5182* (SSR marker) *kwm864* and *kwm929* (SNP markers), 0.8 cM distal to gene *kwm71* and *kwm217* (SNP markers), 2.27 cM proximal to gene | SSR and SNP | Hiebert et al. (2016) |
| 8. | Peace, DHs AC foremost, $F_6$ RILs AC Cadillac, $F_7$ RILs | *SrCad* | Contig11536236_557_ kwm999 and Contig11536236_558_ kwm1000 | SNP | Kassa et al. (2016) |
| 9. | Gabo 56, $F_{4:5}$ RILs | *Sr11* | KASP_6BL_IWB10724 and KASP_6BL_IWB72471 | 90 K Infinium iSelect Custom BeadChip | Nirmala et al. (2016) |

(continued)

**Table 2** (continued)

| S. no. | Donor (its description)/ cross-type/mapping population | QTL/gene | Associated markers | Genotyping technology | Reference |
|---|---|---|---|---|---|
| 10. | U6897-1 to 6; Six $BC_3F_2$ families | *SrTA10187* | *6DS0039* (KASP); 0.2 cM *6DS0050* (STS marker) | SNP and STS | Wiersma et al. (2016) |
| 11. | Harvest (Canadian cultivar resistant to TRTTF), DHs | *Sr_TRTTF* | gwm459 and gwm334 | 90 K iSelect SNP genotyping assay | Hiebert et al. (2017) |
| 12. | RWG35 RWG36 and RWG37, $BC_2F_2$ | Sr47 | Xmag1729 Xwmc41(SSR) Xrwgsnp1 Xrwgsnp4 (STARP) Xrwgs38a Xtnac3119 | SSR and five new STARP (semi-thermal asymmetric reverse PCR) | Klindworth et al. (2017) |
| 13. | PI 306540 (diploid wheat *Triticum monococcum*), $F_2$ | *Sr60* | GH724575 and CJ942731, 0.44 cM | SSR, 90 K SNP iSelect Illumina platform | Chen et al. (2018) |
| 14. | 'Aus27969' landrace x 'Avocet S' cultivar, $F_{5:7}$ RILs | *Sr26* | *sunKASP_224* and *sunKASP_225* | GBS | Qureshi et al. (2018b) |
| 15. | PI 192051 (Portuguese durum landrace), RILs | *QSr.ace-4A* | *IWA603* and *IWA4657* (flanking), 0–15 cM *IWA7521* (closest), 5.9 cM | iSelect 9 K SNP array | Aoun et al. (2019) |
| | | *QSr.ace-7A* | *IWA8390* and *IWA1805*, 1.5 cM | | |
| 16. | TA7682 (disomic addition line) TA5617 (translocation line) | *Sr52* | CINAU1532 6L-4 and 6L11/*Mbo*I | EST STS | Li et al. (2019) |
| 17. | PI193883 (cultivated emmer wheat accession) | *QSr.fcu-2B* | *IWB56465* and *IWB55767* | SSR | Sharma et al. (2019) |
| | | *QSr.fcu-6A* | *IWB3057* and *barc104 rwgsnp7* (STARP marker) | | |

**Table 3** Molecular markers closely associated with wheat stripe rust resistance for their utilization in MAB

| S. no. | Donor (its description)/ cross-type/mapping population | QTL/gene | Associated markers | Genotyping technology | Reference |
|---|---|---|---|---|---|
| 1. | Kenya Kongoni (Kenyan wheat), $F_5$ RILs | *QYr.cim-2AS* | *Xbarc124* and *wPt-1722* (DArT markers) | DArT and SSR | Calvo-Salazar et al. (2015) |
| | | *QYr.cim-2BS*<br>*QYr.cim-5BL* | *wPt-1813* and *wPt-8072* (DArT markers)<br>*Xbarc59 and wPt-3922* (DArT markers) | | |
| 2. | Druchamp (winter wheat), $F_8$ RILs | *QYrdr.wgp-5BL.1*<br>*QYrdr.wgp-5DL*<br>*QYrdr.wgp-6BL.1* | *IWA6271*<br>*IWA8331*<br>*IWA3297* | SSR SNP | Hou et al. (2015) |
| 3. | Sujata (bread wheat cultivar), $F_{4:5}$ RILs | *QLr.cim-1AS/QYr.cim-1AS* | *wPt-9752* and *Xgdm33* | DArT and SSR | Lan et al. (2015) |
| | | *Lr46/Yr29* PAPR gene | *csLV46G22 wPt-8168 Xwmc216* | | |
| | | *Lr67/Yr46* | *Xgwm192* | | |
| | | *QLr.cim-7BL/YrSuj* | *Xcfa2040 and Xwmc526*; 6.0 cM | | |
| 4. | Linmai 2 × Zhong 892; $F_{2:6}$ RILs | *QYr.caas-3AS* | *Kukri_rep_c102131_891* and *Kukri_c96747_274*, 3.7 cM | Illumina 90 K iSelect assay | Liu et al. (2015) |
| | | *QYr.caas-3BS* | *IAAV5662 and BS00056257_51*, 2.3 cM | | |
| | | *QYr.caas-7AL* | *Kukri_c41603_111 and Excalibur_c25335_306*, 3.2 cM | | |
| | | *QYr.caas-7DS.1* | *tplb0024a09_2369 and RAC875_c29314_291*, 12.3 cM | | |
| | | *QYr.caas-2AL* | *wsnp_Ex_c16627_25162391* and *BS00092550_51*, 8.1 cM | | |
| | | *QYr.caas-2BL.3* | *Ra_c21099_1781* and *IACX8602*, 11.0 cM | | |
| | | *QYr.caas-6AL* | *Ku_c45494_267* and *BS00040166_51*, 8.1 cM | | |
| | | *QYr.caas-5DL* | *wsnp_Ex_c508_1008029 and wsnp_Ex_c22984_32207214*, 5.8 cM | | |

(continued)

**Table 3** (continued)

| S. no. | Donor (its description)/ cross-type/mapping population | QTL/gene | Associated markers | Genotyping technology | Reference |
|---|---|---|---|---|---|
| 5. | *Triticum dicoccoides* (wild emmer wheat), $F_2$ | *Yr15* | *Xbarc8* and *Xgwm493* | SSR | Yaniv et al. (2015) |
| 6. | W195 × BTSS; $F_7$ RILs | *QYr.cim-2AL* | *3064488_30:T>G* and *1106314_11:A>G* | DArTseq | Chhetri et al. (2016) |
| | | *QYr.sun-3BS* | *3023704* | | |
| | | *QYr.sun-4DL* | *1052292* | | |
| | | *QYr.sun-7AS* | *1218068* | | |
| 7. | PI 480035 (hexaploid spring wheat landrace), $F_6$ RILs | QYr.wrsggl1-1BS | Xgwm273 Xgwm11 and Xbarc187 1.01 cM Xcfd59 0.59 cM proximal distal to QTL | GBS | Kandel et al. (2017) |
| 8. | Kundan, $F_{5:6}$ RILs | *QYr.cim-2AL* *QYr.cim-3DS* | *3064488_30:T>G* and *1106314_11:A>G* *3021242* and *2243560* | DArT-Seq and SSR | Ren et al. (2017) |
| 9. | PI 182103 (spring wheat landrace) | *QyrPI182103.wgp-7BL*, *Yr79* gene | Xbarc72 and Xwmc335 (SSR markers) | SSR SNP | Feng et al. (2018) |
| 10. | Chilero, $F_{4:5}$ RILs | *QYr.cim-6BS* *QYr.cim-7BL* | 4396419 and 1209575 100006719 and 1112830 | DArT, GBS and SSR | Ponce-Molina et al. (2018) |
| 11. | P10103 (CYMMIT-derived wheat line), RIL | QYr.nwafu-4BL | AX-110963704 and AX-110519862 (SNP markers) | Wheat55 K SNP array | Wu et al. (2018a) |
| 12. | 03031-1-5 H62 wheat line, $F_{2:3}$ | *YrH62* | AX-109352427 and AX-109862469 (KASP markers), 1.0 cM | SNP KASP SSR | Wu et al. (2018b) |
| 13. | 20828 (winter wheat line), RIL | QYr.sicau-1B.1 | *Xwmc216 and Xwmc156* (SSR markers), 1.76-cM | Wheat55K SNP array and SSR | Ma et al. (2019) |
| 14. | Aus27969 (landrace), $F_7$ RILs | *Yr82* | *sun KASP_300 and KASP_8775* (KASP markers), 2 cM | iSelect wheat 90 K Infinium SNP array | Pakeerathan et al. (2019) |
| 15. | Soru#1 (synthetically derived wheat line), $F_6$ RILs | *Yr28* | *BS00108770_51* (KASP marker) | 90 K SNP STS SSR | Zhang et al. (2019a, b, c) |

<table>
<tr><td>16.</td><td>Mianyang351-15, $F_6$ RIL</td><td>QYr.hbau-1BL</td><td>AX-110946149 – Lr46/Yr29, 1.1 cM</td><td>Wheat 55 K SNP array and SSR</td><td>Gebrewahid et al. (2020)</td></tr>
<tr><td></td><td></td><td>QYr.hbau-1DL</td><td>AX-111135128 – AX-111085909, 1.1 cM</td><td></td><td></td></tr>
<tr><td></td><td></td><td>QYr.hbau-2AS</td><td>Lr37/Yr17 – AX-111661031, 2.2 cM</td><td></td><td></td></tr>
<tr><td></td><td></td><td>QYr.hbau-2DS</td><td>AX-109486828 – AX-109403444, 1.1 cM</td><td></td><td></td></tr>
<tr><td></td><td></td><td>QYr.hbau-3AS</td><td>AX-111491666 – AX-110551014, 1.1 cM</td><td></td><td></td></tr>
<tr><td></td><td></td><td>QYr.hbau-3DL</td><td>AX-109333872 – AX-108884636, 0.8 cM</td><td></td><td></td></tr>
<tr><td></td><td></td><td>QYr.hbau-7BL</td><td>AX-108959083 – Lr68, 0.4 cM</td><td></td><td></td></tr>
<tr><td>17.</td><td>Arableu#1 (CIMMYT spring wheat line), $F_5$ RILs</td><td>QYr.cim-1BL.1, Yr29</td><td>5411162|F|0-35:A>C-35:A>C---- 1132278|F|0-20:C>T-20:C>T</td><td>GBS</td><td>Yuan et al. (2020)</td></tr>
<tr><td></td><td></td><td>QYr.cim-4BL</td><td>1235862|F|0-42:C>T-42:C>T---- 1220607|F|0-68:T>C-68:T>C</td><td></td><td></td></tr>
<tr><td></td><td></td><td>QYr.cim-7DS</td><td>1117156----987784|F|0-55:T>G-5 5:T>G</td><td></td><td></td></tr>
</table>

**Table 4** Successful examples of MAB for rust resistance in wheat

| S. no. | Source of gene | Improved cultivar | Approach | Reference |
|---|---|---|---|---|
| *Leaf rust resistance* | | | | |
| 1. | *Lr24 Lr28* and *Yr15* genes from TR380-14*7/3Ag#14 CS2A/2M#4/2 and Avocet*6/*Yr15*, respectively | HD 2687 | Marker-assisted backcross breeding | Tiwari et al. (2014) |
| 2. | *Lr24* and *Lr28* from NIL PBW-343 pyramided line | MP 3299 | Marker-assisted selection | Savitha et al. (2016) |
| 3. | *Lr75* from Forno: Swiss winter wheat cultivar | Arina | Marker-assisted backcrossing | Singla et al. (2017) |
| 4. | *Lr19* and *Lr24* from *Thinopyrum* (syn. *Agropyron*) | HD 2733 (bread wheat variety) | Marker-assisted pyramiding of *Lr19* and *Lr24* | Singh et al. (2017) |
| 5. | NIL PBW 343 introgressed with *Lr24* and *Lr28* | DWR 162 | Marker-assisted backcrossing | Yadawad et al. (2017) |
| 6. | *Lr24* and *Lr28* from MP3299 | MP 3299 | Marker-assisted introgression | Koujalagi et al. (2019) |
| *Stem rust resistance* | | | | |
| 7. | *Sr36* from *Triticum timopheevii* | HI 8498 | Marker-assisted backcross breeding | Prasad et al. (2014) |
| 8. | Three independent *Sr* genes (*Sr25*, *SrWeb* and *Sr50*) from CIMMYT line PMBWIR4 | HUW234 | Marker-assisted backcrossing | Yadav et al. (2015) |
| *Stripe rust resistance* | | | | |
| 9. | *Yr64* from RIL-Yr64 and *Yr15* from AvSYr15NIL | Avocet S (AvS) | Gene pyramiding | Qie et al. (2018) |
| 10. | *Yr10 Yr30* and *Yr48* from Chuanyu12; *Yr10*, *Yr15*, *Yr30*, *Yr48*, *Yr62*, *Yr65* and *YrSP* from 04G368; and *Yr15*, *Yr30* and *Yr65* from Yumai35 | Chuanyu12 | Gene pyramiding strategy and marker-assisted selection | Liu et al. (2020) |
| *Stripe and leaf rust resistance* | | | | |
| 11. | *Yr15* from Avocet/Yr15 and *Lr19/Sr25* and *Lr24/Sr24* from FLW 8 and FLW 21 | UP 2338 | Marker-assisted backcross breeding | Singh et al. (2018) |
| *Stripe and stem rust resistance* | | | | |
| 12. | *Yr51* from AUS91456, *Yr57* from AUS91463, Sr22 from Sr22/3*K441, Sr26 from Sr26 WA1 and Sr50 from Dra-1/Chinese Spring ph1b/2/3* Gabo | Gladius Livingston PBW550 and DBW17 | Marker-assisted selection | Randhawa et al. (2019) |
| *Leaf, stem and stripe rust resistance* | | | | |
| 13. | *Lr24/Sr24* + *Yr36* from Rye Selection 111 Yecora Rojo | PBW343 | Marker-assisted pyramiding | Tyagi et al. (2014) |
| 14. | *Lr19 Sr26* and *Yr10* from NIL of Indian variety HD2687 Eagle and a NIL of exotic variety Avocet, respectively | HD2932 | Marker-assisted backcross breeding | Mallick et al. (2015) |

(continued)

**Table 4** (continued)

| S. no. | Source of gene | Improved cultivar | Approach | Reference |
|---|---|---|---|---|
| 15. | *Yr70/Lr76* + *Lr37/Yr17/Sr38* from P1 line and *Gpc-B1/Yr36* + *QPhs.ccsu-3A.1* + *QGw.ccsu-1A.3* + *Lr24/Sr24* + *Glu-A1-1/Glu-A1-2* from P2 line<br>P1 and P2 are improved PBW343 lines | PBW343 | Marker-assisted pyramiding | Gautam et al. (2020) |

# References


Ali S, Gladieux P, Leconte M, Gautier A, Justesen AF, Hovmøller MS, Enjalbert J, de Vallavieille-Pope C (2014) Origin, migration routes and worldwide population genetic structure of the wheat yellow rust pathogen *Puccinia striiformis* f sp *tritici*. PLoS Pathog 10:e1003903

Aoun M, Kolmer JA, Rouse MN, Elias EM, Breiland M, Bulbula WD, Chao S, Acevedo M (2019) Mapping of novel leaf rust and stem rust resistance genes in the Portuguese durum wheat landrace PI 192051. G3-Genes Genomes Genet 9(8):2535–2547

Arora S, Steuernagel B, Gaurav K, Chandramohan S, Long Y, Matny O, Johnson R, Enk J, Periyannan S, Singh N, Hatta MA (2019) Resistance gene cloning from a wild crop relative by sequence capture and association genetics. Nat Biotechnol 37:139–143

Bansal UK, Muhammad S, Forrest KL, Hayden MJ, Bariana HS (2015) Mapping of a new stem rust resistance gene Sr49 in chromosome 5B of wheat. Theor Appl Genet 128(10):2113–2119

Bariana HS, Bansal UK (2017) Breeding for disease resistance. In: Kole B, Murray GB, Murphy JD (eds) Encyclopedia of applied plant sciences, vol 3. Academic, Waltham, pp 69–76

Bariana HS, Brown GN, Bansal UK, Miah H, Standen GE, Lu M (2007) Breeding triple rust resistant wheat cultivars for Australia using conventional and marker assisted selection technologies. Aust J Agric Res 58:576–587

Basnet BR, Singh S, Lopez-Vera EE, Huerta-Espino J, Bhavani S, Jin Y, Singh RP (2015) Molecular mapping and validation of SrND643: a new wheat gene for resistance to the stem rust pathogen Ug99 race group. Phytopathology 105(4):470–476

Beddow JM, Pardey PG, Chai Y, Hurley TM, Kriticos DJ, Braun H-J, Park RF, Cuddy WS, Yonow T (2015) Research investment implications of shifts in the global geography of wheat stripe rust. Nat Plants 1(10):1–5

Berlin A, Djurle A, Samils B, Yuen J (2012) Genetic variation in *Puccinia graminis* collected from oats, rye, and barberry. Phytopathology 102:1006–1012

Berlin A, Rahmatov M, Muminjanov H, Yuen J (2015) Sexual population contributes to genotypic variation in the population of *Puccinia graminis* in Tajikistan. Eur J Plant Pathol 141:159–168

Bhardwaj SC, Prasad P, Gangwar OP, Khan H, Kumar S (2016) Wheat rust research-then and now. Indian J Agric Sci 86(10):1231–1244

Bhardwaj SC, Gyanendra PS, Om PG, Pramod P, Subodh K (2019) Status of wheat rust research and progress in rust management-Indian context. Agronomy 9:892. https://doi.org/10.3390/agronomy9120892

Bhawar KB, Sharma JB, Singh AK, Sivasamy M, Singh M, Prabhu KV, Singh B (2011) Molecular marker assisted pyramiding of leaf rust resistance genes *Lr19* and *Lr28* in bread wheat (*Triticum aestivum* L) variety HD2687. Indian J Genet Plant Breed 71(4):304

Biffen RH (1905) Mendel's law of inheritance and wheat breeding. J Agric Sci 1:4–48

Bolton MD, Kolmer JA, Garvin DF (2008) Wheat leaf rust caused by *Puccinia triticina*. Mol Plant Pathol 9:563–575

Briggs J, Chen S, Zhang W, Nelson S, Dubcovsky J, Rouse MN (2015) Mapping of SrTm4, a recessive stem rust resistance gene from diploid wheat effective to *Ug99*. Phytopathology 105(10):1347–1354
Browder LE (1973) Probable genotype of some *Triticum aestivum* 'Agent' derivatives for reaction to *Puccinia recondita* f sp *tritici*. Crop Sci 13:203–206
Burdon JJ, Barrett LG, Rebetzke G, Thrall PH (2014) Guiding deployment of resistance in cereals using evolutionary principles. Evol Appl 7(6):609–624
Bux H, Rasheed A, Siyal MA, Mujeeb K (2012) An overview of stripe rust of wheat (*Puccinia striiformis* f sp *tritici*) in Pakistan. Arch Phytopathol Plant Protect 45(19):2278–2289
Calvo-Salazar V, Singh RP, Huerta-Espino J, Cruz-Izquierdo S, Lobato-Ortiz R, Sandoval-Islas S, Lan CX (2015) Genetic analysis of resistance to leaf rust and yellow rust in spring wheat cultivar Kenya Kongoni. Plant Dis 99(8):1153–1160
Charpe A, Koul S, Gupta SK, Singh A, Pallavi JK, Prabhu KV (2012) Marker assisted gene pyramiding of leaf rust resistance genes *Lr 9, Lr 24* and *Lr 28* in a bread wheat cultivar HD 2329. J Wheat Res 4(1)
Chen W, Wellings C, Chen X, Kang Z, Liu T (2014) Wheat stripe (yellow) rust caused by *Puccinia striiformis* f sp *tritici*. Mol Plant Pathol 15:433–446
Chen S, Rouse MN, Zhang W, Jin Y, Akhunov E, Wei Y, Dubcovsky J (2015) Fine mapping and characterization of Sr21, a temperature-sensitive diploid wheat resistance gene effective against the *Puccinia graminis* f sp *tritici* Ug99 race group. Theor Appl Genet 128(4):645–656
Chen S, Guo Y, Briggs J, Dubach F, Chao S, Zhang W, Rouse MN, Dubcovsky J (2018) Mapping and characterization of wheat stem rust resistance genes *SrTm5* and *Sr60* from *Triticum monococcum*. Theor Appl Genet 131:625–635
Chen S, Rouse MN, Zhang W, Zhang X, Guo Y, Briggs J, Dubcovsky J (2020) Wheat gene *Sr60* encodes a protein with two putative kinase domains that confers resistance to stem rust. New Phytol 225:948–959
Chhetri M, Bansal U, Toor A, Lagudah E, Bariana H (2016) Genomic regions conferring resistance to rust diseases of wheat in a W195/BTSS mapping population. Euphytica 209(3):637–649
Chhuneja P, Kaur S, Garg T, Ghai M, Kaur S, Prashar M, Singh K (2008) Mapping of adult plant stripe rust resistance genes in diploid A genome wheat species and their transfer to bread wheat. Theor Appl Genet 116(3):313–324
Chhuneja P, Vikal Y, Kaur S, Singh R, Juneja NS, Bains S, Berry O, Dhaliwal HS (2011) Marker-assisted pyramiding of leaf rust resistance genes Lr24 and Lr28 in wheat (*Triticum aestivum*). Indian J Agric Sci 81(3):214
Draz IS, Mohammed SA, Elmageed MK, Omaima AE, Ahmed FB (2015) Screening of wheat genotypes for leaf rust resistance along with grain yield. Ann Agric Sci 60(1):29–39
Duplessis S, Joly DL, Dodds PN (2012) Rust effectors. In: Effectors in plant-microbe interactions. Wiley-Blackwell, Oxford, pp 155–193
Ellis JG, Lagudah ES, Spielmeyer W, Dodds PN (2014) The past, present and future of breeding rust resistant wheat. Front Plant Sci 5:641
Fang T, Lei L, Li G, Powers C, Hunger RM, Carver BF, Yan L (2020) Development and deployment of KASP markers for multiple alleles of *Lr34* in wheat. Theor Appl Genet:1–13
Feng J, Wang M, See DR, Chao S, Zheng Y, Chen X (2018) Characterization of novel gene Yr79 and four additional quantitative trait loci for all-stage and high-temperature adult-plant resistance to stripe rust in spring wheat PI 182103. Phytopathology 108(6):737–747
Figueroa M, Upadhyaya NM, Sperschneider J, Park RF, Szabo LJ, Steffenson B, Ellis JG, Dodds PN (2016) Changing the game: using integrative genomics to probe virulence mechanisms of the stem rust pathogen *Puccinia graminis* f sp *tritici*. Front Plant Sci 7:205
Figueroa M, Kim EH, Peter SS (2018) A review of wheat diseases-a field perspective. Mol Plant Pathol 19(6):1523–1536
Flor HH (1971) Current status of the gene-for-gene concept. Annu Rev Phytopathol 9:275–296
Flor HH (1995) Host parasite in flax rust- its genetics and other implications. Phytopathology 45:680–685

Gautam T, Dhillon GS, Saripalli G, Singh VP, Prasad P, Kaur S, Gupta PK (2020) Marker-assisted pyramiding of genes/QTL for grain quality and rust resistance in wheat *(Triticum aestivum* L). Mol Breed 40:1–14

Gebrewahid TW, Zhou Y, Zhang P, Ren Y, Pu G, Xia X, Liu D (2020) Mapping of Stripe Rust and Leaf Rust Resistance Quantitative Trait Loci in the Chinese Spring Wheat Line Mianyang351-15. [published online ahead of print, 2020 Apr 6] Phytopathology, PHYTO08190316R. https://doi.org/10.1094/PHYTO-08-19-0316-R

Gessese MK (2019) Description of wheat rusts and their virulence variations determined through annual pathotype surveys and controlled multi-Pathotype tests. Adv Agric. https://doi.org/10.1155/2019/2673706

Gill HS, Li C, Sidhu JS, Liu W, Wilson D, Bai G, Sehgal SK (2019) Fine mapping of the wheat leaf rust resistance gene *Lr42*. Int J Mol Sci 20(10):2445

Gupta SK, Charpe A, Koul S, Prabhu KV, Haq QMR (2005) Development and validation of molecular markers linked to an *Aegilops umbellulata*–derived leaf-rust-resistance gene, *Lr9*, for marker-assisted selection in bread wheat. Genome 48(5):823–830

Gupta PK, Kumar J, Mir RR, Kumar A (2010) Marker-assisted selection as a component of conventional Plant Breed. Plant Breed Rev 33:145–217

Gupta N, Batra N, Bhardwaj SC (2017) Wheat rust research-Status, efforts and way ahead. J Wheat Res 9(2):72–86

Hall D, Tegström C, Ingvarsson PK (2010) Using association mapping to dissect the genetic 1030 basis of complex traits in plants. Brief Funct Genomic Proteomic 9:157–165

Hiebert CW, Kassa MT, McCartney CA, You FM, Rouse MN, Fobert P, Fetch TG (2016) Genetics and mapping of seedling resistance to Ug99 stem rust in winter wheat cultivar Triumph 64 and differentiation of SrTmp, SrCad, and Sr42. Theor Appl Genet 129(11):2171–2177. https://doi.org/10.1007/s00122-016-2765-4

Hiebert CW, Rouse MN, Nirmala J, Fetch T (2017) Genetic mapping of stem rust resistance to *Puccinia graminis* f sp *tritici* race TRTTF in the Canadian wheat cultivar Harvest. Phytopathology 107(2):192–197

Hou L, Chen X, Wang M, See DR, Chao S, Bulli P, Jing J (2015) Mapping a large number of QTL for durable resistance to stripe rust in winter wheat Druchamp using SSR and SNP markers. PLoS One 10(5):e0126794

Hubbard A, Lewis CM, Yoshida K, Ramirez-Gonzalez RH, de Vallavieille-Pope C, Thomas J, Kamoun S, Bayles R, Uauy C, Saunders DGO (2015) Field pathogenomics reveals the emergence of a diverse wheat yellow rust population. Genome Biol 16:23

Johnson R (1983) Genetic background of durable resistance. In: Lamberti F, Waller JM, Van der Graaff NA (eds) Durable resistance in crops. Plenum Press, New York, pp 5–26

Johnson R, Law CN (1973) Cytogenetic studies of the resistance of the wheat variety Bersée to *Puccinia striiformis*. Cereal Rusts Bull 1:38–43

Kandel JS, Krishnan V, Jiwan D, Chen X, Skinner DZ, See DR (2017) Mapping genes for resistance to stripe rust in spring wheat landrace PI 480035. PloS One 12(5):e0177898

Kassa MT, You FM, Fetch TG, Fobert P, Sharpe A, Pozniak CJ, Luo MC (2016) Genetic mapping of SrCad and SNP marker development for marker-assisted selection of Ug99 stem rust resistance in wheat. Theor Appl Genet 129(7):1373–1382

Kassa MT, You FM, Hiebert CW, Pozniak CJ, Fobert PR, Sharpe AG, McCallum BD (2017) Highly predictive SNP markers for efficient selection of the wheat leaf rust resistance gene *Lr16*. BMC Plant Biol 17(1):45

Klindworth DL, Saini J, Long Y, Rouse MN, Faris JD, Jin Y, Xu SS (2017) Physical mapping of DNA markers linked to stem rust resistance gene Sr47 in durum wheat. Theor Appl Genet 130(6):1135–1154

Kolmer JA (2015) A QTL on chromosome 5BL in wheat enhances leaf rust resistance of *Lr46*. Mol Breed 35(2):74

Kolmer A, Ordonez ME, Manisterski J, Anikster Y (2011) Genetic differentiation of *Puccinia triticina* populations in the middle east and genetic similarity with population in Central Asia. Phytopathology 101:870–877

Kolmer JA, Bernardo A, Bai G, Hayden MJ, Chao S (2018a) Adult plant leaf rust resistance derived from Toropi wheat is conditioned by *Lr78* and three minor QTL. Phytopathology 108(2):246–253

Kolmer JA, Chao S, Brown-Guedira G, Bansal U, Bariana H (2018b) Adult plant leaf rust resistance derived from the soft red winter wheat cultivar 'Caldwell' maps to chromosome 3BS. Crop Sci 58(1):152–158

Kolmer JA, Su Z, Bernardo A, Bai G, Chao S (2018c) Mapping and characterization of the new adult plant leaf rust resistance gene *Lr77* derived from Santa Fe winter wheat. Theor Appl Genet 131(7):1553–1560

Kolmer JA, Su Z, Bernardo A, Bai G, Chao S (2019) A backcross line of thatcher wheat with adult plant leaf rust resistance derived from duster wheat has *Lr46* and *Lr77*. Phytopathology 109(1):127–132

Koujalagi D, Desai SA, Biradar S, Savitha BS, Kumar KJ, Naik R, Satish TN (2019) Genetic enhancement of leaf rust resistance in bread wheat (*Triticum aestivum* L) through marker-assisted introgression of *Lr24* and *Lr28*. Appl Biol Res 21(1):1–9

Kthiri D, Loladze A, N'Diaye A, Nilsen KT, Walkowiak S, Dreisigacker S, Pozniak CJ (2019) Mapping of genetic loci conferring resistance to leaf rust from three globally resistant durum wheat sources. Front Plant Sci 10:1247

Kumar J, Mir RR, Kumar N, Kumar A, Mohan A, Prabhu KV, Gupta PK (2010) Marker-assisted selection for pre-harvest sprouting tolerance and leaf rust resistance in bread wheat. Plant Breed 129(6):617–621

Kuraparthy V, Sood S, See DR, Gill BS (2009) Development of a PCR assay and marker-assisted transfer of leaf rust and stripe rust resistance genes *Lr57* and *Yr40* into hard red winter wheats. Crop Sci 49(1):120–126

Kuraparthy V, Sood S, Guedira GB, Gill BS (2011) Development of a PCR assay and marker-assisted transfer of leaf rust resistance gene *Lr58* into adapted winter wheats. Euphytica 180(2):227–234

Lagudah ES (2011) Molecular genetics of race non-specific rust resistance in wheat. Euphytica 179(1):81–91

Lan C, Zhang Y, Herrera-Foessel SA, Basnet BR, Huerta-Espino J, Lagudah ES, Singh RP (2015) Identification and characterization of pleiotropic and co-located resistance loci to leaf rust and stripe rust in bread wheat cultivar Sujata. Theor Appl Genet 128(3):549–561

Lan C, Basnet BR, Singh RP, Huerta-Espino J, Herrera-Foessel SA, Ren Y, Randhawa MS (2017a) Genetic analysis and mapping of adult plant resistance loci to leaf rust in durum wheat cultivar Bairds. Theor Appl Genet 130(3):609–619

Lan C, Hale IL, Herrera-Foessel SA, Basnet BR, Randhawa MS, Huerta-Espino J, Singh RP (2017b) Characterization and mapping of leaf rust and stripe rust resistance loci in hexaploid wheat lines UC1110 and PI610750 under Mexican environments. Front Plant Sci 8:1450

Leonard KJ, Szabo LJ (2005) Stem rust of small grains and grasses caused by *Puccinia graminis*. Mol Plant Pathol 6(2):99–111

Li C, Wang Z, Li C, Bowden R, Bai G, Li C, Carver BF (2017) Mapping of quantitative trait loci for leaf rust resistance in the wheat population Ning7840× Clark. Plant Dis 101(12):1974–1979

Li H, Dong Z, Ma C, Tian X, Qi Z, Wu N, Li T (2019) Physical mapping of stem rust resistance gene *Sr52* from Dasypyrum villosum based on ph1b-induced homoeologous recombination. Int J Mol Sci 20(19):4887

Liu J, He Z, Wu L, Bai B, Wen W, Xie C, Xia X (2015) Genome-wide linkage mapping of QTL for adult-plant resistance to stripe rust in a Chinese wheat population Linmai 2× Zhong 892. PLoS One 10(12):e0145462

Liu R, Lu J, Zhou M, Zheng S, Liu Z, Zhang C, Zhang L (2020) Developing stripe rust resistant wheat (*Triticum aestivum* L) lines with gene pyramiding strategy and marker-assisted selection. Genet Resour Crop Evol 67(2):381–391

Lu Y, Bowden RL, Zhang G, Xu X, Fritz AK, Bai G (2017) Quantitative trait loci for slow-rusting resistance to leaf rust in doubled-haploid wheat population CI13227× Lakin. Phytopathology 107(11):1372–1380

Ma J, Qin N, Cai B, Chen G, Ding P, Zhang H, Liu Y (2019) Identification and validation of a novel major QTL for all-stage stripe rust resistance on 1BL in the winter wheat line 20828. Theor Appl Genet 132(5):1363–1373

Mallick N, Sharma JB, Tomar RS, Sivasamy M, Prabhu KV (2015) Marker-assisted backcross breeding to combine multiple rust resistance in wheat. Plant Breed 134(2):172–177

McCallum BD, Hiebert CW, Cloutier S, Bakkeren G, Rosa SB, Humphreys DG, Marais GF, McCartney CA, Panwar V, Rampitsch C, Saville BJ, Wang X (2016) A review of wheat leaf rust research and the development of resistant cultivars in Canada. Can J Plant Pathol 38:1–18

McIntosh RA, Wellings RC, Park R (1995) Wheat rusts an atlas of resistance genes CSIRO East Melbourne 200pp

McIntosh RA, Dubcovsky J, Rogers WJ, Morris C, Xia XC (2017) Catalogue of gene 1155 symbols for wheat: 2017 supplement *Komugi Wheat Genet Resour Database* Available at: 1156 https://shigennigacjp/wheat/komugi/genes/macgene/ supplement2017pdf

Mondal S, Rutkoski JE, Velu G, Singh PK, Crespo-Herrera LA, Guzman C, Singh RP (2016) Harnessing diversity in wheat to enhance grain yield climate resilience disease and insect pest resistance and nutrition through conventional and modern breeding approaches. Front Plant Sci 7:991

Mujeeb-Kazi A, Kazi AG, Dundas I, Rasheed A, Ogbonnaya F, Kishii M, Bonnett D, Wang RR, Xu S, Chen P, Mahmood T (2013) Genetic diversity for wheat improvement as a conduit to food security. In: Advances in agronomy, vol 122. Academic, pp 179–257

Nirmala J, Chao S, Olivera P, Babiker EM, Abeyo B, Tadesse Z, Pumphrey MO (2016) Markers linked to wheat stem rust resistance gene *Sr11* effective to *Puccinia graminis* f sp *tritici* race TKTTF. Phytopathology 106(11):1352–1358

Nocente F, Gazza L, Pasquini M (2007) Evaluation of leaf rust resistance genes *Lr1 Lr9 Lr24 Lr47* and their introgression into common wheat cultivars by marker-assisted selection. Euphytica 155(3):329–336

Nsabiyera V, Qureshi N, Bariana HS, Wong D, Forrest KL, Hayden MJ, Bansal UK (2016) Molecular markers for adult plant leaf rust resistance gene *Lr48* in wheat. Mol Breed 36(6):65

Nsabiyera V, Bariana HS, Qureshi N, Wong D, Hayden MJ, Bansal UK (2018) Characterisation and mapping of adult plant stripe rust resistance in wheat accession Aus27284. Theor Appl Genet 131:1459–1467. https://doi.org/10.1007/s00122-018-3090-x

Nsabiyera V, Baranwal D, Qureshi N, Kay P, Forrest K, Valárik M, Bansal UK (2020) Fine mapping of *Lr49* using 90K SNP Chip array and flow-sorted chromosome sequencing in wheat. Front Plant Sci 10:1787

Olivera Firpo P, Newcomb M, Flath K, Sommerfeldt-Impe N, Szabo L, Carter M, Luster D, Jin Y (2017) Characterization of *Puccinia graminis* f sp *tritici* isolates derived from an unusual wheat stem rust outbreak in Germany in 2013. Plant Pathol 66:1258–1266. https://doi.org/10.1111/ppa12674

Pakeerathan K, Bariana H, Qureshi N, Wong D, Hayden M, Bansal U (2019) Identification of a new source of stripe rust resistance *Yr82* in wheat. Theor Appl Genet 132(11):3169–3176

Park R, Fetch T, Hodson D, Jin Y, Nazari K, Prashar M, Pretorius Z (2011) International surveillance of wheat rust pathogens progress and challenges. Euphytica 179:109–117

Parlevliet JE (1993) What is durable resistance a general outline. In: Jacobs T, Parlevliet JE (eds) Durability of disease resistance. Kluwer Academic Publishers, Dordrecht, pp 23–29

Parlevliet JE (2002) Durability of resistance against fungal bacterial and viral pathogens; present situation. Euphytica 124:147–156

Pietrusińska A, Czembor JH, Czembor PC (2011) Pyramiding two genes for leaf rust and powdery mildew resistance in common wheat. Cereal Res Commun 39(4):577–588

Ponce-Molina LJ, Huerta-Espino J, Singh RP, Basnet BR, Alvarado G, Randhawa MS, García-Zavala JJ (2018) Characterization of leaf rust and stripe rust resistance in spring wheat 'Chilero'. Plant Dis 102(2):421–427

Prasad SV, Singh SK, Ambati VD, Prakasha TL, Singh JB, Dubey VG, Mishra AN (2014) Introgression of stem rust resistance gene *Sr36* into durum wheat back ground using marker assisted backcross breeding. J Wheat Res 6(1):21–24

Qie Y, Liu Y, Wang M, Li X, See DR, An D, Chen X (2018) Development validation and re-selection of wheat lines with pyramided genes *Yr64* and *Yr15* linked on the short arm of chromosome 1B for resistance to stripe rust. Plant Dis 103(1):51–58

Qureshi N, Bariana H, Forrest K, Hayden M, Keller B, Wicker T, Bansal U (2017a) Fine mapping of the chromosome 5B region carrying closely linked rust resistance genes *Yr47* and *Lr52* in wheat. Theor Appl Genet 130(3):495–504

Qureshi N, Bariana H, Kolmer JA, Miah H, Bansal U (2017b) Genetic and molecular characterization of leaf rust resistance in two durum wheat landraces. Phytopathology 107(11):1381–1387

Qureshi N, Bariana H, Kumran VV, Muruga S, Forrest KL, Hayden MJ, Bansal U (2018a) A new leaf rust resistance gene *Lr79* mapped in chromosome 3BL from the durum wheat landrace Aus26582. Theor Appl Genet 131(5):1091–1098

Qureshi N, Kandiah P, Gessese MK, Nsabiyera V, Wells V, Babu P, Bansal U (2018b) Development of co-dominant KASP markers co-segregating with Ug99 effective stem rust resistance gene *Sr26* in wheat. Mol Breed 38(8):97

Rana M, Sood A, Hussain W, Kaldate R, Sharma TR, Gill RK, Kumar S, Singh S (2019) Gene pyramiding and multiple character breeding. In: Singh M (ed) Lentils potential resources for enhancing genetic gains. Academic, pp 83–124

Randhawa HS, Mutti JS, Kidwell K, Morris CF, Chen X, Gill KS (2009) Rapid and targeted introgression of genes into popular wheat cultivars using marker-assisted background selection. PLoS One 4(6)

Randhawa MS, Bains NS, Sohu VS, Chhuneja P, Trethowan RM, Bariana HS, Bansal U (2019) Marker assisted transfer of stripe rust and stem rust resistance genes into four wheat cultivars. Agronomy 9(9):497

Ren Y, Singh RP, Basnet BR, Lan CX, Huerta-Espino J, Lagudah ES, Ponce-Molina LJ (2017) Identification and mapping of adult plant resistance loci to leaf rust and stripe rust in common wheat cultivar Kundan. Plant Dis 101(3):456–463

Revathi P, Tomar SMS, Singh NK (2010) Marker assisted gene pyramiding of leaf rust resistance genes Lr24 Lr28 along with stripe rust resistance gene *Yr15* in wheat (*Triticum aestivum* L). Indian J Genet Plant Breed 70(4):349–354

Riar AK, Kaur S, Dhaliwal HS, Singh K, Chhuneja P (2012) Introgression of a leaf rust resistance gene from *Aegilops caudata* to bread wheat. J Genet 91(2):155–161

Riaz A, Periyannan S, Hickey L (2016) A rapid phenotyping method for adult plant resistance to leaf rust in wheat. Plant Methods 12:17

Sadeghabad AA, Dadkhodaie A, Heidari B, Razi H, Mostowfizadeh-Ghalamfarsa R (2017) Microsatellite markers for the *Triticum timopheevi*-derived leaf rust resistance gene *Lr18* on wheat 5BL chromosome. Breed Sci 67(2):129–134

Samsampour D, Zanjani BM, Singh A, Pallavi JK, Prabhu KV (2009) Marker assisted selection to pyramid seedling resistance gene Lr24 and adult plant resistance gene *Lr48* for leaf rust resistance in wheat. Indian J Genet Plant Breed 69(1):1

Savitha BS, Desai SA, Biradar SS, Naik R, Lohithaswa HC (2016) Identification of SSR marker linked to leaf rust resistant gene *Lr24* and marker assisted transfer of leaf rust resistance genes into bread wheat. J Agric Sci Technol 18:545–552

Schumann GL, D'Arcy CJ (2006) Essential plant pathology. APS Press, American Phytopathological Society

Sharma JS, Zhang Q, Rouse MN, Klindworth DL, Friesen TL, Long Y, Faris JD (2019) Mapping and characterization of two stem rust resistance genes derived from cultivated emmer wheat accession PI 193883. Theor Appl Genet 132(11):3177–3189

Singh R, Datta D, Singh S, Tiwari R (2004) Marker-assisted selection for leaf rust resistance genes *Lr19* and *Lr24* in wheat (*Triticum aestivum* L). J Appl Genet 45(4):399–404

Singh RP, Hodson DP, Huerta-Espino J, Jin Y, Bhavani S, Njau P, Govindan V (2011a) The emergence of Ug99 races of the stem rust fungus is a threat to world wheat production. Annu Rev Phytopathol 49:465–481

Singh RP, Huerta-Espino J, Bhavani S, Herrera-Foessel SA, Singh D, Singh PK, Velu G, Mason RE, Jin Y, Njau P, Crossa J (2011b) Race non-specific resistance to rust diseases in CIMMYT spring wheats. Euphytica 179(1):175–186

Singh RP, Hodson DP, Jin Y, Lagudah ES, Ayliffe MA, Bhavani S, Rouse MN, Pretorius ZA, Szabo LJ, Huerta-Espino J, Basnet BR (2015) Emergence and spread of new races of wheat stem rust fungus: continued threat to food security and prospects of genetic control. Phytopathology 105:872–884

Singh M, Mallick N, Chand S, Kumari P, Sharma JB, Sivasamy M, Jha SK (2017) Marker-assisted pyramiding of Thinopyrum-derived leaf rust resistance genes *Lr19* and *Lr24* in bread wheat variety HD2733. J Genet 96(6):951–957

Singh A, Jaiswal JP, Badoni S (2018) Enhancing rust resistance in wheat through marker assisted backcross breeding. Indian J Genet 78(1):19–25

Singla J, Lüthi L, Wicker T, Bansal U, Krattinger SG, Keller B (2017) Characterization of *Lr75*: a partial broad-spectrum leaf rust resistance gene in wheat. Theor Appl Genet 130(1):1–12

Sivasamy M, Tiwari S, Tomar RS, Singh B, Sharma JB, Tomar SMS, Chand S (2009) Introgression of useful linked genes for resistance to stem rust leaf rust and powdery mildew and their molecular validation in wheat (*Triticum aestivum* L). Indian J Genet Plant Breed 69(1):17

Thach T, Ali S, de Vallavieille-Pope C, Justesen AF, Hovmøller MS (2016) Worldwide population structure of the wheat rust fungus *Puccinia striiformis* in the past. Fungal Genet Biol 87:1–8

Tian Y, Zhan G, Chen X, Tungruentragoon A, Lu X, Zhao J, Huang L, Kang Z (2016) Virulence and simple sequence repeat marker segregation in a *Puccinia striiformis* f. sp. *tritici* population produced by selfing a Chinese isolate on *Berberis shensiana*. Phytopathology 106(2):185–191

Tiwari S, Tomar RS, Chand S, Singh NK (2014) Combining multiple rust resistance genes by phenotypic and marker assisted selection in wheat (*Triticum aestivum* L). Indian J Genet Plant Breed 74:181–188

Tomar SMS, Singh SK, Sivasamy M, Vinod (2014) Wheat rusts in India: resistance breeding and gene deployment-a review. Indian J Genet 74(2):129–156

Tyagi S, Mir RR, Kaur H, Chhuneja P, Ramesh B, Balyan HS, Gupta PK (2014) Marker-assisted pyramiding of eight QTLs/genes for seven different traits in common wheat (*Triticum aestivum* L). Mol Breed 34(1):167–175

VanderPlank JE (1975) Horizontal resistance: six suggested projects in relation to blast disease of rice. In: Horizontal resistant to blast disease in rice, CIAT series, CE-9, Cali, Colombia, pp 21–26

Vanderplank JE (2012) Disease resistance in plants. Elsevier

Vida G, Gál M, Uhrin A, Veisz O, Syed NH, Flavell AJ, Bedő Z (2009) Molecular markers for the identification of resistance genes and marker-assisted selection in breeding wheat for leaf rust resistance. Euphytica 170(1–2):67–76

Walter S, Ali S, Kemen E, Nazari K, Bahri BA, Enjalbert J, Hansen JG, Brown JK, Sicheritz-Pontén T, Jones J, de Vallavieille-Pope C (2016) Molecular markers for tracking the origin and worldwide distribution of invasive strains of *Puccinia striiformis*. Ecol Evol 9:2790–2804

Wang J, Li Z, Shi L, Zhu L, Ren Z, Li X, Shah SJA (2015) QTL mapping for adult-plant leaf rust resistance genes in Chinese wheat cultivar Weimai 8. Czech J Genet Plant Breed 51(3):79–85

Wang Z, Zhao J, Chen X, Peng Y, Ji J, Zhao S, Lv Y, Huang L, Kang Z (2016) Virulence variations of *Puccinia striiformis* f sp *tritici* isolates collected from *Berberis spp* in China. Plant Dis 100:131–138

Wang C, Hu S, Gardner C, Lübberstedt T (2017) Emerging avenues for utilization of exotic germplasm. Trends Plant Sci 22(7):624–637

Wiersma AT, Brown LK, Brisco EI, Liu TL, Childs KL, Poland JA, Sehgal SK, Olson, EL (2016) Fine mapping of the stem rust resistance gene SrTA10187. Theor Appl Genet 129(12):2369–2378. https://doi.org/10.1007/s00122-016-2776-1

Wu J, Huang S, Zeng Q, Liu S, Wang Q, Mu J, Kang Z (2018a) Comparative genome-wide mapping versus extreme pool-genotyping and development of diagnostic SNP markers linked to QTL for adult plant resistance to stripe rust in common wheat. Theor Appl Genet 131(8):1777–1792

Wu J, Wang Q, Xu L, Chen X, Li B, Mu J, Kang Z (2018b) Combining single nucleotide polymorphism genotyping array with bulked segregant analysis to map a gene controlling adult plant resistance to stripe rust in wheat line 03031-1-5 H62. Phytopathology 108(1)1:03-113

Yadav PS, Mishra VK, Arun B, Chand R, Vishwakarma MK, Vasistha NK, Joshi AK (2015) Enhanced resistance in wheat against stem rust achieved by marker assisted backcrossing involving three independent *Sr* genes. Curr Plant Biol 2:25–33

Yadawad A, Hanchinal RR, Nadaf HL, Desai SA, Biradar S, Rudranaik V (2015) Improvement of leaf rust resistance in bread wheat variety DWR162 (*Triticum aestivum* L) through marker assisted selection. Indian J Genet 75(2):256–259

Yadawad A, Gadpale A, Hanchinal RR, Nadaf HL, Desai SA, Biradar S, Naik VR (2017) Pyramiding of leaf rust resistance genes in bread wheat variety DWR 162 through marker assisted backcrossing. Indian J Genet 77(2):251–257

Yaniv E, Raats D, Ronin Y, Korol AB, Grama A, Bariana H, Fahima T (2015) Evaluation of marker-assisted selection for the stripe rust resistance gene *Yr15* introgressed from wild emmer wheat. Mol Breed 35(1):43

Yu G, Zhang Q, Friesen TL, Rouse MN, Jin Y, Zhong S, Xu SS (2015) Identification and mapping of *Sr46* from *Aegilops tauschii* accession CIae 25 conferring resistance to race TTKSK (Ug99) of wheat stem rust pathogen. Theor Appl Genet 128(3):431–443

Yuan C, Singh RP, Liu D, Randhawa MS, Huerta-Espino J, Lan C (2020) Genome-wide mapping of adult plant resistance to leaf rust and stripe rust in CIMMYT wheat line Arableu# 1. [published online ahead of print 2020 Mar 20] *Plant Dis* 2020;PDIS10192198RE

Zhang P, Yin G, Zhou Y, Qi A, Gao F, Xia X, Liu D (2017) QTL mapping of adult-plant resistance to leaf rust in the wheat cross Zhou 8425B/Chinese Spring using high-density SNP markers. Front Plant Sci 8:793

Zhang P, Lan C, Asad MA, Gebrewahid TW, Xia X, He Z, Liu D (2019a) QTL mapping of adult-plant resistance to leaf rust in the Chinese landraces Pingyuan 50/Mingxian 169 using the wheat 55K SNP array. Mol Breed 39(7):98

Zhang R, Singh RP, Lillemo M, He X, Randhawa MS, Huerta-Espino J, Lan C (2019b) Two main stripe rust resistance genes identified in synthetic-derived wheat line Soru#1. Phytopathology 109(1):120–126

Zhang P, Li X, Gebrewahid TW, Liu H, Xia X, He Z, Liu D (2019c) QTL mapping of adult-plant resistance to leaf and stripe rust in wheat cross SW 8588/Thatcher using the wheat 55K SNP Array. Plant Dis 103(12):3041–3049

Zhao J, Meinan W, Xianming C, Zhensheng K (2016) Role of alternate hosts in epidemiology and pathogen variation of cereal rusts. Annu Rev Phytopathol 54:207–228

# Genome Editing and Trait Improvement in Wheat

**Monika Bansal, Suruchi Jindal, Shabir H. Wani, Showkat Ahmad Ganie, and Ravinder Singh**

**Abstract** Wheat is a major food source for people around the world. During the last decade, an increase in wheat productivity has been observed with the development of novel varieties by using a combination of mutational and molecular breeding approaches. Despite this progress, several environmental factors including biotic and abiotic stresses negatively affect wheat productivity; these include the emergence of new pests and pathogens, global climate change and multiple environmental issues. Keeping all these challenges in mind, there is an urgent need to produce more amount of wheat to alleviate hunger of large and rapidly growing population. In the recent past, advances in genomics and genome editing technologies by the use of engineered nucleases have brought revolution in the field of agriculture. Among several different genome-modifying tools, the CRISPR/Cas9 system is the recent and widely used genome modification tool because it is simple and highly efficient technology. CRISPR/CAS9 along with its variants has immense potential to develop new wheat varieties with higher yield potential. In the present review, we will shed light on the application of genome editing to overcome major challenges and the assessment of its future implications for the improvement of wheat grains, both qualitatively and quantitatively.

**Keywords** Wheat · Genetic engineering · Genome editing · Zinc-finger nucleases · TALENs · CRISPR/Cas

M. Bansal · S. Jindal
School of Agricultural Biotechnology, Punjab Agricultural University, Ludhiana, Punjab, India

S. H. Wani
Mountain Research Centre for Field Crops, Khudwani, Anantnag, Jammu and Kashmir, India

Sher-e-Kashmir University of Agricultural Sciences and Technology of Kashmir, Shalimar, Badgam, Jammu and Kashmir, India

S. A. Ganie
Department of Biotechnology, Visva-Bharati, Santiniketan, West Bengal, India

R. Singh (✉)
School of Biotechnology, Sher-e-Kashmir University of Agricultural Sciences and Technology of Jammu, Jammu, India

© The Editor(s) (if applicable) and The Author(s), under exclusive license to Springer Nature Switzerland AG 2021

S. H. Wani et al. (eds.), *Physiological, Molecular, and Genetic Perspectives of Wheat Improvement*, https://doi.org/10.1007/978-3-030-59577-7_12

# 1 Introduction

Wheat is one of the major food crops to achieve world food security. Wheat alone fulfils about one-fifth of the daily calories and protein demand for the human population; therefore, making it one of the most important protein sources in the world (Shiferaw et al. 2013). However, the ever expanding global human population would need wheat yield increased to about 5 tonnes/ha from its present status of 3.3 tonnes by 2050. Understanding and manipulating the genetic architecture of wheat would help in the development of better yielding varieties. The whole genome sequence coupled with gene mapping studies helps to identify large number of new genes related to (a) resistance to both biotic and abiotic stresses, (b) yield contributing traits, and (c) improved grain quality in wheat. The analysis of sequences of specific key genes involved in growth and development of wheat among large number of wheat accessions would help to unravel new alleles for breeding of promising varieties. Despite the availability of huge amount of genomics information and resources, wheat still lags behind in terms of application of genomic/genetic engineering tools for its improvement as compared to other cereals like maize and rice (Uauy 2017).

# 2 History of Plant Genetic Engineering

The plants have evolved naturally through the process of polyploidization and associated changes both at the genome and the chromosomal levels over tens of thousands of years. Humans have also contributed towards domestication of crops through artificial selection over the past nearly 10,000 years. The crops being grown even today have accumulated considerable genetic variation which is key to crop improvement in the course of evolution. Researchers have also used mutagens for inducing mutations in DNA and then screen the populations for the variation in phenotypes (Shu et al. 2012). The concept of mutation breeding introduced during the 1940s has shown huge success, as in case of wheat varieties with remarkably improved yield that was crucial for bringing about the Green Revolution during the 1970s.

The discovery of *Agrobacterium tumefaciens* responsible for crown gall disease and restriction enzymes laid the foundation of recombinant DNA (rDNA) technology in plants. These tools made it possible to transfer genes even from distantly related organisms leading to the development of transgenics. Despite having huge potential, the rDNA has many drawbacks including the insertion of gene of interest at random places within genome, therefore resulting into a possible disruption of non-target genes.

Later in the 1980s, a number of methods were proposed for the targeted gene editing based on the application of double-strand DNA breaks (DSBs) to target loci in genomes by generating specific DSBs at desired sites (Capecchi 1980; Jasin and

Liang 1996). This method employs innate DNA repair mechanism of cells to undertake targeted gene editing either by not so precise non-homologous end joining (NHEJ) repair or by highly precise repair by homology-directed repair (HDR) (Trevino and Zhang 2014; Baltes and Voytas 2015; Bortesi and Fischer 2015; Schaart et al. 2016).

In protein-DNA interactions, the DNA-binding domains bind specifically to desired sequence in the target genome. This binding domain is fused with another domain with a cleavage activity to cut the DNA at specific location on the target site. These fusion proteins are responsible for precise genome editing, as the site-specific binding proteins are capable of regulating transcription and epigenetic traits and induce base editing changes (Komor et al. 2016; Puchta 2016). These genome editing toolboxes are zinc-finger nucleases (ZFNs), TALEN's and CRISPR/Cas systems.

## 3 Nucleases and Genome Modifications

Nucleases are a class of endonucleases that have been modified to use in genetic engineering. Most of these endonucleases have DNA non-specific activity, but when modified by fusing them to another protein(s) containing DNA-binding sites (DBDs), these can bring about specific nicks in the DNA. These nicks in the DNA are repaired by endogenous DNA repair mechanism either by NHEJ or HDR to induce sequence specific mutation. These mutations can lead to the insertion or deletion of nucleotides, thereby resulting into gene knockouts through NHEJ or cause gene replacements and insertions through HDR (Fig. 1).

These nucleases facilitated successful application of genome editing in diverse fields of life sciences. They have been able to disrupt specific sequence in genes by adding either single or few exogenous bases into intended genomic sites, thereby exhibiting potential to improve performance of agricultural products. The site-specific nucleases used for gene editing in plants can be classified into four major classes: meganucleases (MNs), zinc-finger nucleases (ZFNs), transcription activator-like effector nucleases (TALENs) and clustered regularly interspaced short palindromic repeats/CRISPR-associated protein 9 (CRISPR/Cas9).

### *Meganucleases*

Meganucleases also known as homing nucleases are DNA-cleaving enzymes that mobilize their own reading frames (mobility of self-splicing introns from an intron containing allele to an intron minus allele) by generating double-strand breaks at specific genomic invasion sites (Fig. 2) of size between 14 and 40 bp. These are highly specific endonucleases that have the ability to induce homologous recombination in both mammalian and plant cells. Meganucleases were among the first reported sequence specific nucleases (SSNs) which were used to induce DSBs at

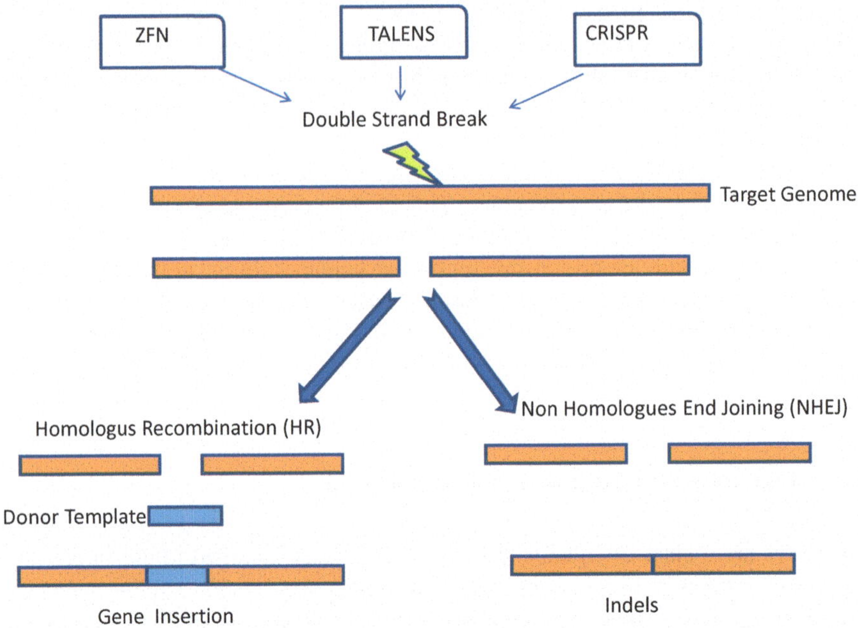


**Fig. 1 Gene editing by site-specific nucleases**: Nucleases (TALENS, zinc-fingers, CRISPR) induce site-specific double-strand breaks of the target DNA. Double-strand breaks are repaired by either non-homologous end joining (NHEJ) or homology-directed repair (HDR). In the NHEJ-mediated repair, there can be either insertion (blue) or deletion of few nucleotides (indel). HDR-mediated repair can introduce mutations, by insertions or replacement on the basis of donor DNA template

precise target location in eukaryotic genomes. The DSBs can be repaired by either NHEJ or HDR (Roth et al. 1992). The meganucleases are difficult to engineer (modify) due to the presence of both DNA recognition and cleavage domains intertwined in a single protein. Meganucleases are challenging to redesign for new target specificity, because redesigning is hindered by the non-modular nature of the protein. As a result, the use of meganucleases has been limited to a few naturally occurring meganucleases (e.g. I-*Sce*I, I-*Cre*I) or to redesigned nucleases made by some research groups with strong expertise in this field.

## *Zinc-Finger Nucleases*

Zinc-finger nucleases are chimeric fusion proteins, which consist of $C_2H_2$ zinc-finger DNA-binding domain fused to *Fok* I nuclease domain. The DNA-binding domain consists of three zinc-fingers; each of them recognizes about 3 bp of target DNA. As a result, ZF domain recognizes a total of 9-bp sequence. Zinc-finger

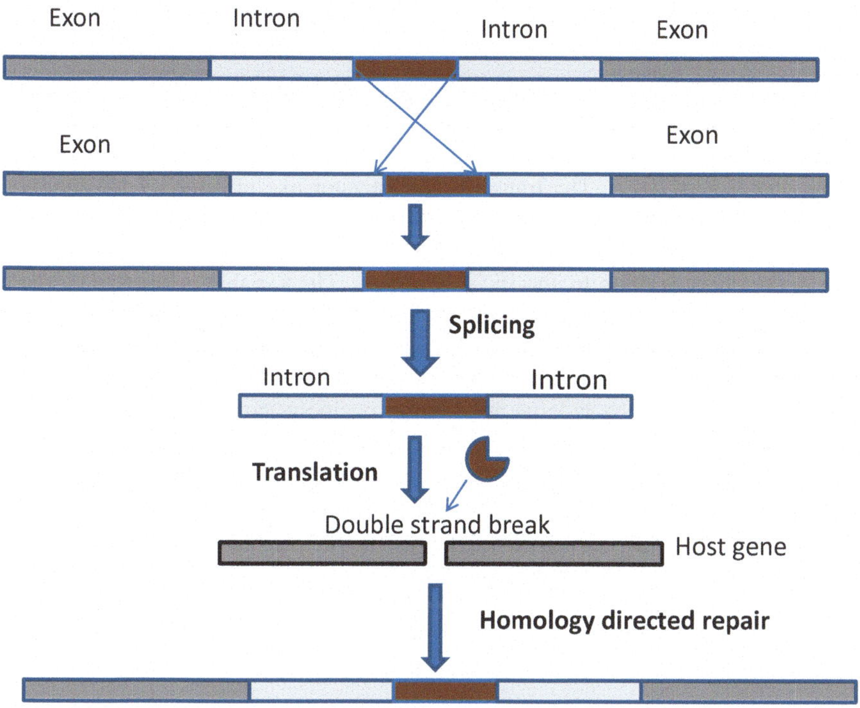


**Fig. 2** Schematic mechanism of homing nucleases: Homing endonuclease gene (maroon bar) is present inside a self-splicing intron within a host gene (grey bars). The homing endonuclease is translated as an independent protein from intron and cleaves a target site found in the homologous allele of the host gene that previously does not contain this sequence and induces repair via homologous recombination (HR)

nucleases always attach to target DNA as a dimer, so a target of three-finger ZFNs dimer becomes 18 bp. DNA-cleavage domain consists of *Fok* I catalytic domain; *Fok* I is a Type IIS restriction enzyme, which has two protein domains one for recognition and other for nuclease activity. *Fok* I domain is the most important component of ZFN which is responsible for targeted cleavage in a complex genome (Vanamee et al. 2001). The major limitations of ZFN-mediated genome modifications are that it is time-consuming and comparatively less efficient and has low reproducibility in germ cells as compared to somatic cells (Ramirez et al. 2008). There are some reports about polyploid crops which were successfully edited with ZFNs including tobacco (Townsend et al. 2009) and rapeseed (Gupta et al. 2012). Some other disadvantages of using ZFNs include its inability to recognize DNA sequence with high specificity, and it may generate many unwanted off targets within host target genome.

## *Transcription Activator-Like Effector Nucleases (TALENs)*

The bacteria Xanthomonas releases certain proteins in plant systems to induce susceptibility genes leading to development of disease. These proteins called as Transcription activator-like effectors (TALEs) often bind to transcription factors (Boch and Bonas 2010). The transcription activator-like effector's DNA-binding domain has a central repeat sequence of ~34 amino acids which has two highly variable residues at 12$^{th}$ and 13$^{th}$ positions (Fig. 3) that form a loop in the proteins and help in target recognition to specific DNA bases. The variable positions (12$^{th}$ and 13$^{th}$) among repeat unit are known as repeat variable di-residues (RVD) recognize one nucleotide. This property of TALEs was used for gene editing by fusing the TALEs to *Fok I* nuclease, leading to formation of TALE nucleases (TALENs). The size of target in TALENs is large and highly specific as compared to other nucleases, because TALEN monomers are designed with 5–20 repeat variable residues (Mussolino et al. 2014). DNA-binding domains of TALENs can easily be engineered to recognize virtually any DNA sequence (Curtin et al. 2012; Sun et al. 2016). One of the major constraints in using TALENs on a large scale is that a new chimeric protein has to be engineered for each target sequence of interest. This is a very time-consuming and complex process to engineer a new protein for each new target. Moreover, TALENs delivery to plant cells is tough and challenging because of their large size of ~950 to ~1900 aa per pair.

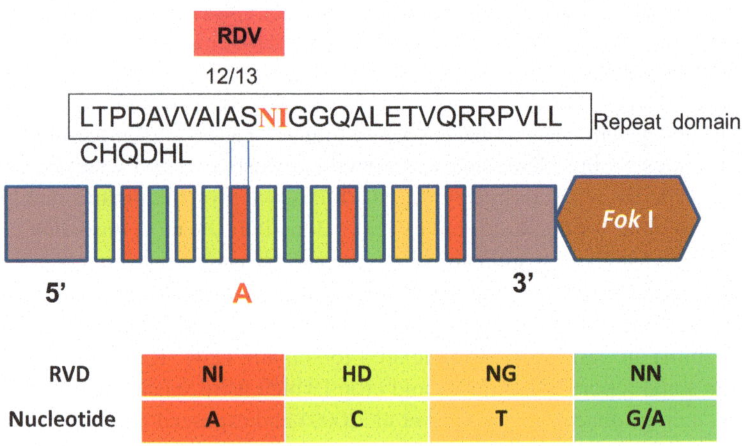


**Fig. 3** TALEN structure: Showing repeat domain of 34 amino acid residue with the repeat-variable di-residue (RVD). This RVD decides which single base the TALE effector will recognize and binds

## *CRISPR/CAS9 System*

CRISPR/Cas is an adaptive immune system in prokaryotes (particularly in bacteria) which confers resistance to invasive viral nucleic acids (Horvath and Barrangou 2010). Immunity is achieved with the help of RNA-guided nucleases, which target and cleave any foreign invading nucleic acids at a specific location in their sequence. A small fragment from the genome of invading bacteriophages is kept in bacteria as a genetic memory in the form of protospacers, therefore leading to the formation of CRISPR arrays, i.e. short palindromic repeats interspersed by spacer sequences of invading bacteriophages. In bacteria, the CRISPR sequences along with many different *Cas* (CRISPR-associated) proteins are used to create CRISPR arrays. The spacer sequences within the CRISPR arrays are transcribed into spacer RNA which along with *Cas* genes leads to catalysis of complimentary sequences found in the nucleic acids of infecting bacteriophages, therefore imparting immunity to the bacterial cells. The genome sequences inserted into CRISPR arrays are excised adjacent to PAM sequences in the bacteriophage DNA. Therefore, these PAM sequence (3–5 nt) are used to induce nicks in the DNA. This ability of *Cas* genes (specifically *Cas*9) to cut (due to DNA endonuclease activity) close to PAM sequences has been utilized in the genetic modification of organisms.

CRISPR/Cas9 system has recently been put to use for genome editing in plant biology (Belhaj et al. 2015; Weeks et al. 2016). This system consists of a guide RNA which is a short RNA molecule associated with *Cas*9. The guide RNA consists of two components: the crRNA (CRISPR-derived RNA) and tracrRNA (trans-activating RNA). The crRNA cuts the target double-stranded DNA and always has some region of homology with tracrRNA. A stem loop is formed by tracrRNA, which interacts with protein *Cas9*. In the experiments involving genome editing system, the crRNA and tracrRNA were combined together into a single-guide RNA (sgRNA) which helps *Cas9*-mediated dsDNA cleavage in a very sequence-specific manner (Jinek et al. 2012). *Cas9*-sgRNA moves along the length of DNA and makes a double-strand break (DSB) near PAM sequence followed by indigenous cell-mediated DNA repair leading to alteration in DNA sequences (Jinek et al. 2014). The successful applications of CRISPR/Cas9 have been reported in plants, animals and human cell lines (Doudna and Charpentier 2014; Khatodia et al. 2016). From 2013 onwards, the CRISPR/Cas9 machinery has been reported for genetic improvement in crops like *Arabidopsis* (Li et al. 2013), tobacco (Nekrasov et al. 2013), rice and wheat (Shan et al. 2013), soybean (Sun et al. 2015; Jacobs et al. 2015; Cai et al. 2015), tomato (Brooks et al. 2014; Pan et al. 2016), potato (Wang et al. 2015), cucumber (Chandrasekaran et al. 2016), maize (Char et al. 2017; Svitashev et al. 2016) and several others. CRISPR technology is useful for regulating the action of both the positive-regulator genes and helps to do away with the detrimental effect of negative regulator genes. Despite much progress of application of CRISPR/Cas system for genetic modification in crop plants, rice and wheat are the two major cereal crops that have been used for modification of genes responsible for biotic and abiotic stresses and other agronomic traits also.

## *Recent Variants Used for Genome Editing*

The identification of variants of CRISPR effectors including *Cpf*I from *Prevotella* and *Francisella* has opened new scope for genome editing to strengthen the research in the field of agriculture (Zetsche et al. 2015). *Cpf*1 is a Class II Type V endonuclease, which has superior and novel features of generating cohesive ends with an overhang of four or five nucleotides in comparison to *SpCas*9, which generates blunt ends only. It is well known that cohesive DNA ends have higher chances of generating InDels at the cleaved site. This insertion follows a mechanism of 'homology-directed repair' (HDR). In case of *Cpf*I, the target DNA molecule is cleaved by a single crRNA which is much smaller than the sgRNA for *SpCas*9, thereby reducing the cost of genome editing with a single and smaller crRNA. Moreover, *Cpf*1 recognizes a T-rich PAM as it requires 5′-TTTN-3′ PAM sequence, whereas *Cas*9 requires a G-rich protospacer sequence. *Cpf*I also carries out RNase III activity for processing of pre-crRNA. This feature can be used efficiently for multiplexing by placing tandem arrayed pre-crRNA-expressing constructs that produce multiple mature crRNAs. The CRISPR *Cpf*I system has been reported for efficient genome engineering by inducing targeted mutagenesis in rice and tobacco (Endo et al. 2016; Begemann et al. 2017). Multiplexed editing of *Os*BEL and *OsPDS* genes in rice reported heritable mutagenesis in rice (Wang et al. 2017; Xu et al. 2017). The CRISPR/*Cpf1* has been reported to induce protoplast-based mutations in soybean and wild tobacco (Kim et al. 2017). As compared to short indels generated by Cas9, majority of *Cpf1*-induced mutations in rice were quite long deletions (Xu et al. 2017).

Another emerging technique in genome editing in plants is base editor, which includes a combination of inactive CRISPR-Cas9 domain and cytosine or adenosine deaminase domain which causes point mutations at the desired target location. This is a new approach for converting a single-base change without a double-strand break in target genome (Komor et al. 2016). Single-base changes can help generate superior or elite trait variations in crop plants. Recently, single-base editing has been achieved in wheat by the fusion of a cytidine deaminase (Zong et al. 2017) or adenosine deaminase (Li et al. 2018a, b) with the *Cas9* nickase for base conversion of C/G to T/A or A/T to G/C. The efficiency of base editing was increased by using a *Cas*9-based nickase rather than inactive Cas9 by using Cas9-APOBEC3A for editing of *TaMTL* (*MATRILINEAL*) which encodes for sperm-specific phospholipase (Zong et al. 2018). Another experiment demonstrated the development of haploids in maize by generating loss of function in *MTL* gene (Kelliher et al. 2017). In wheat, similar approach using *TaMTL* was used to develop knockout mutants, out of which three were homozygous for all six alleles of this gene (Borisjuk et al. 2019).

## 4 Genetics and Genomics of Wheat

Grain quality improvement along with yield is a major objective for wheat breeders around the world. During the last few decades, most efforts have been made for developing semi-dwarf cultivars and also generation of hybrids in wheat, the former undoubtedly resulted in achieving high yield demands. Some level of success has been achieved to improve quality and yield traits in wheat by the application of molecular markers but researchers always tried for novel strategies in research. Dissecting the genetic basis for grain quality using fine mapping and cloning of quantitative trait loci (QTLs) for improvement of grain quality have been achieved in the past in wheat (Cabral et al. 2018; Cheng et al. 2017; Jahani et al. 2019; Li et al. 2018a, b). The biochemical composition of wheat grain determines the nutritional quality of wheat. More than 55–75% of total dry grain weight is formed by starch only, while storage protein accounts for 10–15% of the main reserves in grains. The quality of wheat grain is determined through rheological properties. High-throughput genotyping technologies have enabled to identify major QTLs for wheat grain quality including grain protein content (Groos et al. 2003; Kumar et al. 2018), kernel hardiness (Sourdille et al. 1996; Turner et al. 2004) and traits for mixing time, extensibility and tenacity of dough (Ma et al. 2007; Li et al. 2009).

### *Grain Yield*

During the past few years, the annual increase of wheat production by only 0.5% is very less as compared to the desired level of 2.4% to meet food demand of ever-increasing human population. Therefore, improved production rate can be obtained by increasing the grain yield per area (Sharma et al. 2012; Crespo-Herrera et al. 2018). Marker-assisted selection (MAS) is considered to be a key method to transfer a gene of interest including for yield contributing traits by combining conventional breeding with molecular techniques. The success of MAS application in plants is mainly influenced by the availability of tightly linked molecular markers with the genes of interest. Till date, nearly 65 genes in wheat have been cloned, and about 40 out of them are related to grain yield and associated traits (Liu et al. 2012; Nadolska-Orczyk et al. 2017; Rasheed et al. 2016). Functional markers for these cloned genes have been designed, and KBioscience's competitive allele-specific PCR (KASP) genotyping assays have been widely used for genes related to grain size and weight (Nadolska-Orczyk et al. 2017). Both QTL mapping and genome-wide association study (GWAS) for yield traits have been reported (Edae et al. 2014; Godoy et al. 2018; Sakuma et al. 2019;Liu et al. 2017; Sun et al. 2017). In wheat, a high-throughput 90 K and 660 K SNP arrays are replacing simple sequence repeat (SSR) markers to investigate genetics of most traits including grain yield and quality, along with traits for disease resistance and stress tolerance (Jin et al. 2016; Liu et al. 2017; Sun et al. 2017; Valluru et al. 2017).

### *Grain Quality*

Biochemical composition of wheat grains determines its nutritional and health-related properties. An adequate level of essential elements like iron, zinc, calcium, phosphorus and antioxidants is also essential for balanced wheat products. Starch quality in grains depends on the ratio of two main macromolecules that are amylose and amylopectin. Amylose-rich starch often called as resistant starch is considered beneficial to human health by protecting humans from several health complications such as diabetes and cardiovascular diseases (Meenu and Xu 2018). Downregulation of starch branching enzymes SBEIIa and SBEIIb that leads to increased level of amylose in wheat has been reported (Regina et al. 2006; Sestili et al. 2010). Nitrogen, apart from contributing to crop yield, plays a significant role in accumulation and composition of storage protein in grains of wheat (Zorb et al. 2018). Nitrogen is supplied to the grain by either remobilization from the leaves or stems and root uptake of nitrogen from soil. The dough quality of wheat is determined by visco-elasticity of wheat dough due to the presence of gluten – a major component of seed storage proteins in wheat (Anjum et al. 2007). Glutens are made up of gliadins and glutenins, which account for 70–80% of the total proteins present in wheat flour. Therefore, genes coding different types of storage proteins are targeted to improve both the nutritional and bread-making quality of wheat. The majority of genes responsible for traits related to quality in wheat were identified by genetic and genomics techniques (Yu et al. 2018; Nadolska-Orczyk et al. 2017).

Modern elite wheat cultivars usually contain suboptimal quantities of micronutrients (Cakmak et al. 2000), and majority of it is accumulated in the outer husk and aleurone, and the micronutrients are lost during milling and polishing (Welch and Graham 1999). Biofortification is a promising approach to reduce micronutrient deficiency in plants. Another problem is that of phytic acid, a major antinutritional factor for iron and zinc uptake in the human digestive tract, which is co-deposited with the minerals in aleurone storage vacuoles. Expression of an *Aspergillus niger* phytase gene, a phytic acid degrading enzyme, is targeted to the wheat aleurone to reduce the level of this antinutritional factor (Holm et al. 2002).

## 5 Genetic Modifications of Wheat for Improvement of Quality and Yield Traits

### *Transgenic Approaches*

In order to improve the performance of wheat crop, genetic manipulations have been carried out by introducing transgenes for all major agronomic traits like yield, quality and improving tolerance to biotic and abiotic stress. During the last 10 years, tremendous progress had been made for the genetic manipulation of wheat (Borisjuk et al. 2019) by the overexpression of endogenous genes already present in wheat or by introduction of foreign genes under the control of a specific promoter, which not only helps to understand the function of many novel genes but also contributes to the generation of improved varieties with economically important traits. *TaGW2*, a homologue of rice gene involved in negative regulation of grain size by regulating cell division within the spikelet, has been identified. However, independent experiments to downregulate T*aGW2* in wheat have shown contradictory results, while RNAi suppression of three *TaGW2* homologs A, B and D of bread wheat reported reduction in size and weight of grain (Bednarek et al. 2012); another study used same approach to target *TaGW2* reported an increase in weight of the grain (Hong et al. 2014). Differences in result may be caused by cultivar-specific response to difference in experimental setup or may be due to difference in application of transformation protocols. A positive effect of PEPC and PPDK genes in transgenics on photosynthetic and yield characteristics were observed when both these genes were used either separately or simultaneously in different studies (Zhang et al. 2014). In another study, maize transcription factor *Dof*1 when expressed in wheat upregulated the level of *PEPC* gene leading to an increase in yield and improved drought tolerance in transgenic lines (Qin et al. 2016; Peña et al. 2017).

The nuclear factor Y(NF-Y) is a class of regulators involved in processes related to development and physiological traits in plants (Myers and Holt 2018). NF-Y transcription factors classified into three families (NF-YA, NF-YB and NF-YC) have multiple members. Factor *NF-YA*, when overexpressed in wheat, leads to enhanced level uptake of nitrogen and phosphorus which ultimately resulted in increased grain yield (Qu et al. 2015). Yadav et al. (2015) reported the effect of the second wheat gene, *TaNF-YB4*, and its positive effect on grain yield. Wheat transcription factor, *TaNAC2*-5A, when overexpressed in wheat resulted in improved signalling of nitrogen, influx rate of nitrate and increased nitrate uptake by roots from soil. Transgenic lines showed increase in grain yield and more accumulation of nitrogen in aerial parts, which ultimately was translocated to grains (He et al. 2015). An endogenous chloroplastic glutamine synthase gene (*TaGS2*) when overexpressed in wheat leads to prolonged photosynthesis and remobilization of nitrogen into grains, which resulted into increased number of spike and grain per spike, thereby increasing the overall yield of plants (Hu et al. 2018). Weichert et al. (2010) overexpressed the gene for barley sucrose transporter gene (*HvSUT1*), which is responsible for enhanced uptake of sucrose and protein content in grains, but there was not much change in the level of starch accumulation and biosynthesis (Weichert et al. 2017). Maize ADPglucose pyrophosphorylase (*ZmAGPase*) when expressed in wheat (Smidansky et al. 2007) resulted in enhanced yield and photosynthetic rates in the transgenic lines. Zhao et al. (2015) identified a novel wheat gene, *TaNAC-S* for stay green phenotype leading to increased protein concentration in grains, no effect was observed on biomass and grain yield. The isolation, characterization and overexpression of two wheat Vacuolar Iron Transporter (*TaVIT*) genes under the control of promoter specific for expression in endosperm in case of barley and wheat resulted in twofold increase in the level of iron in the transgenic wheat (Connorton et al. 2017).

## *Application of RNA Interference*

RNA interference (RNAi) present in eukaryotic is a regulatory tool for controlling expression of genes in cells and has now become very popular not only for analysis of a function of a gene but also for producing novel phenotypes. Using this technique, antisense or hairpin RNAi construct molecules were introduced to accomplish post-transcriptional silencing of genes. Application of RNAi has made a strong contribution for helping to make manipulations in size of wheat grains (Uauy et al. 2006; Li et al. 2018a, b) and quality (Barro et al. 2016; Yue et al. 2008). Altenbach and Allen (2011) used RNAi-based silencing for suppressing the expression of $\omega$-gliadins leading to, increased stability of proteins along with improvement in the properties of dough in wheat. Gil-Humanes et al. (2012) used RNAi approach to downregulate the expression of alpha gliadin genes in the wheat *cv.* Bobwhite, and then this trait was transferred by conventional breeding into three other commercial cultivars of wheat. Barro et al. (2016) used seven RNAi plasmids to target $\alpha$-, $\gamma$- and $\omega$-gliadins. Low molecular weight (LMW) glutenins were also targeted to reduce the level of toxic epitopic regions which are responsible for celiac disease. Further analyses of proteins showed absence of toxic epitopes from the $\alpha$- and $\omega$-gliadins in the transgenic wheat.

## *Application of CRISPR /CAS9 for Wheat Grain Quality Improvement*

Improvement of qualitative traits in grains would be cost-effective and sustainable approach. Conventional breeding mainly depends on natural or induced variation for desirable traits in germplasm. The desirable traits are introgressed in the genetic background of locally adapted varieties through repeated backcrossing followed by labour- and time-intensive population screening. CRISPR-Cas system has opened new scope in wheat research for grain quality improvement. The *TaGASR7 locus* associated with grain length belongs to the *Snakin/GASA* class of gene family. A CRISPR/Cas9 system designed to target *TaGASR7* through shoot apical meristem generated 11 mutant plants with desirable alleles, and three plants carried mutation to the next generation (Hamada et al. 2018). The CRISPR/Cas9 system was used to target *TaGW2* gene that encodes RING-type E3 ubiquitin ligase, and is reported to be a, negative regulator of size of wheat grains and thousand grain weight. The $T_1$ knockout plants carried mutations in all three copies of the *TaGW2* gene (Wang et al. 2018a, b). As a result, mutants showed significantly improved characteristics like thousand grain weight, grain area, grain width and grain length in comparison to wild-type plants. CRISPR/Cas9 technology was applied to obtain wheat grains with less immunogenic response. CRISPR/Cas9 technology had been used efficiently to reduce the number of alpha-gliadins giving rise to wheat lines with less immunoreactivity for people suffering from celiac disease (Sanchez-Leon et al. 2018). As a result, a total of 21 lines were generated with desired mutation having a strong reduction in number of copies of $\alpha$-gliadins. Another possible target for manipulation using CRISPR/Cas system is Dense and Erect Panicle 1 (*DEP1*) gene. *DEP1* codes for a subunit of G protein in rice and is responsible for regulation of erect panicles, number of grains in each panicle, nitrogen uptake, and stress tolerance through protein signalling pathway. *TaDEP1*(wheat orthologus gene) mutants in wheat were obtained by CRISPR/CAS9, and frameshift mutations in all six alleles exhibited dwarf phenotype ascompared to wild-type plants (Zhang et al. 2016). The results proved the role of *TaDEP1* as an important regulator of growth and development. *TaGW7,* a homolog of rice *OsGW7* encoding a TONNEAU1-recruiting motif (TRM) protein, affect grain shape and weight in allohexaploid wheat. Editing of *TaGW7* homoeologs by CRISPR/CAS9 in the B and D genomes of wheat resulted in mutations of this gene in both the genomes leading to an increase in grain width and weight but at the cost of reduction in grain length (Wang et al. 2019).

## *Application of CRISPR/CAS9 For Biotic and Abiotic Stress Tolerance*

Plant diseases caused by microorganisms are major factors that negatively affect the quality and yield in wheat. CRISPR-Cas9 system works as a promising tool to engineer genes related to traits of agronomic importance (Table 1) and also provides enhanced tolerance to diseases, pests and abiotic stress in case of wheat (Table 2). The disease powdery mildew in wheat is caused by *Blumeria graminis* f. sp. *tritici* which is responsible for significant losses. Gene knockouts for *TaMLO* gene were reported for the first time in wheat for resistance to powdery mildew (Shan et al. 2013). Zhang et al. (2017) used CRISPR/Cas9 for knock-down of three homologs of *TaEDR*, which is a negative regulator of powdery mildew resistance. The deadly fungus *Fusarium* is responsible for huge yield losses in wheat and silencing of two lipoxygenase genes, *TaLpx1*and *TaLox2* provided resistance to fusarium in wheat. *TaLpx1* and *TaLox2* genes were edited by CRISPR/CAS9, and a mutation frequency of 9 and 45% was achieved, respectively. The CRISPR/Cas9 system was also experimented for editing a wheat homolog of *TaCer9* (*ECERIFERUM9*) which is responsible for improved tolerance to drought stress with better water use efficiency (Liang et al. 2017). The use of Cas9 nickase fused with a human cytidine deaminase, *APOBEC3A*, gave rise to herbicide resistant plants in wheat (Zong et al. 2018) .

**Table 1** Genome editing in wheat for improvement of grains related traits

| Sr. no. | Target gene | Trait effected | Type of editing | Efficiency of mutation | References |
|---|---|---|---|---|---|
| 1 | α-Gliadins | Allergic reaction to gluten reduced | Knockout | Not mentioned | Sanchez-Leon et al. (2018) |
| 2 | *TaGW2* | Negative regulator grain weight | Knockout | 5% | Zhang et al. (2016) |
| 3 | *TaGASR7* | Length and width of grains | Knockout | 1.8% | Liang et al. (2017) |
| 4 | *TaNAC2* | Regulate shoot branching | Knockout | 2% | Zhang et al. (2016) |
| 5 | *TaDEP1* | Inflorescence architecture, grain yield | Knockout | 2% | Zhang et al. (2016) |
| 6 | *TaCKX2–1, TaGW2, TaGW2, TaGLW7 and TaGW8* | Grain number per spikelet | Knockout | 10% | Zhang et al. (2019) |
| 7 | *α–/γ-gliadins* | CRISPR/Cas9 can edit multiple genes simultaneously | Indels | Not specified | Jouanin et al. (2019) |
| 8 | *TaPIN1* | Emergence of adventitious root and tillering | Knockout | 1% | Zhang et al. (2016) |
| 10 | *TaPinB* | Seeds softness | Knockout | Not specified | Brandt et al. (2017) |
| 11 | *TaLOX2* | Grain development and affect the storability of wheat grains | Knockout | 9.5% | Zhang et al. (2016) |

**Table 2** Genome editing in wheat for stress-related traits

| Sr. no. | Target gene | Traits effected | Efficiency (%) | Type | References |
|---|---|---|---|---|---|
| 1 | *TaNFXL1* | Fusarium head blight susceptibility | 42.2 | Knockout | Cui et al. (2019) |
| 2 | *TaMLO* | Repress resistance to powdery mildew | 28 | Knockout | Shan et al. (2014) |
| 3 | *TaMLO-A1* | Powdery mildew resistance | 5.6 | Indel | Wang et al. (2018a, b) |
| 4 | *TaEDR1* | Resistance to powdery mildew | 5 (mutants) | Knockout | Zhang et al. (2017) |
| 5 | *TaDREB2* | Tolerance for drought | 6.7 | Knockout | Kim et al. (2017) |
| 6 | *TaABCC6* | Fusarium head blight (FHB) susceptibility | 6.6 | Knockout | Cui et al. (2019) |
| 7 | *TansLTP9.4* | Fusarium blight resistance | 11.9 | Knockout | Cui et al. (2019) |

### *Application of CRISPR /CAS9 for Other Agronomic Traits*

Male sterility and the haploid induction are effective tools to carry out genetic analysis during wheat breeding program. Male-sterile lines and doubled haploid plants help to the formation of hybrid seeds in wheat. Genetic analysis of mutated plants (Singh et al. 2018) obtained by CRISPR/Cas9 technology showed that all three homeologues of *Ms45* present in wheat contribute for providing male fertility. Mutant plants, *Tams45-abd*, showed absence of pollen development and ultimately provide male sterility.

## 6 Limitations and Bottlenecks

It was earlier known that wheat contained near about 128,000 genes (Montenegro et al. 2017), and most of its genome is highly repetitive sequences (Bhalla et al. 2017). However, recent estimates have reported a total of 107,891 high-confidence genes with threefold redundancy because of it being an allohexaploid genome (Appels et al. 2018). As a result it is challenging to target three or multiple copies of a gene simultaneously for wheat genome editing, and if all copies of a gene are not knocked down, desired phenotype may not be achieved due to some genome buffering. Moreover, regeneration of plants from CRISPR-edited protoplast has been difficult. Despite the considerable efforts of the researchers around the globe, progress in genetic engineering in wheat lacks when compared to other crops like rice and maize; because of its large genome size and highly redundant genome, most of the wheat varieties are recalcitrant for in vitro culture and regeneration (Shrawat and Armstrong 2018). One of the biggest challenges for wheat is the difficulty experienced in the transformation of genotypes, which further reduce scope for applications of CRISPR/cas9 in wheat, although there are some reports where genetic transformation in wheat had been achieved successfully using biolistics transformation (Tassy et al. 2014). However, modification of multiple targets by simultaneous transforming multiple sgRNAs using CRISPR-based multiplex genome editing toolkits has been reported in wheat (Ismagul et al. 2018; Yao et al. 2007) .

## 7 Conclusions and Future Perspectives

The reference genome of wheat IWGSC (http://www.wheatgenome.org/) will speed up the development of CRISPR-based better wheat varieties to meet the global challenge of food security in future. This technology holds promise to meet the increasing food demand world population. As compared to conventional breeding methods, the genome editing tools would equip scientists to target and edit genes for desired traits precisely in shorter duration of time. It would be used to target genes responsible for increased crop productivity, improved nutritional value and enhance abiotic stress tolerance in the crops. Even till now, there have been very few studies that are related to targeted mutagenesis in wheat for improvement of quality and yield in grains. The rapid shift of research interest towards the utilization of CRISPR/Cas9 systems and its variants for targeted mutagenesis could be a promising technique to overcome barriers to breeding improved quality wheat.

## References


Altenbach SB, Allen PV (2011) Transformation of the US bread wheat 'Butte 86' and silencing of omega-5 gliadin genes. GM Crops 2:67–74

Anjum FM, Khan MR, Din A, Saeed M, Pasha I, Arshad MU (2007) Wheat gluten: high molecular weight glutenin subunits structure, genetics, and relation to dough elasticity. J Food Sci 72(3):R56–R63

Appels R, Eversole K, Feuillet C et al (2018) Shifting the limits in wheat research and breeding using a fully annotated reference genome. Science 361(6403). https://doi.org/10.1126/science.aar7191

Baltes NJ, Voytas DF (2015) Enabling plant synthetic biology through genome engineering. Trends Biotechnol 33(2):120–131

Barro F, Iehisa JC, Gimenez MJ et al (2016) Targeting of prolamins by RNAi in bread wheat: effectiveness of seven silencing-fragment combinations for obtaining lines devoid of coeliac disease epitopes from highly immunogenic gliadins. Plant Biotechnol J 14(3):986–996

Bednarek J, Boulafous A, Girousse et al (2012) Down-regulation of the TaGW2 gene by RNA interference results in decreased grain size and weight in wheat. J Exp Bot 63(16):5945–5955

Begemann MB, Gray BN, January E et al (2017) Precise insertion and guided editing of higher plant genomes using Cpf1 CRISPR nucleases. Sci Rep 7:11606

Belhaj K, Chaparro-Garcia A, Kamoun S, Patron NJ, Nekrasov V (2015) Editing plant genomes with CRISPR/Cas9. Curr Opin Biotechnol 32:76–84

Bhalla PL, Sharma A, Singh MB (2017) Enabling molecular technologies for trait improvement in wheat in wheat biotechnology. Method Mol Biol:3–24

Boch J, Bonas U (2010) *Xanthomonas* AvrBs3 family-type III effectors: discovery and function. Annu Rev Phytopathol 48:419–436

Borisjuk N, Kishchenko O, Eliby S, Schramm C, Anderson P, Jatayev S, Kurishbayev A, Shavrukov Y(2019) Genetic modification for wheat improvement: from transgenesis to genome editing. BioMed Res Int Article ID 6216304|18 pages. https://doi.org/10.1155/2019/6216304

Bortesi L, Fischer R (2015) The CRISPR/Cas9 system for plant genome editing and beyond. Biotechnol Adv 33:41–52

Brandt KM, Gunn HM, Buschke BL, Heesacker A, Moretti N, Karasev A, Zemetra RS (2017) Testing non-transgenic CRISPR technology for wheat improvement. Presentation. 13th IWGS – Tulln, Austria

Brooks C, Nekrasov V, Lippman ZB, Van Eck J (2014) Efficient gene editing in tomato in the first generation using the clustered regularly interspaced short palindromic repeats/CRISPRassociated9 system. Plant Physiol 166(3):1292–1297

Cabral AL, Jordan MC, Larson G, Somers DJ, Humphreys DG, McCartney CA (2018) Relationship between QTL for grain shape, grain weight, test weight, milling yield, and plant height in the spring wheat cross RL4452/'AC domain'. PLoS One 13(1):e0190681

Cai Y, Chen L, Liu X, Sun S, Wu C, Jiang B et al (2015) CRISPR/Cas9-mediated genome editing in soybean hairy roots. PLoS One 10(8):e0136064. https://doi.org/10.1371/journal.pone.0136064

Cakmak I, Ozkan H, Braun HJ, Welch RM, Romheld V (2000) Zinc and iron concentrations in seeds of wild, primitive, and modern wheat. Food Nutr Bull 21(4):401–403

Capecchi MR (1980) High efficiency transformation by direct microinjection of DNA into cultured mammalian cells. Cell 22:479–488

Chandrasekaran J, Brumin M, Wolf D, Leibman D, Klap C, Pearlsman M et al (2016) Development of broad virus resistance in non-transgenic cucumber using CRISPR/Cas9 technology. Mol Plant Pathol 17(7):1140–1153

Char SN, Neelakandan AK, Nahampun H, Frame B, Main M, Spalding MH et al (2017) An agrobacterium delivered CRISPR/Cas9 system for high-frequency targeted mutagenesis in maize. Plant Biotechnol J 15(2):257–268

Cheng RR, Kong ZX, Zhang LW, Xie Q, Jia HY, Yu D, Huang YL, Ma ZQ (2017) Mapping QTLs controlling kernel dimensions in a wheat intervarietal RIL mapping population. Theor Appl Genet 130:1405–1414

Connorton JM, Jones ER, Rodríguez-Ramiro I, Fairweather-Tait S, Uauy C, Balk J (2017) Wheat vacuolar iron transporter TaVIT2 transports Fe and Mn and is effective for biofortification. Plant Physiol 174:2434–2444

Crespo-Herrera LA, Crossa J, Huerta-Espino J, Vargas M, Monda S, Velu G et al (2018) Genetic gains for grain yield in CIMMYT's semi-arid wheat yield trials grown in suboptimal environments. Crop Sci 58:1890–1189

Cui X, Balcerzak M, Schernthaner J, Babic V, Datla R, Brauer EK, Labbé N, Subramaniam R, Ouelle T (2019) An optimized CRISPR/Cas9 protocol to create targeted mutations in homoeologous genes and an efficient genotyping protocol to identify edited events in wheat. Plant Methods 15:119

Curtin SJ, Voytas DF, Stupar RM (2012) Genome engineering of crops with designer nucleases. Plant Genome 5:42–50

Doudna JA, Charpentier E (2014) Genome editing. The new frontier of genome engineering With CRISPR-Cas9. Science 346(6213):1258096. https://doi.org/10.1126/science.1258096

Edae EA, Byrne PF, Haley SD, Lopes MS, Reynolds MP (2014) Genome-wide association mapping of yield and yield components of spring wheat under contrasting moisture regimes. Theor Appl Genet 127:791–807

Endo A et al (2016) Efficient targeted mutagenesis of rice and tobacco genomes using Cpf1 from *Francisella novicida*. Sci Rep 6:38169

Gil-Humanes J, Piston F, Gimenez MJ et al (2012) The introgression of RNAi silencing of $\gamma$-gliadins into commercial lines of bread wheat changes the mixing and technological properties. PLoS One 7(9):e45937. https://doi.org/10.1371/journal.pone.0045937

Godoy J, Gizaw S, Chao S, Blake N, Carter A, Cuthbert R, Dubcovsky J, Hucl P, Kephart K, Pozniak C et al (2018) Genome-wide association study of agronomic traits in a spring-planted north American elite hard red spring wheat panel. Crop Sci 58(5):1838–1852

Groos C, Robert N, Bervas E, Charmet G (2003) Genetic analysis of grain protein-content, grain yield and thousand-kernel weight in bread wheat. Theor Appl Genet 106(6):1032–1040

Gupta M, Dekelver RC, Palta A, Clifford C, Gopalan S, Miller JC, Novak S, Desloover D, Gachott E, Connell J (2012) Transcriptional activation of *Brassica napus* B-ketoacyl-ACP synthase ll with an engineered zinc finger protein transcription factor. Plant Biotechnol J 10:783–791

Hamada H, Liu Y, Nagira Y, Miki R, Taoka N, Imai R (2018) Biolistic-delivery- based transient CRISPR/CAS9 expression enables in planta genome editing in wheat. Sci Rep 8:14422

He X, Qu B, Li W et al (2015) The nitrate-inducible NAC transcription factor TaNAC2-5A controls nitrate response and increases wheat yield. Plant Physiol 169(3):1991–2005

Holm PB, Kristiansen KN, Pedersen HB (2002) Transgenic approaches in commonly consumed cereals to improve iron and zinc content and bioavailability. J Nutr 132(3):514S–516S

Hong Y, Chen L, Du L et al (2014) Transcript suppression of TaGW2 increased grain width and weight in bread wheat. Funct Integr Genomics 14(2):341–349

Horvath P, Barrangou R (2010) CRISPR/Cas, the immune system of bacteria and archaea. Science 327:167–170

Hu M, Zhao X, Liu Q et al (2018) Transgenic expression of plastidic glutamine synthetase increases nitrogen uptake and yield in wheat. Plant Biotechnol J 6(11):1858–1867

Ismagul A, Yang N, Maltseva E et al (2018) A biolistic method for high-throughput production of transgenic wheat plants with single gene insertions. BMC Plant Biol 18:135–143

Jacobs TB, LaFayette PR, Schmitz RJ, Parrott WA (2015) Targeted genome modifications in soybean with CRISPR/Cas9. BMC Biotechnol 15(1):16

Jahani M, Mohammadi-Nejad G, Nakhoda B, Rieseberg LH (2019) Genetic dissection of epistatic and QTL by environment interaction effects in three bread wheat genetic backgrounds for yield related traits under saline conditions. Euphytica 215:103

Jasin M, Liang F (1996) Ku80-deficient cells exhibit excess degradation of extrachromosomal DNA. J Biol Chem 271:14405–14411

Jin H, Wen WE, Liu JD, Zhai SN, Zhang Y, Yan J, Liu ZY, Xia XC, He ZH (2016) Genome wide QTL mapping for wheat processing quality parameters in a Gaocheng 8901/ Zhoumai 16 recombinant inbred line population. Front Plant Sci 7:1032

Jinek M, Chylinski K, Fonfara I, Hauer M, Doudna JA, Charpentier E (2012) A programmable dual-RNA-guided DNA endonuclease in adaptive bacterial immunity. Science 337:816–821

Jinek M, Jiang F, Taylor DW, Sternberg SH, Kaya E, Ma E, Anders C, Hauer M, Zhou K, Lin S et al (2014) Structures of Cas9 endonucleases reveal RNA-mediated conformational activation. Science 343(6176):1247997. https://doi.org/10.1126/science.1247997

Jouanin A, Schaart G, Boyd LA, Cockram J, Leigh FJ, Bates R, Wallington EJ, Visser RGF, Smulders MJM (2019) Outlook for coeliac disease patients: towards bread wheat with hypoimmunogenic gluten by gene editing of α- and γ-gliadin gene families. BMC Plant Biol 19:333

Kelliher T, Starr D, Richbourg L et al (2017) MATRILINEAL, a sperm-specific phospholipase, triggers maize haploid induction. Nature 542(7639):105–109

Khatodia S, Bhatotia K, Passricha N, Khurana SM, Tuteja N (2016) The CRISPR/Cas genome-editing tool: application in improvement of crops. Front Plant Sci 7:506

Kim D, Alptekin B, Budak H (2017) CRISPR/Cas9 genome editing in wheat. Funct Integr Genomics 18:31–34

Komor AC, Kim YB, Packer MS, Zuris JA, Liu DR (2016) Programmable editing of a target base in genomic DNA without double stranded DNA cleavage. Nature 533:420–424

Kumar A, Jain S, Elias EM, Ibrahim M, Sharma LK. (2018) An Overview of QTL Identification and Marker-Assisted Selection for Grain Protein Content in Wheat. In: Sengar R, Singh A. (eds) Eco-friendly Agro-biological Techniques for Enhancing Crop Productivity. Springer, Singapore. https://doi.org/10.1007/978-981-10-6934-5_11

Li Y, Song Y, Zhou R, Branlard G, Jia J (2009) Detection of QTLs for bread-making quality in wheat using a recombinant inbred line population. Plant Breed 128(3):235–243

Li J, Norville J, Aach J et al (2013) Multiplex and homologous recombination–mediated genome editing in Arabidopsis and *Nicotiana benthamiana* using guide RNA and Cas9. Nat Biotechnol 31(8):688–691. https://doi.org/10.1038/nbt.2654

Li FJ, Wen WE, He ZH, Liu JD, Jin H, Geng HW, Yan J, Zhang PZ, Wan YX, Xia XC (2018a) Genome-wide linkage mapping of yield related traits in three Chinese bread wheat populations using high-density SNP markers. Theor Appl Genet 131:1903–1924

Li Y, Song G, Gao J et al (2018b) Enhancement of grain number per spike by RNA interference of cytokinin oxidase 2 gene in bread wheat. Hereditas 155(33)

Liang Z, Chen K, Li T et al (2017) Efficient DNA-free genome editing of bread wheat using CRISPR/Cas9 ribonucleoprotein complexes. Nat Commun 8:14261

Liu YN, He ZH, Appels R, Xia XC (2012) Functional markers in wheat: current status and future prospects. Theor Appl Genet 125:1–10

Liu JD, He ZH, Rasheed A, Wen WE, Yan J, Zhang PZ, Wan YX, Zhang Y, Xie CJ, Xia XC (2017) Genome-wide association mapping of black point reaction in common wheat (Triticum aestivum L.). BMC Plant Biol 17:220

Ma W, Sutherland MW, Kammholz S, Banks P, Brennan P, Bovill W et al (2007) Wheat flour protein content and water absorption analysis in a doubled haploid population. J Cereal Sci 45(3):302–308

Meenu M, Xu B (2018) A critical review on anti-diabetic and anti-obesity effects of dietary resistant starch. Crit Rev Food Sci Nutr 30:1–13

Montenegro JD, Golicz AA, Bayer PE et al (2017) The pan genome of hexaploid bread wheat. Plant J 90(5):1007–1013

Mussolino C, Alzubi J, Fine EJ, Morbitzer R, Cradick TJ, Lahaye T, Bao G, Cathomen T (2014) TALENS facilitate targeted genome editing in human cells with high specificity and low cytotoxicity. Nucleic Acid Res 42:6762–6773

Myers ZA, Holt BF III (2018) Nuclear factor-Y: still complex after all these years? Curr Opin Plant Biol 45:96–102

Nadolska-Orczyk A, Rajchel IK, Orczyk W, Gasparis S (2017) Major genes determining yield-related traits in wheat and barley. Theor Appl Genet 130(6):1081–1098

Nekrasov V, Staskawicz B, Weigel D, Jones JD, Kamoun S (2013) Targeted mutagenesis in the model plant Nicotiana benthamiana using Cas9 RNA-guided endonuclease. Nat Biotechnol 31(8):691

Pan C, Ye L, Qin L, Liu X, He Y, Wang J et al (2016) CRISPR/Cas9-mediated efficient and heritable targeted mutagenesis in tomato plants in the first and later generations. Sci Rep 6:24765

Peña PA, Quach T, Sato S et al (2017) Expression of the maize Dof1 transcription factor in wheat and sorghum. Front Plant Sci 8(434)

Puchta H (2016) Using CRISPR/Cas in three dimensions: towards synthetic plant genomes, transcriptomes and epigenomes. Plant J Cell Mol Biol 87:5–15

Qin N, Xu W, Hu L et al (2016) Drought tolerance and proteomics studies of transgenic wheat containing the maize C4 phosphoenolpyruvate carboxylase (PEPC) gene. Protoplasma 253(6):1503–1512

Qu B, He X, Wang J et al (2015) A wheat CCAAT box-binding transcription factor increases the grain yield of wheat with less fertilizer input. Plant Physiol 167(2):411–423

Ramirez CL, Foley JE, Wright DA, Muller F, Lerch SH, Rahman SH, Cornu TI, Winfrey RJ, Fu F, Townsend JA (2008) Unexpected failure rates for modular assembly of engineered zinc fingers. Nat Methods 5:374–376

Rasheed A, Wen WE, Gao FM, Zhai SN, Jin H, Liu JD, Guo Q, Zhang Y, Dreisigacker S, Xia XC, He ZH (2016) Development and validation of KASP assays for genes underpinning key economic traits in bread wheat. Theor Appl Genet 129:1843–1860

Regina A, Bird A, Topping D et al (2006) High-amylose wheat generated by RNA interference improves indices of large-bowel health in rats. Proc Natl Acad Sci U S A 103(10):3546–3551

Roth DB, Menetski JP, Nakajima PB, Bosma MJ, Gellert M (1992) V(D)J recombination: broken DNA molecules with covalently sealed (hairpin) coding ends in scid mouse thymocytes. Cell 70(6):983–991

Sakuma S, Golan G, Guo ZF et al (2019) Unleashing floret fertility in wheat through the mutation of a homeobox gene. PNAS 116:5182–5187

Sanchez-Leon S, Gil-Humanes J, Ozuna CV et al (2018) Low gluten, non transgenic wheat engineered with CRISPR/Cas9. Plant Biotechnol J 16(4):902–910

Schaart JG, van de Wiel CC, Lotz LA, Smulders MJ (2016) Opportunities for products of new plant breeding techniques. Trends Plant Sci 21:438–449

Sestili F, Janni M, Doherty A et al (2010) Increasing the amylose content of durum wheat through silencing of the SBEIIa genes. BMC Plant Biol 10(144)

Shan Q, Wang Y, Li J, Zhang Y, Chen K, Liang Z et al (2013) Targeted genome modification of crop plants using a CRISPR-Cas system. Nat Biotechnol 31(8):686

Shan Q, Wang Y, Li J, Gao C (2014) Genome editing in rice and wheat using the CRISPR/Cas system. Nat Protoc 9:2395–2410
Sharma RC, Crossa J, Velu G, Huerta-Espino J, Vargas M, Payne TS et al (2012) Genetic gains for grain yield in CIMMYT spring bread wheat across international environments. Crop Sci 52:1522–1533
Shiferaw B, Braun HJ, Duveiller E, Duveiller E, Reynolds MP, Muricho G (2013) Crops that feed the world 10. Past successes and future challenges to the role played by wheat in global food security. Food Sec 5(3):291–317
Shrawat AK, Armstrong CL (2018) Development and application of genetic engineering for wheat improvement. Crit Rev Plant Sci 37(5):335–421
Shu QY, Forster BP, Nakagawa H (eds) (2012) Plant mutation breeding and biotechnology. CAB International, Cambridge, MA, pp 453–463
Singh M, Kumar M, Albertsen MC, Young JK, Cigan M (2018) Concurrent modifications in the three homeologs of Ms45 gene with CRISPR-Cas9 lead to rapid generation of male sterile bread wheat (*Triticum aestivum* L.). Plant Mol Biol 97(4–5):371–383
Smidansky ED, Meyer FD, Blakeslee B, Weglarz TE, Greene TW, Giroux MJ (2007) Expression of a modified ADP-glucose pyrophosphorylase large subunit in wheat seeds stimulates photosynthesis and carbon metabolism. Planta 225(4):965–976
Sourdille P, Perretant MR, Charmet G, Leroy P, Gautier MF, Joudrier P et al (1996) Linkage between RFLP markers and genes affecting kernel hardness in wheat. Theor Appl Genet 93(4):580–586
Sun X, Hu Z, Chen R, Jiang Q, Song G, Zhang H et al (2015) Targeted mutagenesis in soybean using the CRISPR-Cas9 system. Sci Rep 5:10342
Sun Y, Li J, Xia L (2016) Precise genome modification via sequence specific nuclease mediated gene targeting for crop improvement. Front Plant Sci 7:1928
Sun C, Zhang F, Yan X, Zhang X, Dong Z, Cui D et al (2017) Genome-wide association study for 13 agronomic traits reveals distribution of superior alleles in bread wheat from the Yellow and Huai Valley of China. Plant Biotechnol J 15(8):953–969
Svitashev S, Schwartz C, Lenderts B, Young JK, Cigan AM (2016) Genome editing in maize directed by CRISPR–Cas9 ribonucleoprotein complexes. Nat Commun 7:13274
Tassy C, Partier A, Beckert M, Feuillet C, Barret P (2014) Biolistic transformation of wheat: increased production of plants with simple insertions and heritable transgene expression. P C TOC 119(1):171–181
Townsend JA, Wright DA, Winfrey RJ, Fu F, Maeder ML, Joung JK, Voytas DF (2009) High frequency modification of plant genes using engineered zinc finger nucleases. Nature 459:442
Trevino F, Zhang F (2014) Genome editing using Cas9 nickases. Methods Enzymol 546C:161–174
Turner AS, Bradburne RP, Fish L, Snape JW (2004) New quantitative trait loci influencing grain texture and protein content in bread wheat. J Cereal Sci 40(1):51–60
Uauy C (2017) Plant genomics: unlocking the genome of wheat's progenitor. Curr Biol 27(20):R1122–R1124
Uauy C, Distelfeld A, Fahima T, Blechl A, Dubcovsky J (2006) A NAC gene regulating senescence improves grain protein, zinc, and iron content in wheat. Science 314(5803):1298–1301
Valluru R, Reynolds MP, Davies WJ, Sukumaran S (2017) Phenotypic and genome wide association analysis of spike ethylene in diverse wheat genotypes under heat stress. New Phytol 214:271–283
Vanamee E, Santagata AK, Aggarwal AK (2001) Fok1 requires two specific sites for cleavage. J Mol Biol 309:69–78
Wang S, Zhang S, Wang W, Xiong X, Meng F, Cui X (2015) Efficient targeted mutagenesis in potato by the CRISPR/Cas9 system. Plant Cell Rep 34(9):1473–1476
Wang M et al (2017) Multiplex gene editing in rice using the CRISPR-Cpf1 system. Mol Plant Published online
Wang W, Pan Q, He F, Akhunova A, Chao S, Trick H, Akhunov E (2018a) Transgenerational CRISPR-Cas9 activity facilitates multiplex gene editing in allopolyploid wheat. CRISPR J 1(1):65–74

Wang W, Simmonds J, Pan Q et al (2018b) Gene editing and mutagenesis reveal inter-cultivar differences and additivity in the contribution of TaGW2 homoeologues to grain size and weight in wheat. Theor Appl Genet 131(11):2463–2475

Wang W, Pan Q, Tian B, He F, Chen Y, Bai G, Akhunova A, Trick HN, Akhunov E (2019) Gene editing of the wheat homologs of TONNEAU1- recruiting motif encoding gene affects grain shape and weight in wheat. Plant J 100(2):251–264

Weeks DP, Spalding MH, Yang B (2016) Use of designer nucleases for targeted gene and genome editing in plants. Plant Biotechnol J 14:483–495

Weichert N, Saalbach I, Weichert H et al (2010) Increasing sucrose uptake capacity of wheat grains stimulates storage protein synthesis. Plant Physiol 152(2):698–710

Weichert H, Högy P, Mora-Ramirez L, Fuchs J, Eggert K, Koehler P, Weschke W, Fangmeier A, Weber H (2017) Grain yield and quality responses of wheat expressing a barley sucrose transporter to combined climate change factors. J Exp Bot 68(20):5511–5525

Welch RM, Graham RD (1999) A new paradigm for world agriculture: meeting human needs: productive, sustainable, nutritious. Field Crops Res 60(1–2):1–10

Xu R, Qin R, Li H, Li D, Li L, Wei P et al (2017) Generation of targeted mutant rice using a CRISPR-Cpf1 system. Plant Biotechnol J 15:713–717

Yadav D, Shavrukov Y, Bazanova N et al (2015) Constitutive overexpression of the TaNF-YB4 gene in transgenic wheat significantly improves grain yield. J Exp Bot 66(21):6635–6650

Yao Q, Cong L, He G et al (2007) Optimization of wheat co transformation procedure with gene cassettes resulted in an improvement in transformation frequency. Mol Biol Rep 34:61–67

Yu Y, Zhu D, Ma C et al (2018) Transcriptome analysis reveals key differentially expressed genes involved in wheat grain development. Crop J 4(2):92–106

Yue S, Li H, Li Y et al (2008) Generation of transgenic wheat lines with altered expression levels of 1Dx5 high-molecular weight glutenin subunit by RNA interference. J Cereal Sci 47(2):153–161

Zetsche B, Gootenberg JS, Abudayyeh OO, Slaymaker IM, Makarova KS, Essletzbichler P et al (2015) Cpf1 is a single RNA-guided endonuclease of a class 2 CRISPR-Cas system. Cell 163:759–771

Zhang H, Xu W, Wang H et al (2014) Pyramiding expression of maize genes encoding phosphoenolpyruvate carboxylase (PEPC) and pyruvate orthophosphate dikinase (PPDK) synergistically improve the photosynthetic characteristics of transgenic wheat. Protoplasma 251(5):1163–1173

Zhang Y, Liang Z, Zong Y et al (2016) Efficient and transgene-free genome editing in wheat through transient expression of CRISPR/Cas9 DNA or RNA. Nat Commun 25(7):12617. https://doi.org/10.1038/ncomms12617

Zhang Y, Bai Y, Wu G, Zou S, Chen Y, Gao C, Tang D (2017) Simultaneous modification of three homoeologs of TaEDR1 by genome editing enhances powdery mildew resistance in wheat. Plant J 91(4):714–724

Zhang Z, Hua L, Gupta A, et al (2019) Development of an Agrobacterium-delivered CRISPR/Cas9 system for wheat genome editing. Plant Biotechnol J, 17: 1623–1635. https://doi.org/10.1111/pbi.13088

Zhao D, Derkx AP, Liu D, Buchner P, Hawkesford MJ, Flemetakis E (2015) Overexpression of a NAC transcription factor delays leaf senescence and increases grain nitrogen concentration in wheat. Electron J Plant Biol 17(4):904–913

Zong Y, Wang Y, Li C et al (2017) Precise base editing in rice, wheat and maize with a Cas9-cytidine deaminase fusion. Nat Biotechnol 35(5):438–440

Zong Y, Song Q, Li C et al (2018) Efficient C-to-T base editing in plants using a fusion of nCas9 and human APOBEC3A. Nat Biotechnol 36:950–953

Zorb C, Ludewig U, Hawkesford MJ (2018) Perspective on wheat yield and quality with reduced nitrogen supply. Trends Plant Sci 23(11):1029–1037

# Index



© The Editor(s) (if applicable) and The Author(s), under exclusive license to Springer Nature Switzerland AG 2021
S. H. Wani et al. (eds.), *Physiological, Molecular, and Genetic Perspectives of Wheat Improvement*, https://doi.org/10.1007/978-3-030-59577-7

The manufacturer's authorised representative in the EU is Springer Nature Customer Service Centre GmbH, Europaplatz 3, 69115 Heidelberg, Germany. If you have any concerns regarding our products, please contact ProductSafety@springernature.com

Printed and bound by CPI Group (UK) Ltd, Croydon, CR0 4YY
30/09/2026
02225508-0010